Artificial Intelligence and Modeling for Water Sustainability

Artificial intelligence and the use of computational methods to extract information from data are providing adequate tools to monitor and predict water pollutants and water quality issues faster and more accurately. Smart sensors and machine learning models help detect and monitor dispersion and leakage of pollutants before they reach groundwater. With contributions from experts in academia and industries who give a unified treatment of AI methods and their applications in water science, this book will help governments, industries, and homeowners not only to address water pollution problems more quickly and efficiently but also to gain better insight into the implementation of more effective remedial measures.

FEATURES

- Provides cutting-edge AI applications in water sector.
- Highlights the environmental models used by experts in different countries.
- Discusses various types of models using AI and its tools for achieving sustainable development in water and groundwater.
- Includes case studies and recent research directions for environmental issues in water sector.
- Addresses future aspects and innovation in AI field related to water sustainability.

This book will appeal to scientists, researchers, and undergraduate and graduate students majoring in environmental or computer science and industry professionals in water science and engineering, environmental management, and governmental sectors. It showcases artificial intelligence applications in detecting environmental issues, with an emphasis on the mitigation and conservation of water and underground resources.

Artificial Intelligence and Modeling for Water Sustainability

Global Challenges

Edited by
Alaa El Din Mahmoud, Manal Fawzy,
and Nadeem Ahmad Khan

CRC Press
Taylor & Francis Group
Boca Raton London New York

CRC Press is an imprint of the
Taylor & Francis Group, an **informa** business

Designed cover image: Dr. Alaa El Din Mahmoud

First edition published 2023
by CRC Press
6000 Broken Sound Parkway NW, Suite 300, Boca Raton, FL 33487–2742

and by CRC Press
4 Park Square, Milton Park, Abingdon, Oxon, OX14 4RN

CRC Press is an imprint of Taylor & Francis Group, LLC

© 2023 selection and editorial matter, Alaa El Din Mahmoud, Manal Fawzy,
and Nadeem Ahmad Khan; individual chapters, the contributors

Library of Congress Cataloging-in-Publication Data
Names: Mahmoud, Alaa El Din, editor.
Title: Artificial Intelligence and modeling for water sustainability : global challenges /
    edited by Alaa El Din Mahmoud, Manal Fawzy, Nadeem Khan.
Description: First edition. | Boca Raton : CRC Press, 2023. | Includes bibliographical references and index.
Summary: "With contributions from experts in academia and industries, who give a unified
    treatment of AI methods and their applications in water science, this book helps governments,
    industries, and homeowners not only address water pollution problems more quickly and
    efficiently, but also gain a better insight into the implementation of more effective remedial
    measures."—Provided by publisher.
Identifiers: LCCN 2022036465 (print) | LCCN 2022036466 (ebook) | ISBN 9781032186993 (hbk) |
    ISBN 9781032197074 (pbk) | ISBN 9781003260455 (ebk)
Subjects: LCSH: Water resources development—Data processing. | Water-supply—Data
    processing. | Artificial intelligence—Industrial applications. | Sustainable engineering.
Classification: LCC TC409 .A77 2023 (print) | LCC TC409 (ebook) | DDC 628.10285/63—dc23/
    eng/20221012
LC record available at https://lccn.loc.gov/2022036465
LC ebook record available at https://lccn.loc.gov/2022036466

ISBN: 978-1-032-18699-3 (hbk)
ISBN: 978-1-032-19707-4 (pbk)
ISBN: 978-1-003-26045-5 (ebk)

DOI: 10.1201/9781003260455

Typeset in Times
by Apex CoVantage, LLC

Contents

Preface

Over the past two decades, there has been significant growth in our understanding of wastewater treatment, with a shift from empirical methods towards a more theoretical, "first principles" approach that incorporates disciplines as diverse as chemistry, microbiology, physical and bioprocess engineering, and mathematics. Many of these advancements have reached a point of maturity, where they may be modeled mathematically and simulated on computers. Especially in developing countries, where advanced-level courses in wastewater treatment are not easily accessible, the quantity, complexity, and diversity of these new developments can be overwhelming for a new generation of young scientists and engineers entering the wastewater treatment profession. This book is an attempt to fill that void. It compiles and synthesizes graduate-level coursework from a half-dozen or so professors worldwide that have made substantial contributions to the improvement of wastewater treatment.

The fundamentals of environmental engineering as it applies to civil engineering include, but are not limited to, the global water crises, the various challenges, integrated water resources, the application of artificial intelligence to sustainability and management in water treatment, its advances, its application towards remediation, the use of sensors and models in water quality monitoring, and its implementation in the context of Industry 4.0 and the circular economy. This book presents the basics of artificial intelligence (AI) that even a novice reader should have no trouble understanding.

Every facet of our society is being revolutionized by the advent of digital technology, from the economy to the classroom to the world's natural resources, to the ways in which we organize our daily lives and interact with one another. In this context, the water industry is increasingly adopting digitalization to meet the expectations of its consumers, remain in compliance with regulations, boost performance, and guarantee the efficiency and dependability of its services. But as global change pressures like urbanization and climate change continue to rise, the water industry will face challenges in effectively managing scarcer and less reliable water resources. To address these issues, we need to rethink how we use (and reuse) water and increase our reliance on natural systems for water and wastewater treatment. These are the core concepts that will drive the necessary shift in water management. From this vantage point, digitalization emerges as a crucial enabler for facilitating this required and significant shift in the water sector. It lays the groundwork for us to think and act more intelligently, as evidenced by the plethora of ingenious digital reactions to the COVID-19 outbreak. The transition to digitalization, however, is an ongoing process. To move from opportunistic deployment of digital solutions to a more revolutionary one, the water sector needs to undergo cultural shifts in tandem with the incorporation of digital technologies. Artificial intelligence (AI) is an important component of this digital transformation of water infrastructure. As an example, it can help with process control, forecasting, and monitoring in real time as well as with predictive maintenance.

We firmly believe that the scientific community will embrace the concepts and findings presented in this book. The current state of affairs is not satisfactory, to say the least, and it needs to be rectified promptly, and I hope that the facts and knowledge offered will serve as a wake-up call to the general public, regulatory agencies, politicians, corporate leaders, and scientists.

Chapter 1 is a general introduction to global water challenges and sustainability. Water, as a priceless and natural resource, is essential for life. Today, the world is facing global water challenges that must be overcome to ensure our security and living conditions and those of future generations. This chapter summarizes the main water problems, such as water scarcity and overconsumption, outlining statistical data and future projections. Chapter 2 predicates that water is a vital and finite natural resource for all species of life, so it needs to be utilized efficiently. Any parameters such as physics, chemistry, or biology can affect the use of water for any purpose, including in aquaculture activities. Good water governance ensures the effective management of water resources. So the integrated water resources management (IWRM) approach aims to manage and develop the water resources to maximize the carrying capacity of waters in fish-farming activities. Monitoring various aspects of water quality in water resource management in real time can take advantage of using the Internet of Things (IoT). The IoT system should enable automated decision-making to control water quality factors measured with sensors set within tolerable limits and store historical monitoring results on a cloud-based platform. Chapter 3 introduces the current status and aims to provide information on the most recent technologies used in agricultural systems, precisely the various types of irrigation techniques, the driving forces behind the search for innovative irrigation techniques, and the artificial intelligence sensors that are commonly used for water preservation and conservation.

Chapter 4 describes research that has improved the accuracy of pollutant removal percentages by 84% to 90% using various models of artificial intelligence (AI) in water treatment, as well as perspectives on the future directions of novel research in the field, with a focus on pollution remediation, cost-effectiveness, energy economy, and water management as proper considerations. Chapter 5 highlights the use of artificial intelligence for water quality monitoring, emphasizing its advantages and the different techniques. The critical problems of using artificial intelligence techniques for intelligent water information extraction are also explored, as well as research goals. Chapter 6 compiles the measuring and monitoring of the critical indicators of water quality to aid researchers and environmentalists in achieving a delicate aquatic ecosystem, which notably affects the underwater flora and fauna, and implements policies to conserve the water domains and encourage biodiversity, the overall protection of the environment. Chapter 7 shows the availability of continuous, dynamic, and verifiable information supplements on the diversity and massive and complex data flow for monitoring the state of the aquatic ecosystem. Storage of these big data requires volumes greater than terabytes, which enables high speed and accuracy in accessing various analytical information in real time. Big data presents a new era that gathers all resources, time-based information, and meta- and deeper analysis. Therefore, it is essential to utilize big data's potential in identifying data gaps, likely vulnerabilities, and accurate predictions to maintain strategies for monitoring and mitigation of aquatic pollution for environmental sustainability. Chapter 8 is critical from a conceptual standpoint: mistakes are continuously made, and people get hurt even when there is enough scientific evidence that could be used to protect the general public adequately. Innovative technologies and approaches such as desalination, rainwater harvesting, water loss reduction, and intelligent water systems, including irrigation, dew harvesting, runoff capturing, wastewater reuse, and application of artificial intelligence (AI), have all been considered. This chapter reviews the status of current approaches to mitigate water shortages, with a significant focus on the role of AI in achieving a sustainable water supply. Chapter 9 discusses Industry 4.0 and its implementation in effective water and wastewater treatment. Chapter 10 deals with some recent experimental works that have been done on various sustainable adsorbents (with a particular focus on applying CS and GO-based materials) for the selective removal of heavy metals (HMs). In addition, it underlines and focuses on the role played by molecular modeling to characterize the efficiency of sustainable materials; in particular, the role played by CS to remove contamination by HMs. Chapter 11 discusses how water can be saved by applying circular economy (CE) models in water and wastewater sectors based on the principles of CE and application models, including those that use artificial intelligence (AI). Various operational benefits can result from implementing CE models, which have a basis of circularity in water, increasing efficiency and providing multiple revenue streams. The waste-to-wealth strategy with a circular water economy is highlighted. In addition to the operational benefits, the carbon footprint associated with water is considered. For further development, investment and a legal framework are needed.

In summary, the implementation of CE has both challenges and opportunities for the water and wastewater sectors. Chapter 12 discusses the state of the circular water economy in Morocco, particularly after regulatory reforms, programs, and the establishment of national strategies by the state. We will provide a comprehensive picture of the state of the circular water economy in Morocco, including the strengths and weaknesses, threats, and opportunities that are driving or hindering the successful transition of the circular water economy model in our kingdom. We will mention avenues of reflection and orientation to establish a model that can be standardized as a circular economy. This study aims to focus on the Moroccan model of the circular water economy that can create added value.

Editors
Dr. Alaa El Din Mahmoud
Prof. Manal Fawzy
Dr. Nadeem Ahmad Khan

Editors

Dr. Alaa El Din Mahmoud is an assistant professor in the Environmental Sciences Department, Faculty of Science, at Alexandria University and the co-founder of Green Technology Lab. Dr. Mahmoud is also the vice-chair of the National Committee of the UNESCO-MAB (Man and Biosphere) program in Egypt and a member of the alumni professional development unit in the Faculty of Science. He received his bachelor's and master's degrees from Alexandria University in Egypt and his PhD degree from Friedrich-Schiller University in Jena, Germany, in 2019. His research focuses on interdisciplinary environmental issues which are related to sustainability, natural resources, water/wastewater treatment, and green nanotechnology. Dr. Mahmoud's teaching activities are oriented to the aspects of environmental sciences (sustainability, remediation, climate change, land degradation, etc.) and environmental chemistry (water and wastewater, integrated waste management, etc.).

Dr. Mahmoud participated in many national and international workshops and conferences as a presenter, a session chair, and a member of the technical committee. He published international peer-reviewed publications with more than 30 scientific articles and conference papers as well as 15 book chapters. Moreover, he is involved in many collaborative funded research projects as a principal investigator, a consultant, and a team member. Currently, he is an editorial member of *Biochar* and *Carbon Research* journals in Springer. Among his current activities, Dr. Mahmoud is a reviewer for several prestigious journals in Elsevier, Springer, RSC society, and MDPI.

In 2022, Dr. Mahmoud was named the "Sustainable Development Ambassador" (top achiever) for the Egyptian Ministry of Planning and Economic Development, as well as a certified trainer. He has been awarded with the Best Young Researcher Award for his scientific contribution from the International Research and Academic Awards of the RanBal Research Institute & Bentham Science. Furthermore, he was awarded the Fulbright Award 2021 for "German Academic Exchange Service, DAAD Award," for conferences and traveling lectures abroad in two consecutive years, 2018–2019; the American Chemical Society Award for Environmental Chemistry Division in 2018; Misr El Kheir Foundation award for postgraduates in 2014; and the "Young Scientists Award" from UNESCO-MAB program in 2013. He was recognized by the Faculty of Science, Alexandria University, during the Day of Excellence for scientific contributions in "teaching and research." During his career, he received fellowship awards from Temple University (2022), University of Texas at Austin (2021), Shanghai University (2019), University of Luxembourg (2018), and TU Berlin Campus El Gouna, Egypt (2015).

Manal Fawzy is one of the founders of the Department of Environmental Sciences, Faculty of Science, Alexandria University, where she is currently acting as a full professor. She is the team leader of Green Technology Lab, chair of the Egyptian National Man and Biosphere Commission to UNESCO, and member of the IUCN (International Union for Conservation of Nature) Commission on Education and Communication (CEC). Prof. Fawzy has obtained her PhD in plant biosystematics from Alexandria University, in conjunction with the Biosystematics Research Centre, Ottawa, Agriculture Canada, in 1991. Afterward, she has dedicated her research to the conservation of natural resources, in the use of indigenous plant species for the remediation of polluted environments, and recently, in the phytosynthesis of nanomaterials for environmental applications. Prof. Fawzy has been awarded the UNESCO Young Scientist Award (1991), Sultan Qaboos Prize for Environmental Preservation (1997), University of Alexandria Prize of Scientific Distinction (2016), and Alexandria University Prize for Recognition (2019). Prof. Fawzy has implemented and participated in many research projects in the field of restoration of degraded ecosystems, biodiversity conservation, and green biotechnology. She has also undertaken environmental consultations for some national, regional, and international authorities or organizations in the fields of biodiversity monitoring and environmental impact assessment. Dr. Fawzy is a reviewer and referee for many internationally sound scientific journals. Prof. Fawzy has published 60 scientific articles and nine book chapters in peer-reviewed journals and national and international conferences.

Dr. Nadeem Ahmad Khan has been an assistant professor in the Department of Civil Engineering since August 1, 2012, in Mewat Engineering College, Nuh. He attained his PhD from Jamia Millia Islamia, New Delhi. He also acted as the academic coordinator of the Civil Engineering Department from 2014 to 2017. He is currently serving as an editorial board member of the Nature journal, Scientific Reports (IF = 4.3), and the Springer journal special issue of Discover Water and is working for the Hindawi journal special issue (IF = 4.2). His recent accomplishments include 15 Coursera course completion certificates. They were awarded the Best Young Scientist Award (Male) by LWT India and GECL for their scientific contribution to environmental engineering. He is also the external examiner and paper setter at Integral University, Lucknow, and Dehradun Technical University since 2013. He got his MTech (civil engineering) degree in 2011 with honors in environmental engineering from Aligarh Muslim University. Dr. Khan was associated with the Ministry of Environment and Forest for two years as a junior research fellow (JRF) in 2009 to 2011 and also worked with UNICEF as a research assistant (RA) on the topic of GIS mapping and water quality parameter estimation, including monitoring of the Agra Region for six months in 2012. He has professional experience in DLF-Laing's Rourke India Ltd. in a SEZ

Project at Gurgaon as a construction engineer during 2008–2009. He qualified for the GATE examinations in 2010. His industry activity areas were postcontract works, reconciliation, planning, and budgeting. Recently, his one chapter was published in Lenin Publication about SBR wastewater treatment technology. He has attended several conferences, workshops, seminars, and research and engaged in research projects. He has published three books with international publishers, 55 papers in SCI, and 15 in national journals. He has also published more than ten patents in civil engineering and water treatment. He is also active in social welfare activities. He is the joint secretary of the Environmental Awareness Society Environsense, which has organized more than 25 events.

Contributors

Abrahan Mora
Tecnológico de Monterrey
Escuela de Ingeniería y Ciencias
Campus Puebla
Puebla, México

Ahmed M. Abdelfatah
Environmental Sciences Department
Faculty of Science
Alexandria University
Alexandria, Egypt

Alaa El Din Mahmoud
Environmental Sciences Department
Faculty of Science
Alexandria University
Alexandria, Egypt

Alan Guajardo
Tecnológico de Monterrey
Escuela de Ingeniería y Ciencias
Campus Monterrey
Nuevo León, México

Asha Ripanda
Chemistry Department
School of Physical Sciences
College of Natural and Mathematical
    Sciences
University of Dodoma
Dodoma, Tanzania

Benton Otieno
Department of Civil Engineering
Vaal University of Technology
Vanderbijlpark, South Africa

Charles Ikenna Osu
Department of Pure and Industrial
    Chemistry
University of Port Harcourt
Rivers State, Nigeria

Chidi Obi
Faculty of Science
Physical Chemistry Unit
Department of Pure and Industrial
    Chemistry
University of Port Harcourt
Rivers State, Nigeria

El-Shimaa A. Rawash
Regional Center for Food and Feed
Agricultural Research Center
Giza, Egypt

Eyas Mahmoud
Department of Chemical and Petroleum
    Engineering
United Arab Emirates University
Al Ain, United Arab Emirates

Faustin Ngassapa
Chemistry Department
College of Natural and Applied
    Sciences
University of Dar es Salaam
Dar es Salaam, Tanzania

Fredrick Ojija
Department of Applied Sciences
Mbeya University of Science and
    Technology
Mbeya, Tanzania

Ghadir A. El-Chaghaby
Regional Center for Food and Feed,
    Agricultural Research Center
Giza, Egypt

Gloria Ukalina Obuzor
Department of Pure and Industrial
    Chemistry
University of Port Harcourt
Rivers State, Nigeria

Hanae Hamidi
Abdelmalek Essaadi University
Faculty of Science, Department of
    Geology
Tetouan, Morocco

Heru Suryanto
Department of Mechanical
    Engineering
Faculty of Engineering
Universitas Negeri Malang
Malang, Indonesia

Houda El Haddad
Abdelmalek Essaadi University
Faculty of Science
Department of Physics
Tetouan, Morocco

Manal Fawzy
Environmental Sciences Department
Faculty of Science
Alexandria University
Alexandria, Egypt

Marco A. Mata-Gómez
Tecnológico de Monterrey
Escuela de Ingeniería y Ciencias
Campus Puebla
Puebla, México

Miraji Hossein
Chemistry Department
School of Physical Sciences
College of Natural and Mathematical
    Sciences
University of Dodoma
Dodoma, Tanzania

Mudiaga Chukunedum Onojake
Centre for Marine Pollution Monitoring
    and Seafood Safety
Faculty of Science
University of Port Harcourt
Rivers State, Nigeria

Muhammad Abdul Moneem
El-Fostat Laboratory
Cairo Water Company
Cairo, Egypt

Mureithi Eunice
MathematicsDepartment
College of Natural and Applied Sciences
University of Dar es Salaam
Dar es Salaam, Tanzania

Murthy Chavali
Department of Science
Faculty of Science & Technology
Alliance University (Central Campus)
Bengaluru, Karnataka, India

Mohamed Hosny
Environmental Sciences Department
Faculty of Science
Alexandria University
Alexandria, Egypt

Nadeem Ahmad Khan
Department of Civil Engineering
Jamia Millia Islamia
New Delhi, India

Nourhan El Maghrabi
Environmental Sciences Department
Faculty of Science
Alexandria University
Alexandria, Egypt

Nico Rahman Caesar
Doctoral Program of Environmental
    Science
University of Brawijaya, Malang, Jl
Malang—East Java, Indonesia

Othman Chande
Chemistry Department
College of Natural and Applied Sciences
University of Dar es Salaam
Dar es Salaam, Tanzania

Pabel Cervantes-Avilés
Tecnológico de Monterrey
Escuela de Ingeniería y Ciencias
Reserva Territorial Atlixcáyotl
Puebla, México

Patrycja Krasucka
Department of Radiochemistry
    and Environmental Chemistry
Faculty of Chemistry, Maria
    Curie—Sklodowska University
Lublin, Poland

S. Abdalla
Department of Materials Science
Institute of Graduate Studies
    and Research
Alexandria University, Alexandria,
    Egypt and
Department of Physics
Faculty of Science
King Abdulaziz University
Jeddah, Saudi Arabia

Sayed Rashad
Regional Center for Food and Feed
Agricultural Research Center
Giza, Egypt

Sherif Kandil
Department of Materials Science
Institute of Graduate Studies and
    Research
Alexandria University, Alexandria, Egypt

Soliman R. Radwan
Regional Center for Food and Feed
Agricultural Research Center
Giza, Egypt

Uun Yanuhar
Aquatic Resources Management
Faculty of Fisheries and Marine Sciences
University of Brawijaya, Malang, Jl
Malang—East Java, Indonesia

Wessam S. Omara
Department of Materials Science
Institute of Graduate Studies and Research
Alexandria University
Alexandria, Egypt

Yung-Tse Hung
Department of Civil and Environmental
    Engineering
Cleveland State University
Cleveland, Ohio, USA

1 Global Water Challenges and Sustainability

Alaa El Din Mahmoud and Patrycja Krasucka

CONTENTS

1.1 INTRODUCTION

The surface of the blue planet Earth is covered by 71% water, but it should be pointed out that it contains only 2.5% fresh water, of which 80% is stored up in glaciers, leaving 20% (or 0.5% of fresh water) available in the world's rivers, lakes, and aquifers (Antoniou et al. 2005). With a rapidly rising and urbanizing population, worldwide water usage is likely to grow; freshwater extraction rates in many regions of the world exceed natural recharge rates. Water consumption is usually different according to the sector type and its needs. Figure 1.1 shows the global water consumption percentage in each sector and the water consumption percentage in Africa and Europe (Gonzalez Sanchez et al. 2020; Statista 2022).

The agricultural sector is considered the primary source of global water consumption so far (Dolan et al. 2021). The increment in food production to meet the requirements of future population will cause having to use the same water resources, which will deteriorate water quality (Mishra et al. 2021; Sawicka et al. 2021; Dotaniya et al. 2022). Around the globe, one-third of utilities reported a loss of >40% of clean water due to leaks in water infrastructure (Mounce et al. 2015). Furthermore, climate change impacts come up with enormous consequences in the water sector, increasing health risks, pollution, and considerable economic impacts (Mansoor et al. 2022; Ratnaweera et al. 2022). Around 83% of disasters are triggered by climate-related events. Aldaya et al. (2022) listed the top challenges that affect the water sector according to category. For instance, (1) climate change, which falls under the environmental category; (2) aging infrastructure, which falls under the technical category; (3) infrastructure capital lack, which falls under the economic category; and (4) population growth and rapid urbanization, which fall under the social category.

Water scarcity as a main water challenge is spreading and worsening around the world as demand for water rises (Mahmoud et al. 2021). It is expected that more

DOI: 10.1201/9781003260455-1

1

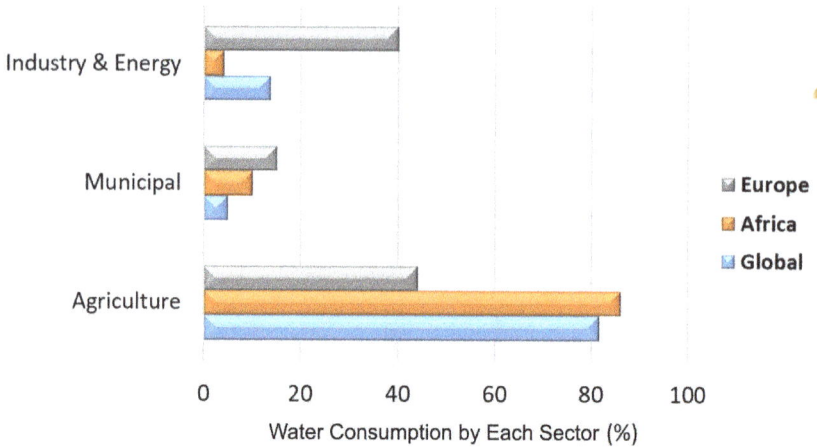

FIGURE 1.1 Water consumption percentage by sector.

humans will face water scarcity, with population expected to rise from 933 million people in 2016 to about 2 billion people in 2050 (He et al. 2021). Such developments in socioeconomic sectors and urbanization would increase domestic and industrial water consumption by 50–80% over the next 30 years (Garrick et al. 2019). *Water scarcity* can be defined as a shortfall of renewable freshwater supply compared to demand, but a more specific definition is needed to create a reliable quantitative indicator. A gauge like this would track and assess efforts in reducing water scarcity, as well as predict where and when it could occur in the future (Damkjaer et al. 2017). Water scarcity is one of the most urgent issues that must be discussed. More than 67% of the world's population face severe water scarcity. In fact, 87 out of 180 countries are expected to be water-scarce by 2050 (Baggio et al. 2021). The global water-scarce areas are identified through the calculations of water scarcity index (WSI) annually and monthly using the following equation, where *WSI* is the total water withdrawal (*TWW* = the water withdrawals sum) from available water resources (*AWR*) (He et al. 2021).

$$WSI = \frac{TWW}{AWR}$$

Water scarcity is exacerbated by the fact that limited natural water resources are becoming polluted. According to projections, worldwide water demand will rise by 20–30% by 2050 unless consumption patterns change substantially (WWAP 2019). More than half of the world's population will be at risk of water scarcity by then. By 2030, severe water scarcity could force 700 million people to relocate. Water scarcity hinders the progress and achievement of the Sustainable Development Goals (SDGs), especially its goal 11, "Sustainable Cities and Communities," and goal 6, "Clean Water and Sanitation" (Larsen et al. 2016). Maintaining water security is also one of major global water challenges. In general, *water security* is defined as:

[Ensuring that freshwater, coastal, and related ecosystems are secured and enhanced, that sustainable development and political stability are supported, that everyone has access to plenty of safe water at an affordable cost to live a healthy and productive life and that the susceptible are safeguarded from the risks of water-related hazards].

(Biswas et al. 2016)

Water security is often understood to imply having and being able to consistently deliver sufficient water of the correct quality where and when you need it for all purposes, including agriculture, but also reasons connected to sustainable natural biodiversity-based Earth system function. It also entails ensuring that your use and management of water in your region does not have a major impact on the water security of regions upstream or downstream from you now or in the future. Water security today entails managing not only the water you always assumed was always accessible but also water amid even greater extremes of excess and shortage than we have seen in the past (Sandford 2017).

Water security includes technical, economic, environmental, social, and legal perspectives. The state of the public field (or the public domain) of water reserves, addressed in most laws, provides public administrations to achieve public benefit aims. This state has been the foundation of water authority consolidation and has advanced mobilization policies that put water at the service of economic and social improvements. This is especially true in water-scarce nations, where water management programs have resulted in near-total control of surface- and groundwater resources in some areas, sometimes with excessive water extraction, posing a danger to the resources' long-term viability (Besbes et al. 2019). Figure 1.2 shows the number of articles published in the last 22 years that are related to water scarcity and water security, which proves the global concern of such issues.

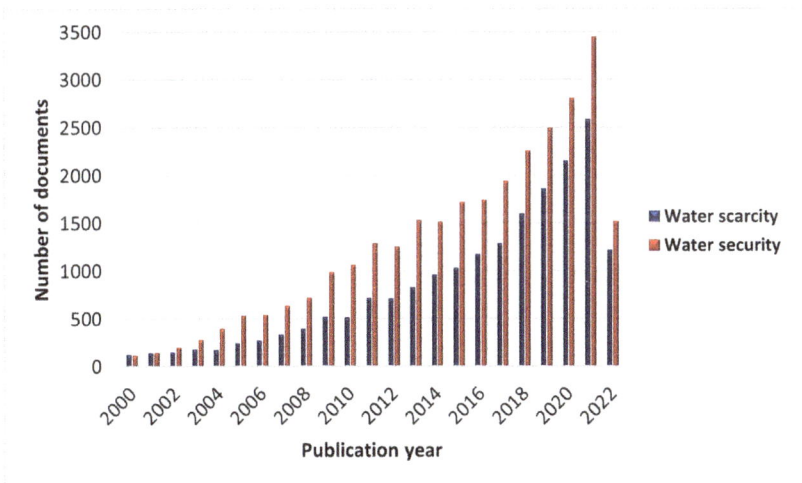

FIGURE 1.2 Water security and water scarcity publications in the last 22 years.

Source: Data retrieved from Scopus.

Human-caused water pollution harms people's health, the economy, and the environment, while also jeopardizing the long-term viability of water sources. Due to poor sanitation, a lack of wastewater treatment by residential and industrial users, and dirty runoff from agriculture and storm drains, water quality in natural bodies is deteriorating. Pollution is exacerbated by unregulated discharges. Around 80% of the world's wastewater, and up to 95% in some underdeveloped nations, is still discharged into the environment. Therefore, "water quality is just as crucial as water quantity," and the spread of water pollution with the search for effective wastewater treatment methods is also an urgent global challenge (Mahmoud et al. 2020). The following sections highlight emerging water pollution and sustainability issues in further details, as well as artificial intelligence and machine learning models that link these two topics in the context of global water challenges.

1.2   WATER POLLUTION OVERVIEW

Water pollution is a well-known problem, and the removal of "common" hazardous pollutants, like dyes, pesticides, heavy metals, etc., is already well recognized (Saravanan et al. 2021; Ahmed et al. 2022; Mahmoud 2022; Mahmoud et al. 2022c); but humans, as well as the entire ecosystem, through different known and unknown sources, are now exposed to different contaminants (Zhu et al. 2019; Chaukura et al. 2022). Emerging contaminants (ECs) have been recognized as major pollutants of water that adversely affect endocrine systems in humans and wildlife (Rodriguez-Narvaez et al. 2017). The term "emerging contaminants" does not only inevitably relate to new chemicals/substances introduced and their degradation products/metabolites or by-products but also refers to compounds with previously unrecognized adverse effects on the ecosystems (Petrovic et al. 2011). There are a number of different types of ECs, like antibiotics (Krasucka et al. 2021; Vambol et al. 2021; Gwenzi et al. 2022), pesticides (Mahmoud et al. 2022b), drugs (Khan et al. 2021a; 2021b), personal care products, and recent nanomaterials (Naidu et al. 2016; Mahmoud and Fawzy 2021). Richardson and Kimura (2020) include a wide variety of compounds in their review to the EC group, including sucralose and other artificial sweeteners, nanomaterials, perfluorinated compounds, drinking water and swimming pool disinfection by-products, sunscreens and UV filters, flame retardants, benzotriazoles and benzothiazoles, siloxanes, naphthenic acids, algal toxins, and ionic liquids, and prions. Many chemical and microbial agents that were not traditionally considered contaminants can be found in various environmental compartments and/or in areas where they were never used, mainly due to their persistence during long-distance transport. The sources and pathways of these emerging contaminants can be increasingly associated with waste and wastewaters resulting from industrial, agricultural, or municipal activities (Badr et al. 2020; Huang et al. 2021; Mahmoud et al. 2022c). With advancements in the chemical industry, the variety of compounds being released to the environment which are potentially harmful to humans and the ecosystem over the long term is expected to grow significantly over the years (Mahmoud et al. 2018b; Fungaro et al. 2021). In recent years, scientists have indicated an intensified interest in monitoring ECs, but little agreement is presented on the list of substances that should be monitored (Rodriguez-Narvaez et al. 2017). The concentrations of most ECs in the Asian region are often higher than in the EU and North America (Tran et al. 2018). Due to the fact that existing wastewater treatment system is very often ineffective at removing such various groups of emerging pollutants, as well, they have not been regulated due to a lack of strict legislation relevant to these contaminants (El Din Mahmoud and Fawzy 2016; Mahmoud et al. 2018a; Mahmoud 2020; Mahmoud et al. 2022d). Although the majority of ECs are still under investigation, with the aim of finding sustainable ways to eliminate or treat them (Richardson and Kimura 2020), ensuring a safe water supply requires efforts not only from science but also from policy and governments.

1.3   WATER AS A KEY TO SUSTAINABLE DEVELOPMENT

Water is critical for socioeconomic development and has a role in practically all the SDGs. Safe drinking water and sanitation are essential for a healthy and productive society. Healthy ecosystems and biodiversity rely on water. Water is also necessary for the production of food and energy, as well as most industrial activities; therefore, a lack of access to it means slower economic growth. Some regions' growth rates could drop by as much as 6% of GDP.

Following are the key points of water relation to sustainable development:

a. Improved sanitation is unavailable to more than half of the world's rural population.
b. By 2035, the amount of water used in energy generation is expected to increase by 85% due to global energy consumption, which is expected to increase 35% between 2010 and 2035 (Figure 1.3).
c. In low-income countries, agriculture consumes 90% of water resources. Keep in mind that more than 70% of fresh water extracted is used for irrigation and agriculture around the world. For further information, refer to Mahmoud and Fawzy (2021).

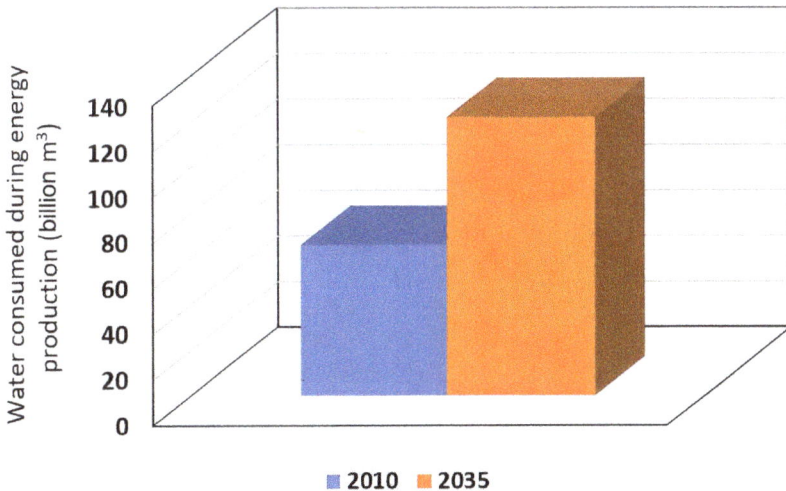

FIGURE 1.3   Water consumption during energy production globally.

Source: Adapted from www.iea.org/.

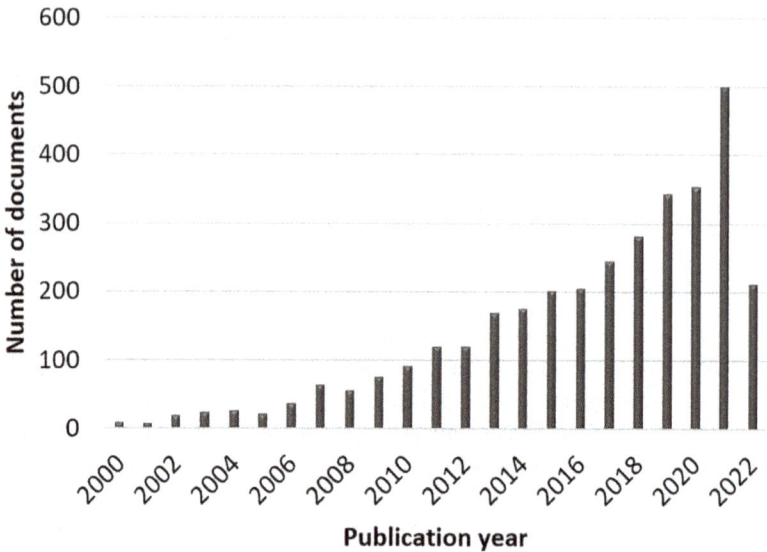

FIGURE 1.4 Number of publications each year from 2000 to 2022 in Scopus database.

Source: Retrieved the published papers using the TITLE-ABS-KEY of "sustainable* AND water global challenges." Publications published in 2022 were included till June 2022.

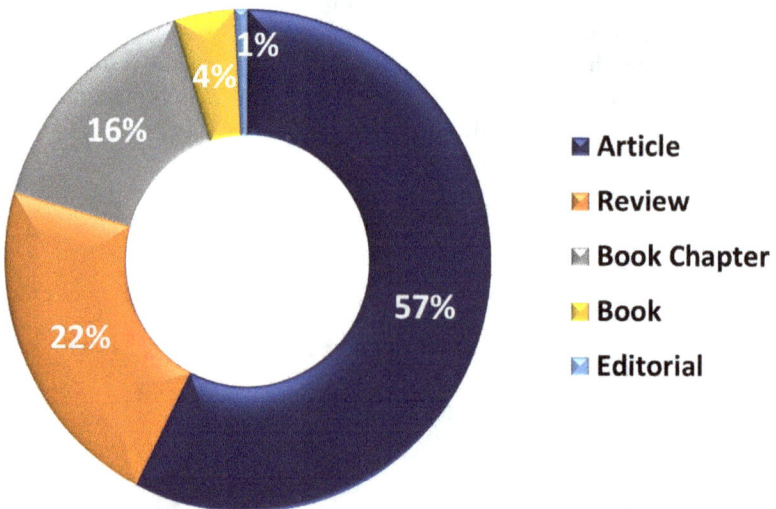

FIGURE 1.5 Distribution of publication types from 2000 to 2022 in Scopus database.

Figure 1.4 illustrates the number of published paper (about 3,379 documents from 2000 to 2022) regarding both sustainability and global water challenges, demonstrating that these issues are interrelated, and that solving SDG 6 problems is the need of the hour and involves many researchers from different fields. In further details, Figure 1.5 demonstrates most publication types are articles (57%), followed

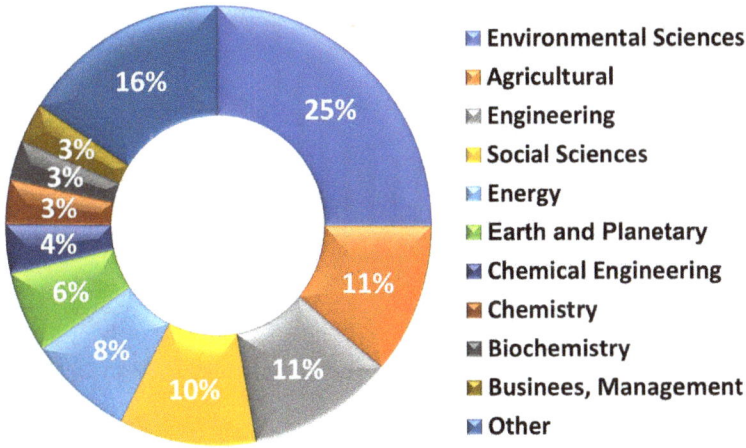

FIGURE 1.6 Percentage of publications in each subject area.

by reviews (22%). Book chapters are ranked third with 16% of all publication types. Most of these publications are related to environmental sciences subject (~ 25%), followed by agricultural (~ 11%) and engineering (~ 11%), as well as social sciences (~ 10%) (Figure 1.6). This indicates the relevance of water sustainability in different perspectives, which requires a holistic approach to find sustainable solutions.

1.4 ARTIFICIAL INTELLIGENCE AND MODELS AS A TOOL IN FIGHT WITH GLOBAL WATER CHALLENGES

In brief, the major global water challenge is to ensure adequate quantity and quality of water for all people—it seems so simple, yet it is so difficult to meet these two issues in real life. As reported in literature (Figure 1.7), science is not helpless in this fight, and thanks to new knowledge provided by information technology, especially artificial intelligence (AI) and machine learning models (MLM), new methods and solutions to these problems are developed. AI based on statistical models that were trained with so-called big data refers to the algorithms with the ability to perform tasks and make inferences, while ML acts as intelligent systems that can learn/adapt during the training stage with newly received data (Doorn 2021; Lowe et al. 2022)

   AI and ML have been used in a wide range of water quality studies (Lowe et al. 2022). From modeling, monitoring, and/or predicting of pollution (El Bilali et al. 2021; Che et al. 2022) and/or composition/properties (salinity, pH, etc. [Liang 2021; Mahmoud et al. 2022a]) of natural waters (e.g., rivers, groundwater) to the investigation of wastewater aimed to control and/or improve optimalization of common wastewater treatment processes, such as a membrane filtration, coagulation, and chlorination, as well as adsorption of various heavy metals and organic water contaminants on diverse adsorbents (carbon, clay, plant waste) (Lowe et al. 2022). By using AI and ML to optimize/design the wastewater treatment process (resulted in, e.g., decrease using chemicals or increase application of waste-derived adsorbent),

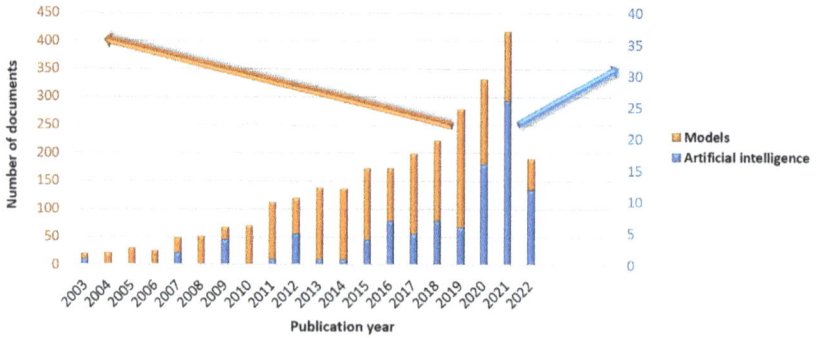

FIGURE 1.7 Number of available publications each year from 2003 to 2022 in Scopus database.

Source: Retrieved the published papers using the TITLE-ABS-KEY of "sustainable* AND water global challenges." Publications published in 2022 were included till June 2022.

it is possible not only to reduce costs and improve safety and efficiency of wastewater treatment plants but also to comply with the idea of sustainability and circular economy. In addition to water quality testing, AI and ML techniques can be successfully used in water quantity control and management. As was mentioned earlier, 70% of the water is used for irrigation in crop production, and due to increasing population and food demand, it is expected that agricultural water consumption will be constantly rising (80% in 2030) (El Bilali et al. 2021). Therefore, ML and AI are involved in the creation of schedules and whole irrigation systems to sustain and manage when, where, and how much to irrigate, by estimating the specified crop-type water demand (depending on the climate, soil conditions, etc.), with maximizing water-use efficiency and crop yield (Sidhu et al. 2020; Glória et al. 2021; Dehghanisanij et al. 2022). Not only farming but also the urban part of water management can be supported by AI and ML. According to the estimation of the World Bank and the International Water Association in developing countries, 45 million m³ of water are lost per day, which costs over US$3 billion per year. To minimalize this problem, ML coupled with sensors (capable of high-resolution monitoring) are involved in leak detection in water supply network, as well as overall water consumption control (Mounce et al. 2015; Alves Coelho et al. 2020; Fan et al. 2021; Bhardwaj et al. 2022). In global context, ML and AI gain importance also in hydrology, starting from groundwater-level prediction, crucial to the management of water resources (Tao et al. 2022), to the estimation of precipitation and related forecasting of natural disasters, such as floods or droughts (Schmidt et al. 2020; Almikaeel et al. 2022; Mosaffa et al. 2022), which is essential not only for the protection of life through early warning of floods but also for the economic aspect, through analysis and prediction of droughts or efficient management of hydropower plants, which is strictly related to the sustainable development strategy. As all aforementioned examples demonstrate, AI and ML are multidisciplinary approaches with the potential to be successfully applied in improving the quality and quantity of global water resources.

1.5 CONCLUSION AND RECOMMENDATIONS

Tackling water scarcity while ensure water safety is a main global problem. Rational water management in all domains (agriculture, industry, and daily life), estimating and caring for natural water resources, improving water quality by reducing pollution and increasing the efficiency of water treatment, and generally following the principles of sustainable development are the most appropriate ways to reduce these problems. Such complex actions require a combination of knowledge and experience from many research fields, which can be challenging. Artificial intelligence and machine learning models as multidisciplinary tools are one of the new proposed solutions to fight global water challenges. However, it should be kept in mind that they have limitations. One of the main problems in AI and ML applications is collecting and selecting the right dataset for the "learning" process. Aquatic/hydrological data are often selected for specific time, conditions, and locations, making it difficult to standardize models and compare received results. Thus, it requires real-time and reliable global database. We recommend interdisciplinary efforts in research, industry, policy, and practice to develop alternatives sustainable solutions.

REFERENCES

Ahmed, M., Mavukkandy, M.O., Giwa, A., Elektorowicz, M., Katsou, E., Khelifi, O., Naddeo, V., Hasan, S.W., 2022. Recent developments in hazardous pollutants removal from wastewater and water reuse within a circular economy. *NPJ Clean Water* 5, 12.

Aldaya, M.M., Sesma-Martín, D., Schyns, J.F., 2022. Advances and challenges in the water footprint assessment research field: Towards a more integrated understanding of the water—energy—food—land nexus in a changing climate. *Water* 14, 1488.

Almikaeel, W., Čubanová, L., Šoltész, A., 2022. Hydrological drought forecasting using machine learning—gidra river case study. *Water* 14, 387.

Alves Coelho, J., Glória, A., Sebastião, P., 2020. Precise water leak detection using machine learning and real-time sensor data. *IoT* 1, 474–493.

Antoniou, M.G., De La Cruz, A.A., Dionysiou, D.D.D., 2005. Cyanotoxins: New generation of water contaminants. *Journal of Environmental Engineering* 131, 1239–1243.

Badr, N.B.E., Al-Qahtani, K.M., Mahmoud, A.E.D., 2020. Factorial experimental design for optimizing selenium sorption on *Cyperus laevigatus* biomass and green-synthesized nano-silver. *Alexandria Engineering Journal* 59, 5219–5229.

Baggio, G., Qadir, M., Smakhtin, V., 2021. Freshwater availability status across countries for human and ecosystem needs. *Science of the Total Environment* 792, 148230.

Besbes, M., Chahed, J., Hamdane, A., 2019. Elements for a conceptual model of national water security. National Water Security: Case Study of an Arid Country: Tunisia. Springer International Publishing, Cham, pp. 257–268.

Bhardwaj, A., Dagar, V., Khan, M.O., Aggarwal, A., Alvarado, R., Kumar, M., Irfan, M., Proshad, R., 2022. Smart IoT and machine learning-based framework for water quality assessment and device component monitoring. *Environmental Science and Pollution Research* 29, 46018–46036.

Biswas, A.K., Tortajada, C., 2016. Water Security, Climate Change and Sustainable Development. Springer, Singapore.

Chaukura, N., Muzawazi, E.S., Katengeza, G., Mahmoud, A.E.D., 2022. Chapter 20—Remediation technologies for contaminated soil systems. In: Gwenzi, W. (Ed.). Emerging Contaminants in the Terrestrial-Aquatic-Atmosphere Continuum. Elsevier, pp. 353–365.

Che, Y., Wei, Z., Wen, J., 2022. Construction of surface water pollution prediction model based on machine learning. *Scientific Programming* 2022, 7211094.

Damkjaer, S., Taylor, R., 2017. The measurement of water scarcity: Defining a meaningful indicator. *Ambio* 46, 513–531.

Dehghanisanij, H., Emami, H., Emami, S., Rezaverdinejad, V., 2022. A hybrid machine learning approach for estimating the water-use efficiency and yield in agriculture. *Scientific Reports* 12, 6728.

Dolan, F., Lamontagne, J., Link, R., Hejazi, M., Reed, P., Edmonds, J., 2021. Evaluating the economic impact of water scarcity in a changing world. *Nature Communications* 12, 1915.

Doorn, N., 2021. Artificial intelligence in the water domain: Opportunities for responsible use. *Science of the Total Environment* 755, 142561.

Dotaniya, M.L., Meena, V.D., Saha, J.K., Dotaniya, C.K., Mahmoud, A.E.D., Meena, B.L., Meena, M.D., Sanwal, R.C., Meena, R.S., Doutaniya, R.K., Solanki, P., Lata, M., Rai, P.K., 2022. Reuse of poor-quality water for sustainable crop production in the changing scenario of climate. *Environment, Development and Sustainability*.

El Bilali, A., Taleb, A., Brouziyne, Y., 2021. Groundwater quality forecasting using machine learning algorithms for irrigation purposes. *Agricultural Water Management* 245, 106625.

El Din Mahmoud, A., Fawzy, M., 2016. Bio-based methods for wastewater treatment: Green sorbents. In: Ansari, A.A., Gill, S.S., Gill, R., Lanza, G.R., Newman, L. (Eds.). Phytoremediation: Management of Environmental Contaminants, Volume 3. Springer International Publishing, Cham, pp. 209–238.

Fan, X., Zhang, X., Yu, X., 2021. Machine learning model and strategy for fast and accurate detection of leaks in water supply network. *Journal of Infrastructure Preservation and Resilience* 2, 10.

Fungaro, D.A., Silva, K.C., Mahmoud, A.E.D., 2021. Aluminium tertiary industry waste and ashes samples for development of zeolitic material synthesis. *Journal of Applied Materials Technology* 2, 66–73.

Garrick, D., De Stefano, L., Yu, W., Jorgensen, I., O'Donnell, E., Turley, L., Aguilar-Barajas, I., Dai, X., de Souza Leão, R., Punjabi, B., 2019. Rural water for thirsty cities: A systematic review of water reallocation from rural to urban regions. *Environmental Research Letters* 14, 043003.

Glória, A., Cardoso, J., Sebastião, P., 2021. Sustainable irrigation system for farming supported by machine learning and real-time sensor data. *Sensors* 21, 3079.

Gonzalez Sanchez, R., Seliger, R., Fahl, F., De Felice, L., Ouarda, T.B.M.J., Farinosi, F., 2020. Freshwater use of the energy sector in Africa. *Applied Energy* 270, 115171.

Gwenzi, W., Selvasembian, R., Offiong, N.-A.O., Mahmoud, A.E.D., Sanganyado, E., Mal, J., 2022. COVID-19 drugs in aquatic systems: A review. *Environmental Chemistry Letters* 20, 1275–1294.

He, C., Liu, Z., Wu, J., Pan, X., Fang, Z., Li, J., Bryan, B.A., 2021. Future global urban water scarcity and potential solutions. *Nature Communications* 12, 4667.

Huang, L., Yan, T., Mahmoud, A.E.D., Li, S., Zhang, J., Shi, L., Zhang, D., 2021. Enhanced water purification via redox interfaces created by an atomic layer deposition strategy. *Environmental Science: Nano* 8, 950–959.

Khan, A.H., Abutaleb, A., Khan, N.A., El Din Mahmoud, A., Khursheed, A., Kumar, M., 2021a. Co-occurring indicator pathogens for SARS-CoV-2: A review with emphasis on exposure rates and treatment technologies. *Case Studies in Chemical and Environmental Engineering* 4, 100113.

Khan, A.H., Tirth, V., Fawzy, M., Mahmoud, A.E.D., Khan, N.A., Ahmed, S., Ali, S.S., Akram, M., Hameed, L., Islam, S., Das, G., Roy, S., Dehghani, M.H., 2021b. COVID-19 transmission, vulnerability, persistence and nanotherapy: A review. *Environmental Chemistry Letters* 9, 2773–2787.

Krasucka, P., Pan, B., Sik Ok, Y., Mohan, D., Sarkar, B., Oleszczuk, P., 2021. Engineered biochar—A sustainable solution for the removal of antibiotics from water. *Chemical Engineering Journal* 405, 126926.

Larsen, T.A., Hoffmann, S., Lüthi, C., Truffer, B., Maurer, M., 2016. Emerging solutions to the water challenges of an urbanizing world. *Science* 352, 928–933.

Liang, L., 2021. Water pollution prediction based on deep belief network in big data of water environment monitoring. *Scientific Programming* 2021, 8271950.

Lowe, M., Qin, R., Mao, X., 2022. A review on machine learning, artificial intelligence, and smart technology in water treatment and monitoring. *Water* 14, 1384.

Mahmoud, A.E.D., 2020. Graphene-based nanomaterials for the removal of organic pollutants: Insights into linear versus nonlinear mathematical models. *Journal of Environmental Management* 270, 110911.

Mahmoud, A.E.D., 2022. Recent advances of TiO2 nanocomposites for photocatalytic degradation of water contaminants and rechargeable sodium ion batteries. In: Shalan, A.E., Hamdy Makhlouf, A.S., Lanceros-Méndez, S. (Eds.). Advances in Nanocomposite Materials for Environmental and Energy Harvesting Applications. Springer International Publishing, Cham, pp. 757–770.

Mahmoud, A.E.D., Fawzy, M., 2021. Nanosensors and nanobiosensors for monitoring the environmental pollutants. In: Makhlouf, A.S.H., Ali, G.A.M. (Eds.). Waste Recycling Technologies for Nanomaterials Manufacturing. Springer International Publishing, Cham, pp. 229–246.

Mahmoud, A.E.D., Fawzy, M., Abdel-Fatah, M.M.A., 2022a. Technical Aspects of nanofiltration for dyes wastewater treatment. In: Muthu, S.S., Khadir, A. (Eds.). Membrane Based Methods for Dye Containing Wastewater: Recent Advances. Springer Singapore, Singapore, pp. 23–35.

Mahmoud, A.E.D., Franke, M., Braeutigam, P., 2022d. Experimental and modeling of fixed-bed column study for phenolic compounds removal by graphite oxide. *Journal of Water Process Engineering* 49, 103085.

Mahmoud, A.E.D., Fawzy, M., Khairy, H., Sorour, A., 2022b. Environmental bioremediation as an eco-sustainable approach for pesticides: A case study of mena region. In: Siddiqui, S., Meghvansi, M.K., Chaudhary, K.K. (Eds.). Pesticides Bioremediation. Springer, Switzerland, p. 479.

Mahmoud, A.E.D., Franke, M., Stelter, M., Braeutigam, P., 2020. Mechanochemical versus chemical routes for graphitic precursors and their performance in micropollutants removal in water. *Powder Technology* 366, 629–640.

Mahmoud, A.E.D., Kathi, S., 2022c. Chapter 7 — Assessment of biochar application in decontamination of water and wastewater. In: Kathi, S., Devipriya, S., Thamaraiselvi, K. (Eds.). Cost Effective Technologies for Solid Waste and Wastewater Treatment. Elsevier, pp. 69–74.

Mahmoud, A.E.D., Stolle, A., Stelter, M., 2018a. Sustainable synthesis of high-surface-area graphite oxide via dry ball milling. *ACS Sustainable Chemistry & Engineering* 6, 6358–6369.

Mahmoud, A.E.D., Stolle, A., Stelter, M., Braeutigam, P., 2018b. Adsorption technique for organic pollutants using different carbon materials. In Abstracts of Papers of the American Chemical Society. American Chemical Society, Washington, DC.

Mahmoud, A.E.D., Umachandran, K., Sawicka, B., Mtewa, T.K., 2021. 26 — Water resources security and management for sustainable communities. In: Mtewa, A.G., Egbuna, C. (Eds.). Phytochemistry, the Military and Health. Elsevier, pp. 509–522.

Mansoor, S., Farooq, I., Kachroo, M.M., Mahmoud, A.E.D., Fawzy, M., Popescu, S.M., Alyemeni, M.N., Sonne, C., Rinklebe, J., Ahmad, P., 2022. Elevation in wildfire frequencies with respect to the climate change. *Journal of Environmental Management* 301, 113769.

Mishra, B.K., Kumar, P., Saraswat, C., Chakraborty, S., Gautam, A., 2021. Water security in a changing environment: Concept, challenges and solutions. *Water* 13, 490.

Mosaffa, H., Sadeghi, M., Mallakpour, I., Naghdyzadegan Jahromi, M., Pourghasemi, H.R., 2022. Chapter 43 — Application of machine learning algorithms in hydrology. In: Pourghasemi, H.R. (Ed.). Computers in Earth and Environmental Sciences. Elsevier, pp. 585–591.

Mounce, S.R., Pedraza, C., Jackson, T., Linford, P., Boxall, J.B., 2015. Cloud based machine learning approaches for leakage assessment and management in smart water networks. *Procedia Engineering* 119, 43–52.

Naidu, R., Arias Espana, V.A., Liu, Y., Jit, J., 2016. Emerging contaminants in the environment: Risk-based analysis for better management. *Chemosphere* 154, 350–357.

Petrovic, M., Ginebreda, A., Acuña, V., Batalla, R.J., Elosegi, A., Guasch, H., de Alda, M.L., Marcé, R., Muñoz, I., Navarro-Ortega, A., Navarro, E., Vericat, D., Sabater, S., Barceló, D., 2011. Combined scenarios of chemical and ecological quality under water scarcity in Mediterranean rivers. *TrAC Trends in Analytical Chemistry* 30, 1269–1278.

Ratnaweera, H., Sætersdal, T., Weerakoon, S.B., Mutua, F.M., 2022. Editorial: Water management addressing societal and climate change challenges. *Journal of Water and Climate Change* 13, v–vii.

Richardson, S.D., Kimura, S.Y., 2020. Water analysis: Emerging contaminants and current issues. *Analytical Chemistry* 92, 473–505.

Rodriguez-Narvaez, O.M., Peralta-Hernandez, J.M., Goonetilleke, A., Bandala, E.R., 2017. Treatment technologies for emerging contaminants in water: A review. *Chemical Engineering Journal* 323, 361–380.

Sandford, R., 2017. The Human Face of Water Security. Springer, pp. 1–24.

Saravanan, A., Senthil Kumar, P., Jeevanantham, S., Karishma, S., Tajsabreen, B., Yaashikaa, P.R., Reshma, B., 2021. Effective water/wastewater treatment methodologies for toxic pollutants removal: Processes and applications towards sustainable development. *Chemosphere* 280, 130595.

Sawicka, B., Umachandran, K., Fawzy, M., Mahmoud, A.E.D., 2021. 27 — Impacts of inorganic/organic pollutants on agroecosystems and eco-friendly solutions. In: Mtewa, A.G., Egbuna, C. (Eds.). Phytochemistry, the Military and Health. Elsevier, pp. 523–552.

Schmidt, L., Heße, F., Attinger, S., Kumar, R., 2020. Challenges in applying machine learning models for hydrological inference: A case study for flooding events across Germany. *Water Resources Research* 56, e2019WR025924.

Sidhu, R.K., Kumar, R., Rana, P.S., 2020. Machine learning based crop water demand forecasting using minimum climatological data. *Multimedia Tools and Applications* 79, 13109–13124.

Statista, 2022. Global consumption for water by sector. www.statista.com/statistics/1012228/global-consumption-for-water-by-sector/.

Tao, H., Hameed, M.M., Marhoon, H.A., Zounemat-Kermani, M., Heddam, S., Kim, S., Sulaiman, S.O., Tan, M.L., Sa'adi, Z., Mehr, A.D., Allawi, M.F., Abba, S.I., Zain, J.M., Falah, M.W., Jamei, M., Bokde, N.D., Bayatvarkeshi, M., Al-Mukhtar, M., Bhagat, S.K., Tiyasha, T., Khedher, K.M., Al-Ansari, N., Shahid, S., Yaseen, Z.M., 2022. Groundwater level prediction using machine learning models: A comprehensive review. *Neurocomputing* 489, 271–308.

Tran, N.H., Reinhard, M., Gin, K.Y.-H., 2018. Occurrence and fate of emerging contaminants in municipal wastewater treatment plants from different geographical regions-a review. *Water Research* 133, 182–207.

Vambol, S., Vambol, V., Mozaffari, N., Mahmoud, A.E.D., Ramsawak, N., Mozaffari, N., Ziarati, P., Khan, N.A., 2021. Comprehensive insights into sources of pharmaceutical wastewater in the biotic systems. In: Khan, N.A., Ahmed, S., Vambol, V., Vambol, S. (Eds.). Pharmaceutical Wastewater Treatment Technologies: Concepts and Implementation Strategies. IWA Publishing, UK. pp. 17–47.

WWAP. 2019. The United Nations World Water Development Report 2019: Leaving No One Behind. UNESCO, Paris.

Zhu, Y.-G., Zhao, Y., Zhu, D., Gillings, M., Penuelas, J., Ok, Y.S., Capon, A., Banwart, S., 2019. Soil biota, antimicrobial resistance and planetary health. *Environment International* 131, 105059.

2 Integrated Water Resources Management
Perspective and Challenges

*Uun Yanuhar, Heru Suryanto,
and Nico Rahman Caesar*

CONTENTS

2.1 INTRODUCTION

Water has a vital role in supporting human productive activities, such as fisheries, agriculture, energy generation, industrial production, sanitation, transportation services, tourism, and many other activities. Water resources management must be designed properly to meet current and future needs by maintaining the various hydrological variations needed to preserve ecological integrity and the aquatic environment (Hussen et al. 2018). According to Hwang et al. (2020), water and soil resources are highly vulnerable, and they exist in organic relationships with significantly reciprocal effects. To manage this vulnerable resource, food supply and sustainability and adequate water availability for public health must be ensured. Peculiarly, water is an important resource needed to sustain animal and plant life. Therefore, water management becomes very important (Detenbeck et al. 2018). Integrated water resources management (IWRM) is one of the views of the Sustainable Development Goals

DOI: 10.1201/9781003260455-2

(SDGs) that are applied to waters (Mahmoud et al. 2021). IWRM has been very clearly defined, which is actually a concept that includes many umbrella principles, with the aim of managing water resources in a more coordinated manner. The principle in question covers the political, economic, and administrative systems. IWRM aims to maximize the economic and social welfare benefits of water use without compromising ecosystem sustainability (Pires et al. 2017).

Aquaculture has considerable benefits beyond its potential to improve the community's economy. One of the most effective ways to do aquaculture is to pay attention to the water quality of fish-farming ponds (Mishra et al. 2022). Fish cultivation media is a place for fish to grow and develop, namely, water. Water that can be used for fish-farming must have quantity and quality standards that are in accordance with the requirements of fish life. Water that can be used as a living medium for fish must be studied so that fish as aquatic organisms can be cultivated according to human needs. Water that can meet good criteria for low-level animals and plants has plankton as the easiest indicator that the water can be used for fish-farming. This is because these organisms are primary producers as supporters of water fertility. Therefore, water/water conditions must ensure the creation of natural food for aquatic organisms, especially in favorable conditions, with a low level of activity (phytoplankton) in the assimilation process into a food source for animals, especially fish.

Good fish-farming activities must be supported by good water quality management as well. This is because water quality is very influential on the life and growth of aquaculture fish organisms. The efficiency and productivity of aquaculture are highly dependent on the development of advanced technology for water quality management systems, namely, controlling the water quality of fish-rearing media. One of the parameters that can be used to monitor water quality is the availability of oxygen (DO) in aquaculture ponds. Generally, DO in aquaculture ponds is very limited. Therefore, it is necessary to apply technology that can increase and maintain the stability of DO availability in the pool, such as microbubble, which is integrated with the IoT. Building an intelligent monitoring system of water quality based on IoT technology requires the integration of various components to monitor and control it in real time (Doss 2018). IoT applications can be used to support life activities in many domains, such as homes and offices, through transportation and resource management.

According to Qaddoori et al. (2021), microbubble generator (MBG) is a technology capable of producing very small bubbles. This technology has characteristics such as very fast bubble production, very wide bubble dispersion, and longer bubble burst resistance, as well as being able to reduce drag friction. The drag friction of gas-liquid two-phase flow is reduced by increasing the microbubble volume fraction in vertical pipe flow. Magallón-Servín et al. (2020) stated that aeration is one of the most important parts of intensive or hyperintensive cultivation, where aeration costs reach 15% of the total production cost. Therefore, the selection of aerators and diffusion systems is very important. Increasing the density of organisms and biomass production, and being in the same volume of water, increases the rate of metabolism and respiration. Efficient use of aeration is very important because aeration requirements vary according to the conditions of the culture pond, where there is a constant need for oxygen associated with that consumed at the surface and bottom areas of the pond water. Therefore, the development of technology-based integrated water management with microbubble system integrated with the Internet of Things (IoT) to support sustainable water quality management automation so as to provide a synergistic effort to monitor the marine environment better.

2.2 INTERNET OF THINGS (IOT) TECHNOLOGIES

IoT is described as an enabler that seamlessly connects objects around an environment and performs a kind of message exchange between them. IoT is a concept where objects can transfer data over a network without requiring human-to-human or human-to-computer interaction. It collects objects that work together to serve consumer tasks in a unified way (Biju and Mathew 2020). The development of IoT starts from internet convergence, wireless technology, and microelectromechanics (MEMS). IoT creates this small network among its system devices. It uses computing power to transmit data to the surrounding environment (Priyanka et al. 2022; Valentini et al. 2019). These devices can be in the form of embedded systems, tools, sensors, and data analysis microchips.

IoT technology allows remote control of objects across existing network infrastructure and is able to create opportunities for integration between the physical world and cyber-based digital systems so as to increase efficiency and economic benefits. Every object or thing can be identified through an embedded computing system and can operate on the internet network (Jian et al. 2021). Figure 2.1 shows the workflow of an IoT system.

The monitoring system in specific environments apply unique architecture system. It uses a specific component for building an automated device that applies sensor, microcontroller (Arduino Uno) for processing the sensor signal until it can be visualized at output display via the internet in the cloud server, and power supply for monitoring the environment condition (Figure 2.1). The sensor consists of many wireless sensing systems specially designed for environmental monitoring, both quality and quantity, based on certain environmental parameters that need to be monitored (Afifi et al. 2018).

The local wireless water quality monitoring unit consists of several wireless data collectors, namely a Wi-Fi gateway and router. A central server is useful in collecting environmental data with a wireless sensing system using the Arduino Uno microcontroller (Kamaruidzaman and Rahmat 2020) equipped with a node that is a responsible for the sensor–microcontroller connection to support the interface. Sensor signal data processing is carried out using the Arduino Uno platform (Encinas et al. 2017). Complete sensing scheduling parameters are performed via a web-based application from the cloud (Afifi et al. 2018). Cloud servers serve to send data from sensors to a central server via an internet base. It provides services as a data storage and analysis tool. The interface communication type in IoT systems are device-to-cloud (Majumdar 2018) and device-to-device (Gismalla and Abdullah 2017). Device-to-device are able to connect directly and communicate with each other through the internet network. In the device-to-cloud model, the device connects directly to the internet cloud via a Wi-Fi connection and an internet network to exchange data and control message traffic. The cloud allows gaining remote access to their devices, for example, via a smartphone or a web interface using a web-based application (Alam 2020).

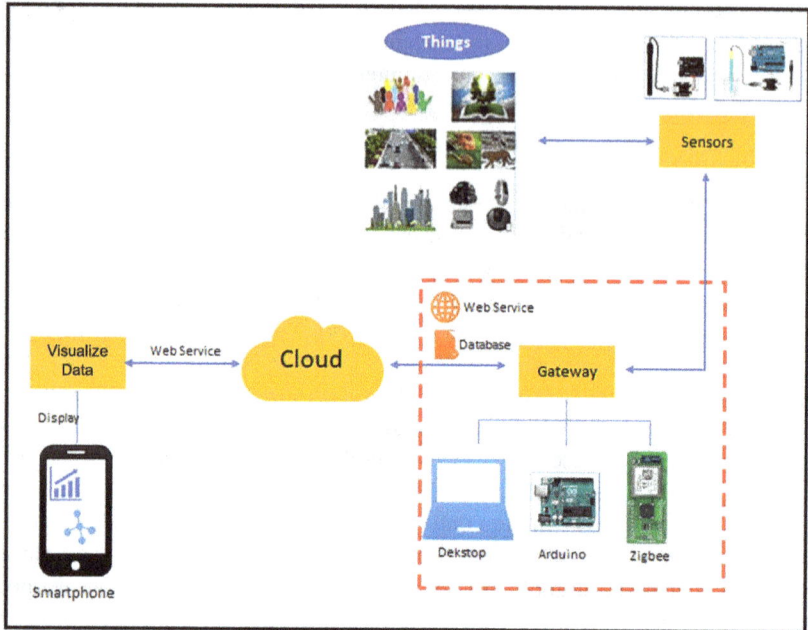

FIGURE 2.1   Workflow in IoT system.

2.3 WATER RESOURCES MANAGEMENT

Management of water quality in aquaculture is important in order to increase production and support the success of aquaculture. The good growth and development of fish can be influenced by external and internal factors. One of the important external factors in aquaculture management is the management of water quality as a living medium for aquatic organisms. Water is the main medium for fish to live; it must be considered both in terms of quantity and quality. Intensive-scale aquaculture relies heavily on water quality checks. Fish infection is very fast and affects crop yields (Robles-Porchas et al. 2020). Low water quality can also affect fish development and production yields. Sustainable fish-farming relies heavily on the physical, synthetic, and natural characteristics of the water (Maulu et al. 2021), regardless of the type of cultivation. Furthermore, water quality is a way to succeed in viable fisheries. It is determined by factors, for example, temperature, turbidity, carbon dioxide, pH, alkalinity, odorous salts, nitrites, nitrates, and so on. Among them, the most basic are temperature, DO, and pH. The ideal temperature depends on the fish species; it is very important that the temperature is controlled and kept within the proper range. What's more, even within the right range, higher temperatures build up the rate of biochemical movement of the microbiota, thereby expanding oxygen demand. To limit disease and oxygen utilization, the temperature must be properly directed (Omer 2019). The ideal DO should always be more than 5 ppm. Fish need enough oxygen in the water to survive; otherwise, they stay on the surface to make up for the lost time, a slower digestion and slower development, and ultimately, a tremendous lack of oxygen (Little et al. 2020). Ahmed and Turchini (2021) proposed the concept of recirculation of aquaculture systems in this study using temperature, DO, and pH scales, which are monitored directly continuously because they have a tendency to change rapidly and have major adverse results in the culture system if left at sub-optimal values. Therefore, these three parameters were chosen to be monitored in this system. Running and maintaining this kind of framework is very costly. Many small-scale cultivators decided not to use it and took risks to get meager results. That is why water quality is the way to achieve success in aquaculture, and improving water quality is a major test, especially in small-fish-farming in developing countries (Hwang et al. 2020).

Management of water quality in aquaculture is important to remove the body's metabolic waste from harmful fish, such as ammonia (NH_3), as well as a substitute for aeration. Good water quality will reduce the level of turbidity in the water so that sunlight can enter to meet the needs of fish productivity. Decreased water quality will cause the accumulation of feed residues, organic matter, toxic compounds or toxins, and other harmful substances (Gitau et al. 2016; Yanuhar and Arfiati 2018). A very dangerous and toxic waste for fish is ammonia. Ammonia can cause toxicity and the emergence of disease in fish. Management of water quality in aquaculture can be done in several ways. One way that is usually done in aquaculture is by changing the pond water periodically. The speed of regular and maximum water change will be able to improve water quality in aquaculture. However, if the water change process is not regular, it will not necessarily be able to eliminate all harmful substances in the water, so the water quality is still not optimal. Another tool that can be used to manage water quality in aquaculture is a paddlewheel or waterwheel. The use of a waterwheel for 24 hours can increase the supply of oxygen in the water.

2.3.1 Recirculation System

The circulation system is one way to improve water quality as a medium for maintaining fish in aquaculture activities. Circulation of water can help the distribution of oxygen in all directions, both in the water and its diffusion or exchange with air, and can keep the accumulation or collect toxic metabolic products so that toxic levels can be reduced. The effectiveness of the circulation system in improving the water quality of the cultivation media is influenced by the rate of water discharge. Water discharge is the amount of water flowing per unit time in rivers, ditches, or pipes. Water discharge is usually also referred to as the quantity of flowing water, the volume of flowing water, or the supply of flowing water, where this water discharge varies in its use. Knowledge of this amount of water will give the advantage of optimizing water use. Problems with the content of phosphate and nitrogen (ammonia, nitrite, nitrate) can be reduced by this circulation system. In addition, water circulation can help the distribution of oxygen in all directions, both in the water and its diffusion or exchange with air, and can keep the accumulation or collect toxic metabolic products so that toxic levels can be reduced (Omer 2019).

Aquaponics is a recirculation system using the principle of integrating vegetable, herbal, and ornamental plants with fish-farming. Ekawati et al. (2021) argue that aquaponics systems can reduce ammonia by absorbing aquaculture wastewater or using plant roots. The absorbed ammonia undergoes an oxidation process assisted by oxygen and bacteria; the ammonia is converted into nitrate. Plants have an important role in reducing carbon because plants are able to use carbon to carry out the process of photosynthesis to produce oxygen (Ziarati et al. 2019). Kangkung is a plant with roots that are not too strong, and in its maintenance, it requires continuous water. According to Setijaningsih and Suryaningrum (2015), this system utilizes a mutualism symbiosis between plants and fish based on the utilization of waste products of fish metabolism by plants, application of polyculture systems, efficient use of water, provision of organic food products, and increased income. Water which is a medium for fish cultivation is used as a source of nutrients for plant maintenance; on the other hand, plants function as a biofilter for water.

2.3.2 FILTRATION SYSTEM APPLICATION

The condition of the water as a living medium for aquatic biota must be adapted to optimal conditions for the biota being maintained. The water quality includes the quality of physics, chemistry, and biology. As we know, water quality also depends on the source of the water. If the quality of the water source is good, it will have a good influence on the cultured organisms, and vice versa. If the water source has poor quality, it will also have a bad effect on the organisms. The water suitability factor is important because it involves whether or not fish-farming can be carried out in an environment. The filtration technique according to Jenti and Nurhayati (2014) is one way to physically manage water. The filtration system is a solid–liquid separation process by flowing liquid through a porous medium or filter to remove as many small particles of suspended solids as possible from the liquid. Usually, this filtration technique uses certain materials, such as sand, gravel, and other materials that have the ability to filter. Purnama and Arief (2018) show that through this filtration method, water quality can be improved by reducing turbidity in the water source. In addition, through this technology, the pH of the water can be lowered, especially water media that comes from calcareous soil or former coal mines. Purwoto et al. (2015) state that the use of this filtration system was chosen because it is practical, cheap, easy to obtain, easy to wash, easy to shape and has high filterability. In addition, the output obtained is the problem of difficulty in cleaning river water, which is only filtered using gravel and sand to become clearer by means of polyaluminum chloride (PAC), filter spoon, MGS, zeolite, silica, anion resin, and cation resin. In addition, the opportunity to increase the processed discharge by adding new processing tub units is more open.

2.4 WATER QUALITY MANAGEMENT WITH MICROBUBBLE IMPLEMENTATION INTEGRATED WITH IOT

Traditionally, water quality monitoring was manually conducted, where samples of water were taken and delivered to the laboratories for examination. It takes many cost, human resources, and time process and also does not provide data in real time (Biju and Mathew 2020). Water quality management activities can be carried out in real time with the help of Internet of Things technology. The application of technology can be utilized to facilitate controlling pool water. Increasing aquaculture productivity can be assisted by the use of technology and the application of technology to manage and monitor sustainable koi pond water quality, such as the microbubble generator (MBG) (Yanuhar et al. 2021). A *microbubble* is an aerator that can produce air bubbles in micro size in pool water. This technology has the advantage of having a larger specific surface area and can last longer in water. Water quality can be monitored periodically, and if the water quality is below the water quality standard, the device will be activated automatically. Microbubbles will stop if the DO content is sufficient for the water quality standard that has been determined. Water quality monitoring can be carried out according to the monitoring time scale, namely, per minute, hourly, and per day, to weekly. The inputted data will always be updated every second and set in manual and automatic modes. The use of manual mode can be essential in the event of an emergency on water quality so that it requires a fast response and must be handled immediately. In automatic mode, it is carried out with an upper and lower benchmark, which means that if the concentration of DO and pH of the water has reached the optimal limit according to the quality standard, the tool will turn off automatically; on the contrary, if the concentration has reached the minimum value, the tool will turn on automatically (Ahmed and Turchini 2021; Hwang et al. 2020; Xiao et al. 2019). Water quality management can be done in real time with the help of the IoT. IoT technology can monitor the quality of existing or set water, such as DO content, pH, and others. IoT technology assistance can help fish farmers control water quality remotely. The condition of poor water quality can be known by farmers through smartphones. Water quality in fish-farming greatly affects the resulting cultivation. Water quality parameters that will be used as fish cultivation media must meet water quality standards. Water quality conditions must support fish growth properly. Physical parameters that need to be considered include temperature, brightness, dissolved solids, turbidity levels, and dissolved oxygen. This parameter needs to be monitored in order to get maximum cultivation results. What is obtained from monitoring activities is that farmers know more about the condition of fishponds and know the effects experienced by fish in real time. Therefore, if there is a problem with the quality of the pool water, it can be handled quickly and correctly (Bekele et al. 2018; Hwang et al. 2020).

In aquaculture or fish-farming, water quality management is very important in increasing aquaculture yields and supporting the success of aquaculture. The good growth and development of fish can be influenced by external and internal factors. One of the important external factors in aquaculture management is the management of water quality as the living environment of aquatic organisms. Water is the main habitat of fish, so both quantity and quality must be considered. Intensive cultivation will reduce water quality as a medium for fish development and growth (Preena et al. 2021). The factors that can reduce water quality in aquaculture are high levels of ammonia and low oxygen content, causing an increase in the content of ammonia (NH_3) and carbon dioxide (CO_2) in the water. High levels of ammonia and carbon dioxide in the water can reduce the quality of pond water. In addition, water temperature can affect the process of digestion of food in fish. At low temperatures, the digestive process of fish is usually very slow, but at high temperatures, the digestive process of fish takes place quickly (Ahmed and Turchini 2021; Martinsen et al. 2019). Water quality management in fish-farming is very important. It plays a role in removing harmful body metabolic waste from fish and other aquatic biotas, such as ammonia (NH_3), as well as, as a substitute for aeration. Good water quality will reduce the level of turbidity in the water so that sunlight can enter to meet the needs of fish productivity. Decreasing water quality will result in the accumulation of feed residues, organic matter, toxic compounds or toxins, and other harmful substances. A very dangerous and toxic waste for fish is ammonia. Ammonia can cause toxins as well as release disease in fish (Deb et al. 2020).

According to Zhang and Li (2020), over the past few decades, effective planning methods for water quality management have become a major research topic because water quality management strategies can significantly affect the living medium of fish, namely, water supply, which is very important to maintain the socioeconomic sustainability of fish-farming activities. Optimization models with useful technology have been widely applied in the field of water quality management. However, the initial study only considered optimization problems with deterministic input parameters. With the application of water quality and disease control technology with IoT-based microbubble (Figure 2.2), it is able to maintain water quality at optimal levels and have a positive impact on water quality and disease management that will be carried out in ponds. Water quality is a very decisive factor for the survival of aquatic organisms. Water quality fluctuations that occur will result in decreased fish growth. In addition, suboptimal water quality will become a habitat for pathogens that cause disease in fish, which in turn will cause losses in fish-farming (Budhijanto et al. 2017; Yanuhar et al. 2019). The presence of microbubbles can also support the process of improving water quality conditions and also the process of water decomposition by probiotic bacteria.

Application of microbubble that is integrated with IoT has been carried out in catfish and koi fish culture ponds (Suryanto et al. 2022; Yanuhar et al. 2022). The results

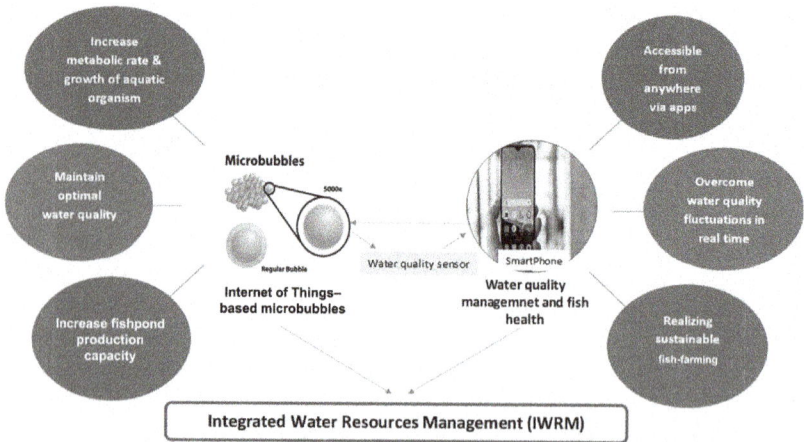

FIGURE 2.2 The basic concept of *integrated water resources management* (IWRM) with *microbubble* integrated with the IoT.

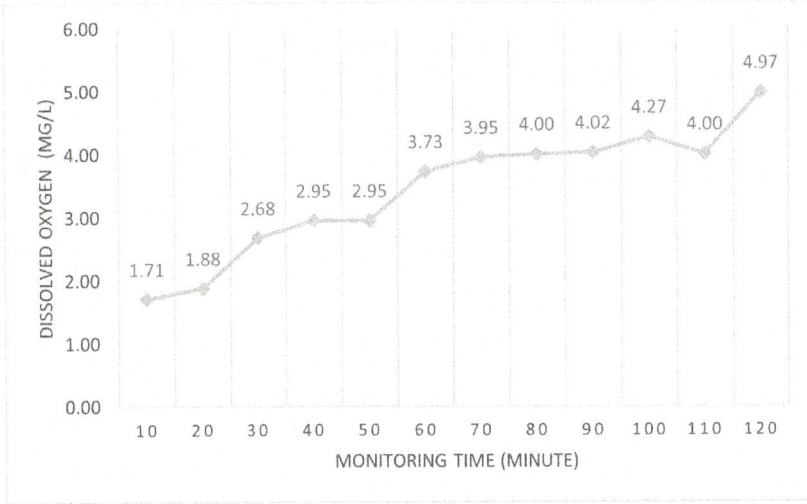

FIGURE 2.3 Monitoring DO levels in fishponds with the application of microbubble integrated with the IoT for two hours (DO data recording time interval is every ten minutes) (Suryanto et al. 2022).

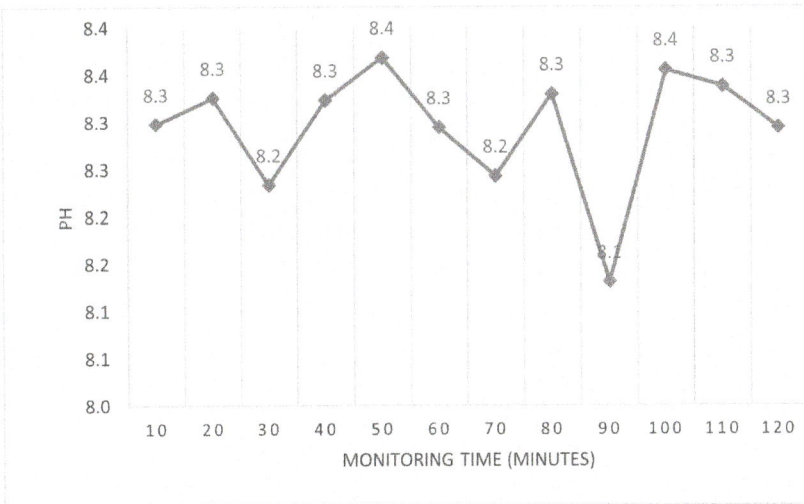

FIGURE 2.4 Monitoring of fishpond pH levels with the application of microbubble integrated with the IoT for two hours (DO data recording time interval is every ten minutes) (Suryanto et al. 2022).

show an increase in DO values and maintained optimal water pH values (Figures 2.3 and 2.4). Water quality really needs special attention in fish-farming business activities (Suryanto et al. 2022; Yanuhar et al. 2019). According to Liu et al. (2019), the parameters of water quality of cultivation media should be in accord with environmental conditions supporting the growth properly. Physical parameters that need to be considered are the presence of dissolved solids, turbidity level, temperature, and pond water depth. Chemical parameters that also need to be controlled are pH value, carbon dioxide, DO, the presence of ammonia, phosphate, and nitrate. The results obtained from water quality monitoring activities are that fish farmers know how the conditions of fishponds are and their effects on fish in real time so that if there is a problem with pond water quality, fast and effective resolution actions can be taken immediately.

According to research conducted by Chang et al. (2017), it is known that water quality management is very important in response to the need for sustainable development, such as with the development of remote sensing-based multiscale modeling systems that integrate multisensor satellite data and image reconstruction algorithms to support feature extraction with machine learning, leading to the automation of sustainable water quality so as to provide a synergistic effort to better monitor the aquatic environment.

According to Bekele et al. (2018), scientific developments, especially in the area of water quality assessment, may include molecular detection methods for microbial pathogens and analytical chemical resolution methods for detecting chemical traces. It provides unprecedented new insight into the "polishing" offered by natural water quality management treatments. So this provides an opportunity to set targets and reduce the risk of failure. Water quality management in aquaculture can be done in several ways. One of the methods commonly used in cultivation is to change the water medium periodically. Regular and maximum water change rates can improve air quality in cultivation. However, if the water exchange process is not regular, it will not be possible to remove all harmful substances in the water so that it affects air quality. The tool that can regulate water quality in aquaculture is a waterwheel. Using a waterwheel for 24 hours can improve air oxygenation and can support water quality. There is a water quality management method, namely, the water treatment water management system.

2.4.1 Smart Management of Water Quality with IoT

The purpose of smart management of water quality with IoT is to integrate the certain sensors for detecting water quality in real time. The system also uses different filters which are used to improve water quality efficiently because the system is only used once needed and not every time. Upon utilization, the system leads to an increase in the health of the waters and a reduction in cost for controlling water quality.

Ramadhan et al. (2020) have conducted research related to water quality measurement systems based on intelligent network sensors for monitoring environmental factors of aquaculture. Huan et al. (2020) also developed a ZigBee-based WSN to look at aquaculture recycling systems. This system can read various water quality parameters, such as temperature, pH, DO, nitrate, brightness, and salinity. Cario et al. (2017) focus on the application of an effective and suitable underwater acoustic network for long-term water quality monitoring in aquaculture activities. All the aforementioned systems have no processing and analysis of stored data. The system predicts changes in water quality for each parameter and controls the performance of various aquaculture instruments according to the predicted results as a forecast to maintain the stability of pond water quality parameters and optimize the control of managed equipment to improve energy conservation.

An IoT-based water quality monitoring approach has also been developed for marine and lake waters. For such needs, a wireless sensor network is needed to monitor water quality parameters in a wider area and send monitoring data directly to available servers. The application usually monitors parameters such as chlorophyll and DO concentration (Prapti et al. 2022). Dissolved oxygen measurement is important for aquaculture activities because it determines whether the media/water can support the survival of aquaculture commodities (Ullo and Sinha 2020). Based on research conducted by Kageyama et al. (2016), a WSN system has been developed for monitoring water quality. The sensors are connected to the transmission module using UART. Communication on the outer sensor node is carried out using a 3G cellular network internet connection.

Communication between the controller and the centralized data store is carried out using long-distance communication standards, such as 3G and the internet. Some previous work aimed to remind users in the form of SMS about water quality. Such a system based on Pasika and Gandla (2020) requires an additional SIM card to connect between the GPRS module and the controller. IoT is a new communication paradigm where objects in everyday life are equipped with microcontrollers and transceivers for digital communication, which will make objects communicate with each other and become an integral part of the internet (Jian et al. 2021; Valentini et al. 2019).

Another system that monitors water quality remotely uses low-power sensors on multiple nodes that enter all data in a database, then compares it with standard values and sends automatic alerts via SMS to each node (Ragavan et al. 2016). Another system provides a low-power model that is used to collect water quality data based on IoT technology. The system then sends an alert to the remote user if any value differs from the default value preloaded in the program (Geetha and Gouthami 2016). Some communication systems use Bluetooth technology with controllers, but this system has the disadvantage of a dropped signal causing some datasets to disappear (Kumar and Samalla 2019). Monitoring water quality in the distribution system is a challenge in managing a distributed wireless sensor network (WSN). A water distribution network for monitoring chlorine concentrations has been presented by Olatinwo and Joubert (2019). Solar-powered distributed WSN has been proposed by Son et al. (2020) to monitor water quality conditions, such as oxygen, temperature, turbidity, and pH. The location of fishponds in different areas can be monitored directly using a system consisting of water quality sensors and a database that is supported by energy from solar panels.

2.5   INTERNET OF THINGS (IOT) WATER QUALITY MONITORING

In the era of mobile technology and interconnectivity, the concept of the IoT has emerged, which has communication and interconnectivity with objects. It provides intelligent services with a combination of the internet and sensor networks. The current procedure for monitoring aquaculture ponds is less efficient. Based on the experience gained by cultivators, this takes a lot of time and operational costs. Quality measurements are usually carried out when pond waters are experiencing problems or there has been a significant decrease in the carrying capacity of water quality. During these conditions, the restoration of the condition of the pool will usually require very expensive costs. This causes the water quality factor to not be monitored properly and efficiently. Of the various problems that occur in aquaculture, one of the main obstacles is a disease (Ighalo et al. 2021). Diseases are caused by a decrease and fluctuation of water quality in fishponds. With the application of IoT, it is expected to be able to implement appropriate, fast, and efficient water quality management at low costs. This system can be implemented in the ponds to monitor in real time the most important water physicochemical variables, by having a faster response with regard to what action to take when conditions arise in the pool water quality. Its contribution is a mobile sensor platform for monitoring ponds (Encinas et al. 2017). The model used mainly focuses on constant monitoring of water quality parameters over time to take preventive measures.

The DO meter and pH meter sensors will send information to the control (Figure 2.5). Information is processed into data that is displayed on the LCD monitor in the form of DO and pH values. The value is also sent to the server using SIM900 communication media, namely, via GPRS communication. The data stored on the

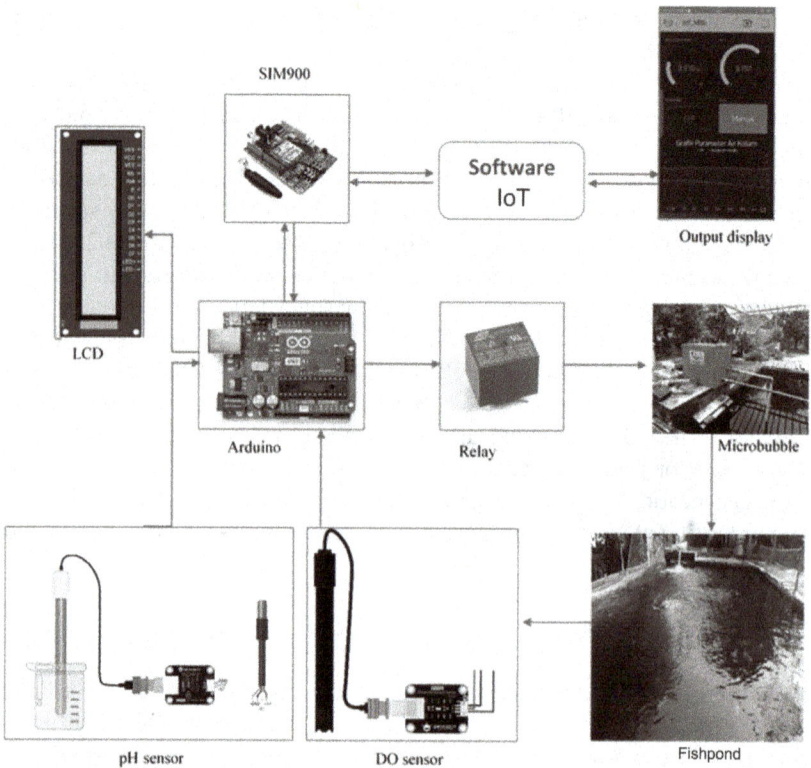

FIGURE 2.5   Microbubble work scheme integrated with IoT

server can then be displayed on a smartphone using the application. The DO and pH values are then used as parameters to activate the microbubble generator engine with the help of a relay. The machine will be active when DO is 5 mg L^{-1} and inactive when DO is 7 mg L^{-1}. This function will run automatically when all devices are started/restarted. This automatic function setting can be deactivated using a smartphone application so that there are two operating modes, namely, automatic and manual. Manual mode requires the user to deactivate and activate the microbubble generator engine manually through the application. DO and pH values can be monitored every second but are stored on the server every 60 seconds (Suryanto et al. 2022; Yanuhar et al. 2022).

The IoT-based fishpond water quality management module is primarily designed to manage water quality so as to support fish-farming. This module consists of five parts: (i) data and database acquisition, (ii) query management, (iii) integrated sensor components, (iv) data analysis and optimization, and (v) control actuator. The module enables the transmission, data acquisition, storage, analysis, and processing of data and supports monitoring and management functions in aquaculture ponds (Gao et al. 2019).

Another example of an IoT model mainly focuses on constant monitoring of water quality parameters over time to take preventive measures. Its model architecture consists of four modules: (1) power module, (2) sensor module, (3) microcontroller module, and (4) output module. The selector module is more environmentally friendly. The sensor module consists of several sensors, such as for nitrate, salt, temperature, pH, ammonia, DO, and carbonate. This sensor is installed on the Raspberry Pi and is used to describe water quality parameters in fishponds over time. It is considered to be the heart of this architecture (Ismail et al. 2020).

2.6 MICROBUBBLE CONCEPT

According to Liew et al. (2020), based on the working principle, the microbubble generator (MBG) group is divided into three groups, namely, pressurized dissolution (decompression), rotating flow (spiral flow), and cavitation for the ejector, or venturi method. The latest MBG developments are systems based on constant flow nozzles and membranes, porous media, or gas spargers combined with a mixer. For the pressurized-type MBG, a highly saturated gas is injected into the tank via nozzles, along with pressurized water, to increase the solubility of the gas. The liquid-gas mixture then forms microsized bubbles as a result of the sudden drop in pressure when flashed by reducing valve at lower pressures. Spiral flow fluids are generally designed in a conical shape, with the aim of facilitating gas-water circulation. Water is fed tangentially into a cylindrical tank to create a hollow spiral pattern that flows like a vortex. Meanwhile, spiral or vortex microbubbles can also work based on a self-suction mechanism for gas supply, such as venturi- or orifice-type MBG. The venturi effect has been exploited to produce microbubbles and factors affecting the formation of the microbubble. In orifice type, gas is sucked from an orifice at the tank base into the central core of a low-pressure whirlpool. Then, the gas-liquid mixture is reduced to microbubbles due to the shear effect of the centrifugation formed by the rapidly rotating liquid stream.

This is also supported by research by Wilson et al. (2021), which states that the way for microbubbles can be generated in three different ways. The first stage is to provide low-pressure air so that the air bubbles burst due to mechanical vibration. Another method that can be used to produce microbubbles is the use of powerful ultrasound waves. The ultrasonic bubbles are used to induce cavitation at tiny points high in the liquid. Next, the gas will be injected into the liquid through a specially designed nozzle. This nozzle is made to produce small bubbles in the nanobubble region.

According to Tsuge (2019), microbubbles have been widely used in various other fields, such as overcoming problems in waste treatment in the environment; MBG is considered to be a modern tool that is very helpful in treating waste. This is based on the fact that the structure of MBG itself is quite simple, installation is easy, time efficiency is quite fast, and it has low energy consumption. MBG is also able to damage the structure of free radicals that enter the waters, which makes the water clearer. MBG can also improve soil layers in soil liquefaction. Soil improvement against soil liquefaction is carried out by utilizing the principle of bubble shrinkage; part of the bubble containing oxygen is introduced into the soil, and an air cushion made of bubbles prevents the increase in pore water pressure during an earthquake. Although this method has not yet become practical, it could have prospects as an effective antiliquefaction method for solving problems such as high costs and very narrow coverage. But in terms of effectiveness, it works very well because the structure of the tool is very simple, so it doesn't require a lot of effort in assembling it, it doesn't require a large area to operate, and the way it works, which only requires water and air, makes MBG very friendly to the environment. The process of reducing the level of soil saturation by using microbubbles is carried out with a mechanism where the soil absorbs bubbles so that high vibration pressures on the soil can be minimized (Brasileiro et al. 2020). The collection of microbubbles can enter the soil and pass through the sand particles because the diameter of the bubbles is ten microns and is smaller than the size of the cavity between the gaps of the sand particles. However, other microbubbles may be trapped in tiny air pockets between the sand particles, and the two of them coalesce to sink into the soil to form a size that is milliliter in diameter so that the trapped microbubble can reduce its surface tension and air pressure. Since the water surrounding the bubbles dissolves the air, the air that settles will increase the size of the bubbles.

The use of MBG in the fisheries sector can improve the quality and quantity of DO, make the pH in the pond stable, make the ammonia content low due to the high organic matter decomposition process, and minimize the occurrence of disease attacks in cultured fish caused by viruses, bacteria, or chemical compound pollutants that enter the pool (Lim et al. 2021). The design of the MBG shape must be adapted to the shape of the pool; this is because each form of the pool has different needs. The design of the MBG will affect the effectiveness of its work, such as if the aquaculture pond is large enough, and in feeding it takes more than 1 MBG to operate and it can be modified by modifying the nozzles to become double chamber so that the MBG has a double chamber with a wider water inlet at the bottom of the outlet to draw more water up the inner chamber and create a double whirlpool inside. Inlet whirlpools can rotate more rapidly than outlet whirlpools and allow the production of a much larger number of microbubbles (Wilson et al. 2021).

The tools needed in fish-farming with the application of the microbubble genera-tor (MBG) consist of several parts in producing microbubbles, which are as follows:

1. Media for keeping fish or ponds.

In this media, fish-farming activities occur, where the longer the cultivation pro-cess, the more the water quality will decrease, in the form of a decrease in DO, and the color of the water becomes cloudy. This turbidity can be caused by the presence of foreign pollutants that mask the pond or excessive feeding so that the remaining food will collect together and will change the color of the water. Steps that can be taken to minimize the turbidity of water in aquaculture ponds are regulation of fish density, use of types and doses in feeding, and water quality management (Yanuhar et al. 2021). The amount of feed given must be adjusted to the number of fish culti-vated in the pond. Excessive feeding will affect the amount of secretion in cultured fish, and the remaining feed that accumulates will cause the ammonia content in the pond to increase so that the cultured fish will be susceptible to disease.

2. Filtration installation.

This filter is useful in filtering water from impurities resulting from fish-farming activities. Filters can be carried out in three forms, namely, physical, chemical, and biological. The physical filter can use the use of stones. Chemical filters can use ginger coral and zeolite, and biological filters can use microbiological assistance in the form of bacteria, where these bacteria can decompose organic matter into simple ones so that phytoplankton can use them in photosynthesis. The use of this filter is expected to reduce the content of ammonia and nitrate in aquaculture ponds.

3. Filtered water reservoir.

The place will then be drained by a pump connected to the pipe. This media container is an indicator of the success of the filtration process that has been carried out. If the water quality in the storage media container is better than the container for biological, physical, and chemical filters, it is declared that the filtering has been successful.

4. Water pump.

The water pump will drain the water in the reservoir to the main container for fish-farming. The use of pumps is very important because they are interrelated with microbubbles and the ongoing filtration process.

5. Microbubble device.

The microbubble device consists of determining the size of the tool, the size of the pipe used, the size of the valve, the size of the water flow meter and airflow meter, the size of the pump, and the size of the hole. Each part of the tool will greatly affect

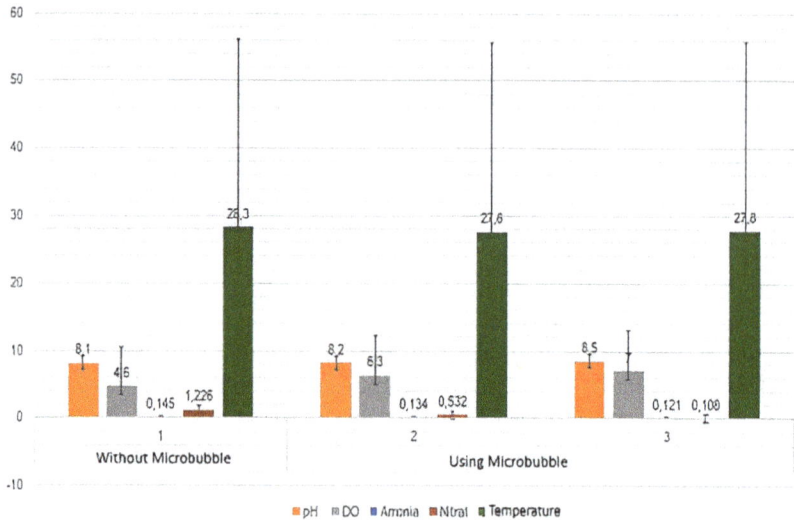

FIGURE 2.6   Results of water quality measurements before and after the application of microbubble (Yanuhar et al. 2021).

the working system. Therefore, accuracy is needed in regulating and measuring the size of each part.

The use of a microbubble generator (MBG) indirectly also increases the production and quality of fish produced. Usually, fish production will increase in terms of health, color, and growth (Tsuge 2019; Yanuhar et al. 2021). MBG is highly recommended for use in fish-farming because it can improve water quality (Figure 2.6) in the form of high and long-lasting DO, stable pH, and lower organic matter content in ponds.

Based on the complete water quality measurement, the results obtained on all parameters indicate the optimal value. Especially the stable and increasing values of temperature, pH, and DO. One of the factors that play a role in determining the success of fish-farming is water quality management, because fish are aquatic animals whose life, health, and growth depend on water quality as their medium of life (Abdel-Tawwab et al. 2019; Yanuhar et al. 2019). Water temperature affects the diet and growth of fish. Fish generally experience stress and disease outbreaks when temperatures chronically approach their maximum tolerances or fluctuate abruptly (Patang et al. 2021). DO is one of the main limiting parameters in aquaculture because it will affect the anaerobic metabolic system, which requires DO at an efficient level. The DO value of waters is usually influenced by several factors, namely, the respiration process of aquatic organisms, photosynthesis, and oxygen diffusion from the atmosphere. The O_2 solubility in water pond depends on environment conditions, such as salinity, water temperature, and O_2 partial pressure. Most fish have some ability to adjust to DO levels fluctuations, but if severe hypoxia persists, the fish will eventually die. DO level of about 5 mg/L is essential for acceptable fish growth, and as DO levels decrease, respiratory and feeding activity also decreases (Aboagye and Allen 2018). Abdel-Tawwab et al. (2019) reported that the differences in DO levels and fish size have impact on the growth, feed utilization, physiological changes, and fish innate immunity.

2.7 PRINCIPLES AND CONCEPTS OF INTEGRATED WATER RESOURCES MANAGEMENT FOR SUSTAINABLE AQUACULTURE

The ecosystem approach is used for the integration of strategies in aquaculture related to the wider ecosystem to realize sustainable aquaculture, equity, and a strong and interrelated socioecological system. It is guided by three strategic principles, namely:

1. Cultivation development and management must pay attention to various ecosystem functions and services so that they do not threaten the sustainability of the community.
2. Aquaculture carried out must realize the welfare of human resources and equality for all parties involved. Cultivation should be further developed with other sectors, such as appropriate goals and policies.
3. An ecosystem approach to aquaculture should be based on the principle of providing a planning and governance framework that integrates all sectors effectively.

This approach provides a clear mechanism for government authorities and producers to engage with each other for the effective and sustainable management of aquaculture operations. Together they embrace the environmental, socioeconomic, and governance objectives of the sector. There is a gap between the two sides in applying this approach to aquaculture, with general considerations regarding the development of aquaculture at the global level and practitioners. Based on this, practical entry points have been found for the application of this approach, including spatial planning, zoning, and biophysical carrying capacity (Brugere et al. 2019).

Good water resource management can be realized by implementing good water management. According to Katusiime and Schütt (2020), IWRM approaches and water governance have a close relationship. Water governance is very important because it can describe the powers, rights, decisions, and priorities related to water resources. Integrated water resources management (IWRM) relies on three foundations, namely, (a) building a supportive environment, (b) institutional framework, and (c) management instruments.

The application of the integrated water resources management approach aims to maximize the carrying capacity of the waters in fish-farming activities that can be done by using the IoT as a form of fish-farming management by utilizing technological advances and the application of the IoT in the management of fish-farming in an integrated manner through monitoring and quality control in real time. Advances in sensor technology, wireless communication, and the IoT have paved the way for effective monitoring and control of water quality in aquaculture, as well as in increasing productivity and efficiency in aquaculture. Fisheries is an important and fast-growing sector that has seen the adoption of advanced technologies, such as the adoption of IoT (Ouyang et al. 2021). According to Dhenuvakonda and Sharma (2020), IoT is about connecting everyday things embedded with electronics, software, and sensors to the internet, enabling them to collect and exchange data. The use of IoT has high potential in the field of fisheries and aquaculture because it has advantages in a competitive market. The application of IoT systems can be used to monitor water quality, such as pH, DO, temperature, ammonia, automatic feed distribution, etc. The monitored parameters are sent to the remote control via the Internet of Things (IoT) and are displayed. IoT is an emerging technology, a technological advancement that provides large-capacity data storage via the internet.

The use of the Internet of Things (IoT) for fish-farming can improve the economic status of the entrepreneurs involved, which will also increase the country's economic growth through the aquaculture industry (Ismail et al. 2019). This application has started to change traditional cultivation into intensive cultivation. However, aquaculture IoT equipment is often used in outdoor ponds located in remote areas. Mistakes are common in this difficult environment. In addition, the use of the IoT requires professional experts because if an error occurs, the expert must check the IoT in the field and carry out maintenance outdoors (Kumar and Samalla 2019).

Fish-farming can provide high profits if water quality management and monitoring are carried out properly. One of the efforts to improve water quality in fish-farming is using microbubble (MBG) to create high levels of DO. In principle, pressurized water is pushed through the orifice to create a lower-pressure area at a certain point. Outside air will be automatically sucked into the pipe to this lower-pressure area through the porous pipe so that microbubbles will be produced downstream of the hole. According to Yun et al. (2019), *microbubbles* are generally defined as small gaseous particles with a diameter of less than 100 μm. Macrobubbles exhibit different characteristics from microbubbles, which are larger than 100 μm. Microbubble generators have been used in aquaculture to improve water quality and reduce mortality, as well as promote growth by increasing available oxygen. These bubbles rise slowly to the surface and remain in solution for some time. In addition, they can assist in dissolving gases into solution. The use of microbubble is emerging as a promising option to improve the fishing industry. The application of microbubble in the fish-farming industry helps increase productivity and water quality. For example, microbubble has been used in oyster culture to increase growth rate, shell opening, and increase oyster blood flow rate. In the intensive cultivation of tilapia, the application of microbubble also encourages the growth rate of fish (both length and weight). A special type of microbubble has also been applied to purify fishery wastewater.

2.8   CHALLENGES/OBSTACLES IN THE FUTURE

Aquaculture is one of the productive sectors that have applied technology in its implementation, both on household scale cultivation and large industrial scale. The application of technology is tailored to the needs of aquaculture ponds, such as real-time monitoring of water quality and recommendations for actions to be taken. The technology is usually based on the Internet of Things (IoT). Several aquaculture units are widespread in almost all areas, both urban and rural. Especially in rural areas, often not equipped with adequate electricity and connections. This is an inhibiting factor in the application of IoT technology in fish-farming so that all monitoring is carried out manually and inefficiently. For these two types of companies, the challenges are different to enable increased productivity, increased yields, and reduced workforce. For the first type, the main challenge is the measurement and communication of devices with unstable power grids and no internet connection. For the second type, the main challenge is direct communication with farmers remotely in places where there are environmental constraints. Regardless of the specificity, all farmers have the same need through IoT integration: real-time and continuous monitoring of aquaculture pond water quality is the basis of the application of IoT technology in aquaculture. This makes it possible to establish the state of the situation in real time in the pool; 24-7 action makes it possible to have a global vision of the grower and then act if necessary. This section covers the development of probes capable of measuring various physico-chemical parameters at a lower cost, that are easy to use, and with very high-frequency measurements and transmission capacities. For data acquisition for big data analysis and prediction, anticipation is at the heart of the problem in aquaculture. The sooner the information arrives at the farmer, the more choices of action can be taken. Therefore, real-time measurement of water quality becomes a parameter for IoT development in relation to activity in and around fishponds through cameras and environmental data (weather, satellite data), which will enable the creation of very large datasets for big data analysis. This is to predict and anticipate risks (Lafont et al. 2019).

The implementation of IWRM requires social participation to monitor and evaluate the implemented water management. A strategy for monitoring capacity development management should be defined, implemented, and continuously updated. Management evaluation should be based on the selected indicators to find out the results of the measures that have been taken and their effect on water quality (Schuler et al. 2020). According to Jazi (2021), IWRM explicitly challenges fragmented conventional water management and development systems and emphasizes an integrated approach with more coordinated decision-making across sectors and scales.

2.9   CONCLUSION

Water quality management using the Internet of Things (IoT)–based microbubble has positive impact on waters; by increasing quality carrying capacity, the distribution of oxygen in the waters is more stable and even. Then, it provides convenience and accuracy in monitoring water quality. The use of IoT has high potential in the field of fisheries and aquaculture in monitoring water quality, such as in measuring pH, dissolved oxygen, temperature, ammonia, automatic feed distribution, etc. This is very helpful for aquaculture actors in maximizing productivity and improving the quality of the products produced and also increasing in a competitive market.

REFERENCES

Abdel-Tawwab, M., Monier, M.N., Hoseinifar, S.H., Faggio, C., 2019. Fish response to hypoxia stress: Growth, physiological, and immunological biomarkers. Fish Physiol. Biochem. 45, 997–1013.

Aboagye, D.L., Allen, P.J., 2018. Effects of acute and chronic hypoxia on acid—base regulation, hematology, ion, and osmoregulation of juvenile American paddlefish. J. Comp. Physiol. B 188, 77–88.

Afifi, M., Abdelkader, M.F., Ghoneim, A., 2018. An IoT system for continuous monitoring and burst detection in intermittent water distribution networks. In: 2018 International Conference on Innovative Trends in Computer Engineering (ITCE). IEEE, Piscataway, NJ, pp. 240–247.

Ahmed, N., Turchini, G.M., 2021. Recirculating aquaculture systems (RAS): Environmental solution and climate change adaptation. J. Clean. Prod. 297, 126604. https://doi.org/10.1016/j.jclepro.2021.126604

Alam, T., 2020. Device-to-device communications in cloud, manet and internet of things integrated architecture. J. Inf. Syst. Eng. Bus. Intell. 18–26. https://doi.org/10.20473/jisebi.6.1.18–26

Bekele, E., Page, D., Vanderzalm, J., Kaksonen, A., Gonzalez, D., 2018. Water recycling via aquifers for sustainable urban water quality management: Current status, challenges and opportunities. Water 10, 457.

Biju, S.M., Mathew, A., 2020. Internet of Things (IoT): Securing the next frontier in connectivity. Intl. J. of Sci. Eng. and Sci. 4(5), 18–21.

Brasileiro, P.P.F., dos Santos, L.B., Chaprão, M.J., de Almeida, D.G., Roque, B.A.C., dos Santos, V.A., Sarubbo, L.A., Benachour, M., 2020. Construction of a microbubble generation and measurement unit for use in flotation systems. Chem. Eng. Res. Des. 153, 212–219.

Brugere, C., Aguilar-Manjarrez, J., Beveridge, M.C.M., Soto, D., 2019. The ecosystem approach to aquaculture 10 years on—a critical review and consideration of its future role in blue growth. Rev. Aquac. 11, 493–514.

Budhijanto, W., Darlianto, D., Pradana, Y.S., Hartono, M., 2017. Application of micro bubble generator as low cost and high efficient aerator for sustainable fresh water fish farming. In: AIP Conference Proceedings. AIP Publishing LLC, New York, p. 110008.

Cario, G., Casavola, A., Gjanci, P., Lupia, M., Petrioli, C., Spaccini, D., 2017. Long lasting underwater wireless sensors network for water quality monitoring in fish farms. In: Oceans 2017-Aberdeen. IEEE, Piscataway, NJ, pp. 1–6.

Chang, N.-B., Bai, K., Chen, C.-F., 2017. Integrating multisensor satellite data merging and image reconstruction in support of machine learning for better water quality management. J. Environ. Manage. 201, 227–240.

Deb, S., Noori, M.T., Rao, P.S., 2020. Application of biofloc technology for Indian major carp culture (polyculture) along with water quality management. Aquac. Eng. 91, 102106.

Detenbeck, N., Le, A., Piscopo, A., Stagnitta, T., White, J., McKenrick, S., Brown, A., ten Brink, M., 2018. User-Friendly Decision Support for Integrated Water Management: EPA's Watershed Management Optimization Support Tool (WMOST). In Proceedings, 9th International Congress on Environmental Modelling and Software (iEMSs), Fort Collins, CO, June 24–28, 2018. Brigham Young University, Provo, UT, 128.

Dhenuvakonda, K., Sharma, A., 2020. Mobile Apps and Internet of things (IoT): A Promising Future for Indian Fisheries and Aquaculture Sector. J. Entomol Zool Stud. 8, 1659–1669.

Doss, P., 2018. Smart Water Conservation and Management System Using IOT. vol 7109, 9–12.

Ekawati, A.W., Ulfa, S.M., Dewi, C.S.U., Amin, A.A., Salamah, L.N., Yanuar, A.T., Kurniawan, A., 2021. Analysis of aquaponic-recirculation aquaculture system (A-Ras) application in the catfish (Clarias gariepinus) aquaculture in Indonesia. Aquac. Stud. 21, 93–100.

Encinas, C., Ruiz, E., Cortez, J., Espinoza, A., 2017. Design and implementation of a distributed IoT system for the monitoring of water quality in aquaculture. In: 2017 Wireless Telecommunications Symposium (WTS). IEEE, Piscataway, NJ, pp. 1–7. https://doi.org/10.1109/WTS.2017.7943540

Gao, G., Xiao, K., Chen, M., 2019. An intelligent IoT-based control and traceability system to forecast and maintain water quality in freshwater fish farms. Comput. Electron. Agric. 166, 105013.

Geetha, S., Gouthami, S., 2016. Internet of things enabled real time water quality monitoring system. Smart Water 2, 1–19.

Gismalla, M.S.M., Abdullah, M.F.L., 2017. Device to device communication for internet of things ecosystem: An overview. Int. J. Integr. Eng. 9, 118–123.

Gitau, M.W., Chen, J., Ma, Z., 2016. Water quality indices as tools for decision making and management. Water Resour. Manag. 30, 2591–2610.

Huan, J., Li, H., Wu, F., Cao, W., 2020. Design of water quality monitoring system for aquaculture ponds based on NB-IoT. Aquac. Eng. 90, 102088.

Hussen, B., Mekonnen, A., Pingale, S.M., 2018. Integrated water resources management under climate change scenarios in the sub-basin of Abaya-Chamo, Ethiopia. Model. Earth Syst. Environ. 4, 221–240.

Hwang, J.H., Park, S.H., Song, C.M., 2020. A study on an integrated water quantity and water quality evaluation method for the implementation of integrated water resource management policies in the republic of Korea. Water 12, 2346.

Ighalo, J.O., Adeniyi, A.G., Marques, G., 2021. Internet of things for water quality monitoring and assessment: A comprehensive review. In: Artificial Intelligence for Sustainable Development: Theory, Practice and Future Applications. Springer Nature, Switzerland, pp. 245–259.

Ismail, R., Shafinah, K., Latif, K., 2020. A proposed model of fishpond water quality measurement and monitoring system based on Internet of Things (IoT). In: IOP Conference Series: Earth and Environmental Science. IOP Publishing, Bristol, p. 12016.

Ismail, W., Shinohara, N., Nameh, S., Boonsong, W., Alifah, S., Kamaludin, K.H., Anwar, T., 2019. Development of Smart Aquaculture Quality Monitoring (AQM) System with Internet of Things (IoT). IEEE, Perak, Malaysia, pp. 1–3.

Jazi, H.H., 2021. Integrated water resources management: A tool for sustainable development. Futur. Eng. J. 2, 1–14.

Jenti, U.B., Nurhayati, I., 2014. The effect of the use of filtration media on the quality of dug well water in Tambak Rejo Waru Village, Sidoarjo Regency. WAKTU J. Tek. UNIPA 12, 34–38.

Jian, Z., Jin, Z., Xie, X., Lu, Y., Li, G., Chen, X., Baker, T., 2021. Sysnif: A log-based workflow construction method and performance measurement in intelligent IoT system. Measurement 186, 110175.

Kageyama, T., Miura, M., Maeda, A., Mori, A., Lee, S.-S., 2016. A wireless sensor network platform for water quality monitoring. In: 2016 IEEE Sensors. IEEE, Piscataway, NJ, pp. 1–3.

Kamaruidzaman, N.S., Rahmat, S.N., 2020. Water monitoring system embedded with internet of things (IoT) device: A review. In: IOP Conference Series: Earth and Environmental Science. IOP Publishing, Bristol, p. 12068.

Katusiime, J., Schütt, B., 2020. Integrated water resources management approaches to improve water resources governance. Water 12, 3424.

Kumar, M.J.V., Samalla, K., 2019. Design and development of water quality monitoring system in IoT. Int. J. Recent Technol. Eng. 7, 527–533.

Lafont, M., Dupont, S., Cousin, P., Vallauri, A., Dupont, C., 2019. Back to the future: IoT to improve aquaculture: Real-time monitoring and algorithmic prediction of water parameters for aquaculture needs. In: 2019 Global IoT Summit (GIoTS). IEEE, Piscataway, NJ, pp. 1–6.

Liew, K.C.S., Rasdi, A., Budhijanto, W., Yusoff, M.H.M., Bilad, M.R., Shamsuddin, N., Md Nordin, N.A.H., Putra, Z.A., 2020. Porous venturi-orifice microbubble generator for oxygen dissolution in water. Processes 8, 1266.

Lim, Y.S., Ganesan, P., Varman, M., Hamad, F.A., Krishnasamy, S., 2021. Effects of microbubble aeration on water quality and growth performance of Litopenaeus vannamei in biofloc system. Aquac. Eng. 93, 102159.

Little, A.G., Loughland, I., Seebacher, F., 2020. What do warming waters mean for fish physiology and fisheries? J. Fish Biol. 97, 328–340.

Liu, H., Li, H., Wei, H., Zhu, X., Han, D., Jin, J., Yang, Y., Xie, S., 2019. Biofloc formation improves water quality and fish yield in a freshwater pond aquaculture system. Aquaculture 506, 256–269.

Magallón-Servín, J.A., Bórquez-López, R.A., Quadros-Seiffert, W., Magallón-Barajas, F.J., Casillas-Hernández, R., 2020. Influencia de la columna de agua y eficiencia energética de dos tipos de generadores de microburbujas en un cultivo hiper-intensivo de camarón. Rev. Latinoam. Recur. Nat. 16, 79–87.

Mahmoud, A.E.D., Umachandran, K., Sawicka, B., Mtewa, T.K., 2021. 26 — Water resources security and management for sustainable communities. In: Mtewa, A.G., Egbuna, C. (Eds.). Phytochemistry, the Military and Health. Elsevier, Amsterdam, pp. 509–522.

Majumdar, A.K., 2018. Optical Wireless Communications for Broadband Global Internet Connectivity: Fundamentals and Potential Applications. Elsevier. Amsterdam, pp. 1–290.

Martinsen, G., Liu, S., Mo, X., Bauer-Gottwein, P., 2019. Joint optimization of water allocation and water quality management in Haihe River basin. Sci. Total Environ. 654, 72–84.

Maulu, S., Hasimuna, O.J., Haambiya, L.H., Monde, C., Musuka, C.G., Makorwa, T.H., Munganga, B.P., Phiri, K.J., Nsekanabo, J.D., 2021. Climate change effects on aquaculture production: Sustainability implications, mitigation, and adaptations. Front. Sustain. Food Syst. 5, 609097. https://doi.org/10.3389/fsufs.2021.609097

Mishra, B., Tiwari, A., Mahmoud, A.E.D., 2022. Microalgal potential for sustainable aquaculture applications: Bioremediation, biocontrol, aquafeed. Clean Technol Environ. Policy, 1–13.

Olatinwo, S.O., Joubert, T.H., 2019. Efficient energy resource utilization in a wireless sensor system for monitoring water quality. EURASIP J. Wirel. Commun. Netw. 2019, 1–22.

Omer, N.H., 2019. Water quality parameters. In: Water Quality Science, Assessments and Policy, IntechOpen, London, pp. 1–34.

Ouyang, B., Wills, P.S., Tang, Y., Hallstrom, J.O., Su, T.-C., Namuduri, K., Mukherjee, S., Rodriguez-Labra, J.I., Li, Y., Den Ouden, C.J., 2021. Initial development of the hybrid aerial underwater robotic system (HAUCS): Internet of things (IOT) for aquaculture farms. IEEE, Piscataway, NJ, Volume 8, p. 14013–14027.

Pasika, S., Gandla, S.T., 2020. Smart water quality monitoring system with cost-effective using IoT. Heliyon 6, e04096.

Patang, P., Nurmila, N., Wahab, I., 2021. Modification of aeration on increased dissolved oxygen affecting growth and survival rates in Tilapia (Oreocrobis Niloticus). J. Pendidik. Teknol. Pertan. 5, 65–72.

Pires, A., Morato, J., Peixoto, H., Botero, V., Zuluaga, L., Figueroa, A., 2017. Sustainability Assessment of indicators for integrated water resources management. Sci. Total Environ. 578, 139–147.

Prapti, D.R., Mohamed Shariff, A.R., Che Man, H., Ramli, N.M., Perumal, T., Shariff, M., 2022. Internet of Things (IoT)-based aquaculture: An overview of IoT application on water quality monitoring. Rev. Aquac. 14, 979–992.

Preena, P.G., Rejish Kumar, V.J., Singh, I.S.B., 2021. Nitrification and denitrification in recirculating aquaculture systems: The processes and players. Rev. Aquac. 13, 2053–2075.

Priyanka, J.S., Kiran, M.S., Nalla, P., 2022. A secured IoT-based health care monitoring System using body sensor network. In: Emergent Converging Technologies and Biomedical Systems. Springer, Singapore, pp. 483–490.

Purnama, J., Arief, Z., 2018. Counseling and training on water purification as a step to minimize the shortage of clean water in Tulung Village, Gresik Regency. J. Abdikarya J. Karya Pengabdi. Dosen Dan Mhs. 1, 72–76.

Purwoto, S., Purwanto, T., Hakim, L., 2015. River water purification by coagulation, filtration, absorption and ion exchange treatment. WAKTU J. Tek. UNIPA 13, 45–53.

Qaddoori, A.S., Saud, J.H., Hamad, F.A., 2021. A classifier design for micro bubble generators based on deep learning technique. Mater. Today Proc. https://doi.org/10.1016/j.matpr.2021.07.013 (Article In Press).

Ragavan, E., Hariharan, C., Aravindraj, N., Manivannan, S.S., 2016. Real time water quality monitoring system. Int. J. Pharm. Technol. 8, 26199–26205.

Ramadhan, A.J., Ali, A.M., Kareem, H.K., 2020. Smart water-quality monitoring system based on enabled real-time internet of things. J. Eng. Sci. Technol. 15, 3514–3527.

Robles-Porchas, G.R., Gollas-Galván, T., Martínez-Porchas, M., Martínez-Cordova, L.R., Miranda-Baeza, A., Vargas-Albores, F., 2020. The nitrification process for nitrogen removal in biofloc system aquaculture. Rev. Aquac. 12, 2228–2249.

Schuler, A.E., LIMA, J., Cruz, M.A.S., 2020. Chapter 5: Integrated water resources management. pp. 49–58. https://ainfo.cnptia.embrapa.br/digital/bitstream/item/222411/1/SDG-6-cap-5-2020.pdf

Setijaningsih, L., Suryaningrum, L.H., 2015. Utilization of catfish (Clarias batrachus) culture waste for tilapia (Oreochromis niloticus) with a recirculation system. Ber. Biol. 14, 287–293.

Son, Y., Kang, M., Kim, Y., Yoon, I., Noh, D.K., 2020. Energy-efficient cluster management using a mobile charger for solar-powered wireless sensor networks. Sensors 20, 3668.

Suryanto, H., Aminuddin, Uun Yanuhar, Muhamad Syaifuddin, Bili Darnanto Susilo;, Nico Rahman Caesar, 2022. Community empowerment through the application of Iot controlled microbubble technology in catfish ponds in pokdakan roi lele, Malang regency. J. Serv. Educ. Technol. 3, 1–9.

Tsuge, H., 2019. Micro-and Nanobubbles: Fundamentals and Applications. CRC Press, Pan Stanford, Boca Raton.

Ullo, S.L., Sinha, G.R., 2020. Advances in smart environment monitoring systems using IoT and sensors. Sensors 20, 3113.

Valentini, R., Belelli Marchesini, L., Gianelle, D., Sala, G., Yarovslavtsev, A., Vasenev, V., Castaldi, S., 2019. New tree monitoring systems: From Industry 4.0 to Nature 4.0. Ann. Silvicul. Res. 43(2), 84–88.

Wilson, D.A., Pun, K., Ganesan, P.B., Hamad, F., 2021. Geometrical Optimization of a Venturi-Type Microbubble Generator Using CFD Simulation and Experimental Measurements. Designs 5, 4.

Xiao, R., Wei, Y., An, D., Li, D., Ta, X., Wu, Y., Ren, Q., 2019. A review on the research status and development trend of equipment in water treatment processes of recirculating aquaculture systems. Rev. Aquac. 11, 863–895.

Yanuhar, U., Anitasari, S., Muslimin, A., Taufiq, A., Junirahma, N. S., Caesar, N. R., 2021. Application of microbubble in koi pools for sustainable water quality management in nglegok village, blitar regency. In Proceedings of the National Seminar on Fisheries and Marine Affairs 9, Faculty of Fisheries and Marine Sciences Universitas Brawijaya Malang, pp. 90–94.

Yanuhar, U., Arfiati, D., 2018. Opportunity plankton as vector transmission of koi herpes virus infection on carp (Cyprinus carpio). Aquac. Aquarium, Conserv. Legis. 11, 1869–1881.

Yanuhar, U., Caesar, N.R., Setiawan, F., Sumsanto, M., Musa, M., Wuragil, D.K., 2019. The aquatic environmental quality of koi fish (Cyprinus carpio) pond infected by Myxobolus sp. based on the biological status of the phytoplankton. In: Journal of Physics: Conference Series. IOP Publishing, London, p. 12017.

Yanuhar, U., Musa, M., Evanuarini, H., Wuragil, D. K., Shodiq, F., Junirahma, N. S., Caesar, N. R., 2022. Increasing the potential of community groups through application of microbubble technology and internet of things at pokdakan rukun abadi makmur. J. Serv. Educ. Technol. 3, 1–7.

Yun, S., Giri, S.S., Kim, H.J., Kim, S.G., Kim, S.W., Kang, J.W., Han, S.J., Kwon, J., Oh, W.T., Chi, C., 2019. Enhanced bath immersion vaccination through microbubble treatment in the cyprinid loach. Fish Shellfish Immunol. 91, 12–18.

Zhang, Q., Li, Z., 2020. Development of an interval quadratic programming water quality management model and its solution algorithms. J. Clean. Prod. 249, 119319.

Ziarati, P., El-Esawi, M., Sawicka, B., Umachandran, K., Mahmoud, A.E.D., Hochwimmer, B., Vambol, S., Vambol, V., 2019. Investigation of prospects for phytoremediation treatment of soils contaminated with heavy metals. J. Med Discov. 4, 1–16.

3 Artificial Intelligence for Sustainable Water Management and Treatment

*Manal Fawzy, Alaa El Din Mahmoud,
Mohamed Hosny, Nourhan El Maghrabi,
and Ahmed M. Abdelfatah*

CONTENTS

3.1 INTRODUCTION

Agriculture expansion has led to a significant increase in per-capita income in rural areas. It makes a substantial contribution to the economic prosperity of industrialized countries, as well as playing an important role in the economies of emerging countries. By 2050, the world's population is projected to reach approximately ten billion, expanding agricultural order in a condition of humble financial development by roughly 50% compared to that in 2013 (Mogili et al. 2018). As a result, putting a greater prominence on the agricultural sector is both rational and feasible.

Water is, without question, a critical resource for all living things (Mahmoud et al. 2016; Mahmoud 2020; Hosny et al. 2021), and its management is now a necessity because of discharge of various contaminants into surface- and groundwater systems (Gwenzi et al. 2022; Mahmoud et al. 2022a, 2022b). Herein, we focus on the agricultural sector, due to water scarcity (Huang et al. 2021; Sawicka et al. 2021). The operations of irrigation systems are still manual in most countries around the world, which reduces water efficiency. Therefore, the current practices and activities in agricultural fields are not sustainable. Agriculture is the world's largest consumer of fresh water, accounting for up to 70% of total use (Kamienski et al. 2019; Mahmoud and Fawzy 2021). Agricultural management practices must reconsider the long-term

DOI: 10.1201/9781003260455-3

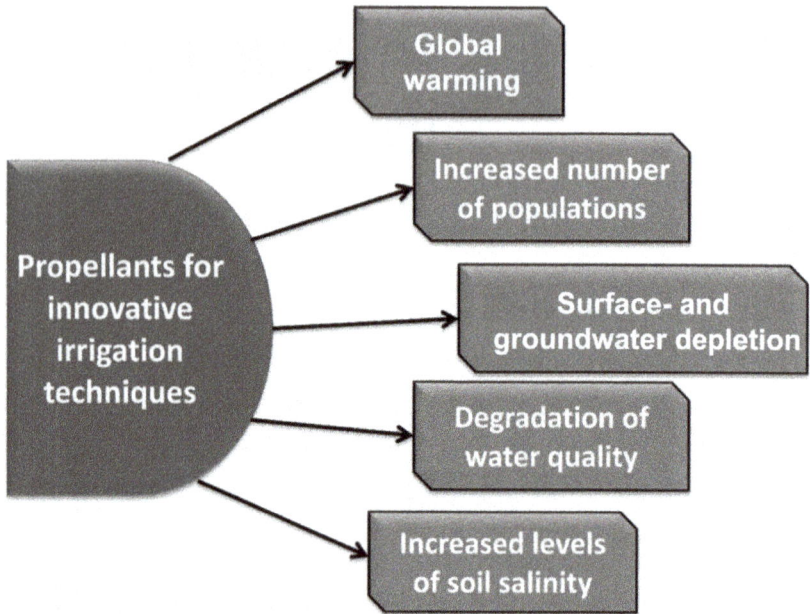

FIGURE 3.1   Different propellants for innovative irrigation techniques.

availability and quality of fresh water while still supplying water for the production of crops in a hot and dry climate (Papadopoulou et al. 2020). However, water scarcity issue is on the edge in Mediterranean nations that efforts should also be focused on water-saving measures (Iglesias et al. 2011).

There are many factors determining such issue—for instance, dramatic increase in the population's number, global warming, increased levels of soil salinity, degradation of the quality of available water, and depletion of surface- and groundwater sources (Figure 3.1). It is obvious that there has been a progressive reduction in groundwater level over the last ten years, as well as bad monsoon conditions. This needs irrigation system automation in order to properly utilize water resources, and most studies are currently focusing on irrigation system automation (Barkunan et al. 2019a). Therefore, innovative techniques have to be applied in the irrigation of agricultural lands.

The agricultural sector is emphasizing the need for effectively managing existing water resources to maintain the economic sector's survival (Monckton 2019; García et al. 2020b; Mahmoud et al. 2021b). It has to be noticed that the number of publications related to the use of water increased substantially, especially in the last decade, as the number reached almost 1,000 in 2020 (Figure 3.2a), and most of this water was employed in agricultural activities, as shown in Figure 3.2b.

Both water and food security depend on sustainable water resource management (Mishra et al. 2022). Besides the shortage of water availability, innovative irrigation techniques have become a quintessential demand in modern life. In an attempt to minimize productivity loss due to water stress, farmers overirrigate (spray more

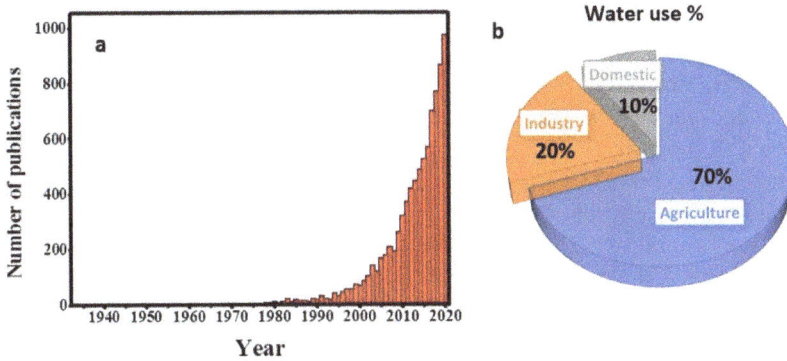

FIGURE 3.2 (a) Number of publications concerning the use of water. (b) Water use percentage by category.

Source: Data obtained from Scopus in June 2021.

water than is required), not only putting production at risk but also wasting water and energy (Venot et al. 2017).

Precision irrigation, on the other hand, may make better use of water by minimizing underirrigation and overirrigation. In agriculture, smart water management for precise irrigation is critical for crop growth (Kamienski et al. 2018). However, it is also critical to establish a link between climate policy and water management. Different variables, such as water demand and the impact of global warming on hydrological resources, can have an impact on water management. Climate change and its consequences are a recurring theme in water resource and agriculture research publications (García et al. 2020a).

As a result of the potential impacts of global warming (Mansoor et al. 2022), water adaptation strategies are being considered to assure water supply for food production, as well as to sustain ecosystems (Iglesias et al. 2018). Climate change will most certainly have an impact on rural society and the environment, posing or exacerbating new water management difficulties (Iglesias et al. 2018; Özerol et al. 2020). To satisfy agricultural demands, increasing water usage efficiency is required, but not sufficient. However, it is conceivable to considerably increase water delivery to agriculture in areas where it has been underperforming, therefore alleviating existing and future imbalances (Sawicka et al. 2021).

The concept of artificial intelligence (AI) is to develop technology that behaves like a human brain (Parekh et al. 2019). AI is the science of creating intelligent machines and programs (Kodali et al. 2016).

AI is built by understanding how the human brain thinks, learns, decides, and works while solving a problem (Parekh et al. 2020). These intelligent systems, like the human brain, are supplied with training data and then provide us with the desired result for every valid input. The core parts of AI are machine learning (ML) and deep learning (DL) (Patel et al. 2020). The ability to learn anything without being explicitly taught is known as ML, and deep neural network learning is known as DL.

AI has a major role in agriculture in numerous areas—for instance, developing methods for harvest planning, identifying flowers and leaves, and integrating in advanced irrigation techniques. It has enhanced real-time monitoring of crop production and various parameters, harvesting, and processing (Liakos et al. 2018). Plessen (2019) presented a harvest planning method based on the connection of crop assignment and vehicle routing. With this new technology, workers who formerly worked in only a few industrial sectors can now work in a variety of fields.

The application of AI to agriculture depends on other technological advances, such as robotics, the availability of cheap sensors, the internet, big data analytics, cameras, and drone technology. AI systems are able to provide predictive insights into the planting of certain crops in a given year, as well as the optimal dates to sow and harvest, by analyzing soil parameters in a specific area, such as moisture, weather, historic crop performance, and temperature; as a result, crop yields will be improved and water, fertilizers, and pesticides will not be overused (Awokuse and Xie 2015). Its technologies applications reduce the adverse impacts on ecosystems and increase the safety environment of the workers, which may lower the prices of food and increase food production to levels that meet the high demand of the increasing populations (Eli-Chukwu 2019).

Thanks to AI-based technology solutions, farmers have been able to produce more with less input, improve the quality of their product, and shorten the time it takes for their produced crops to reach market. It is predicted that farmers will create 4.1 million data points by 2050 using millions of linked devices.

This chapter aims to provide information about the most recent technologies utilized in agricultural systems, particularly the various types of irrigation techniques, the propellants behind searching for innovative irrigation techniques, and the artificial intelligence sensors that are commonly used for the sake of water preservation and conservation. Furthermore, we highlight on multi-intelligent control system of water management in the agricultural sector.

3.2 IRRIGATION TECHNIQUES IN AGRICULTURE

Various techniques were utilized in irrigation of agricultural lands, including surface and subsurface irrigation, drip, sprinkler, center pivot, and lateral move irrigation, as illustrated in Figure 3.3. Among these numerous irrigation techniques, drip irrigation was deemed to be the most efficient technique, as it could be used to apply water at the time when plants and agricultural crops need it most and in rates needed for proper crop production, as elaborated by many workers (Barkunan et al. 2019b; Wang et al. 2020; Zou et al. 2020). Drip irrigation (microirrigation) is a method of supplying water to the soil surface or root zone using pressurized pipes and drippers to drop water gently (Barkunan et al. 2019b). When compared to typical flood irrigation, drip irrigation saves about 40–80% of water (Barman et al. 2020; Tokarev et al. 2020).

Figure 3.4 shows different numbers of publications concerning the irrigation techniques collected from Scopus data, and it was found that the drip irrigation technique garnered most of the researchers' interest, as indicated by the number of publications that significantly rose from just 1 in 1966 to 431 in 2020, as shown in Figure 3.4b.

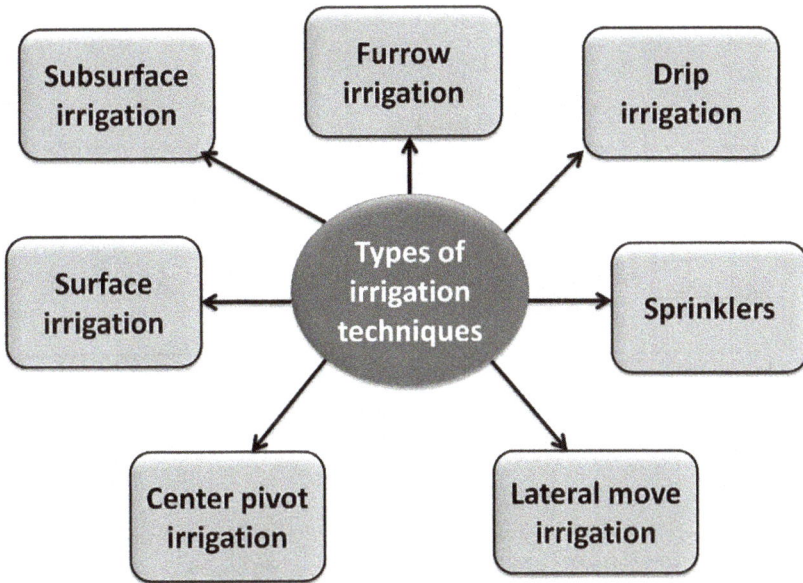

FIGURE 3.3   Types of irrigation techniques.

On the other hand, lateral irrigation received the least attention, as the number of publications in this point varied from 1 in 1990 to only 2 publications in 2020, as illustrated by Figure 3.4e. Similarly, the number of publications on sprayer irrigation technique reached only 6 publications in 2019 (Figure 3.4f). Regarding other irrigation techniques, the increasing trend in their publication numbers was quite similar, and they all were recorded to be more than 20 during 2020 (Figures 3.4a, c, d, and g).

3.3   ARTIFICIAL INTELLIGENCE FOR OPTIMIZATION OF IRRIGATION

With increasing food demand and the population growth, the available freshwater resources will not be adequate to cope up with this high demand. That's why more efficient technologies are urgently needed to guarantee proper usage of water resources (Mahmoud et al. 2020; Mahmoud et al. 2021a; Vambol et al. 2021).

In recent years, AI has begun to play a larger part in many fields, and one of them is the agricultural sector. Recent advancements in offering automation between diverse equipment using sensory applications by artificial intelligence in the agro-industrial sector indicate that operation accuracy improves while prices reduce (Tran and Ha 2015; Muangprathub et al. 2019). To reduce the switching number of pump operations, Chunduri and Menaka (2019) suggested a genetic control method with different threshold levels for a medium-sized city. The results showed that the methods could be utilized to optimize a nonlinear multivariable cost function satisfactorily. Muangprathub et al. (2019) demonstrated a crop watering control system that uses a wireless sensor network to control environmental conditions. Temperature

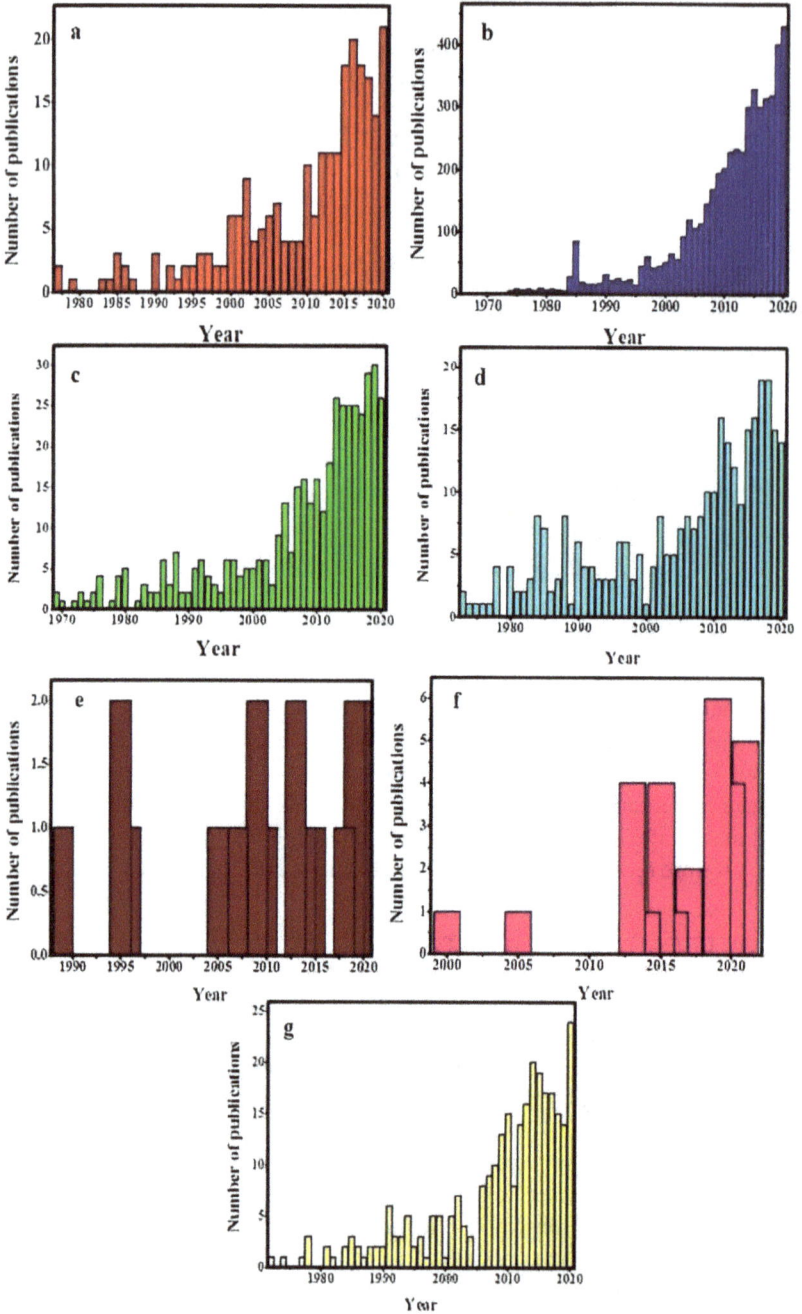

FIGURE 3.4   Number of publications for (a) center pivot irrigation, (b) drip irrigation, (c) flood irrigation, (d) furrow irrigation, (e) lateral irrigation, (f) sprayers, and (g) subsurface irrigation.

Source: Data obtained from Scopus in June 2021.

and humidity were discovered to be important factors in production. The ideal temperature was found to be between 29 and 32°C for increasing the productivity of home-grown crops. Furthermore, the author discovered that the maximum output for vegetables and lemons may be attained around 79–81% humidity.

Automatic irrigation scheduling techniques have replaced manual irrigation independent of soil water measurement, but with taking plant evapotranspiration into consideration, which is dependent on crop factors, including plant density, the growth stage, and soil properties and other different atmospheric parameters, such as solar radiations, humidity, and wind speed (Talaviya et al. 2020).

Smart irrigation technology is primarily utilized to enhance production by detecting water levels, nutrient content, weather predictions, and soil temperature without the use of a huge number of manpower. Automatic plant irrigators are used to ensure soil fertility and efficient use of water resources. They are installed on the field using wireless technology for drip irrigation.

Researchers created soil moisture sensors and raindrop sensors that are controlled by a wireless broadband network and powered by solar panels to identify the demand for water supply to the crops. Using GSM module, the farmer would be notified of the moisture content in the soil using these sensors through SMS on their cell phone. As a result, the farmer can use SMS to turn on and off the water supply. This technology allows farmers to discover spots in their crops that demand more water and may prevent them from overwatering if it is raining.

Furthermore, the microcontroller controls the irrigator pump's on/off operation, and machine-to-machine (M2M) technology has been developed to facilitate communication and data sharing among nodes in the agricultural field, as well as to the server or cloud, via the main network (Shekhar et al. 2017). They created an automated robotic model to detect moisture on the Arduino's and Raspberry Pi 3's moisture content and temperature. Data is collected at regular intervals and transferred to the Arduino microcontroller (which is coupled to edge-level hardware), which converts the analogue input to digital. The signal is transferred to the Raspberry Pi 3 (which is embedded with the KNN algorithm), which then transmits it to Arduino to activate the irrigation water source.

Remote sensors were developed by (Savitha and UmaMaheswari 2018) using the technology of Arduino in order to increase production up to 40%, which encourages the idea of efficient and automated irrigation system. The moisture content and temperature of the Arduino and Raspberry Pi 3 were detected using automated robotic model that was developed by Shekhar et al. (Shekhar et al. 2017). The technology of Arduino was also used by Jha et al. (Jha et al. 2019) in a developed automated irrigation system for decreasing human power and time consumption in the irrigation process.

In another approach by Varatharajalu and Ramprabu (2018), different sensors were designed for automated irrigation, each with its own purpose, such as the temperature sensor to detect temperature, the soil moisture sensor to detect soil moisture content, the pressure regulator sensor to regulate pressure, and the molecular sensor to improve crop growth. All these devices' output is then transformed to a digital signal.

Kuyper and Balendonck (1997) presented an automated system for real-time irrigation control that leverages dielectric soil moisture sensors. The soil is made up of a variety of elements, including air, water, and minerals (Ziarati et al. 2019). Water-on-demand irrigation is a moisture sensor technique where the threshold is based on the soil's field capacity. When the planned time comes, the sensor reads the moisture content or level for that zone, and then watering happens if the moisture content is below the threshold. For each zone, the start time and duration are required (Yong et al. 2018).

Kumar (2014) discusses the various irrigation systems with the goal of developing a system that uses less resources and is more efficient. pH and fertility meters are set up to detect the proportion of the primary ingredients of the soil, such as potassium, phosphorous, and nitrogen, to measure the fertility of the soil. Wireless technology is used for drip irrigation to automatic plant irrigators on the field. This technology ensures the soil's fertility as well as the efficient use of water resources.

Another automated irrigation system was given by Savitha and UmaMaheswari (2018). Diverse sensors were constructed for different reasons in this technique. For instance, a sensor for soil moisture, a sensor to detect temperature, a sensor to maintain and regulate pressure, and a molecular sensor to improve crop growth.

Sabitha et al. (2017) elaborated that work after the sensor responds will be controlled by the controller, and the soil water threshold will be chosen. After that, water is applied for a day to the area where the sensor is buried. The microcontrollers are the major components of the automated process. Therefore, the irrigation process could be automated using them.

3.4 THE APPLICATION OF WIRELESS AND SMART SENSORS IN AGRICULTURE

Wireless sensor network (WSN) is attaining popularity as a platform for altering our interactions with the physical environment, and with the progress of sensor technology and the technology of network-based information, WSN is considered to play a key role in lots of different applications, including the monitoring of wildlife habitat, deforestation and desertification rates, intelligent home monitoring, and with a special focus on precision agriculture (Hempstead et al. 2008; Ramson and Moni 2017).

Pumping systems are typically deployed in the agricultural sectors on farmlands where environmental conditions are harsh. This system not only increases cable usage and expenses but also risks destroying electrical and mechanical equipment. In hostile settings, lengthy distances, and inaccessible locations, cabling has some constraints. These circumstances will raise the expense of cabling, labor, system maintenance, and most critically, water waste. As a result, WSN can be considered one of the most acceptable alternatives in severe environments, and AI can be used to overcome the constraints that have just been mentioned (Gungor and Lambert 2006; Ali et al. 2009). Due to high installation costs, data transfer and wireless technology might play a role in the selection of some control systems, such as programmable logic controllers (PLC). Wireless networks have a number of limitations, including security concerns, coverage area, and fault tolerance (Robles et al. 2015).

The ability of WSN devices to measure values across a wide variety of operating time scales and environmental circumstances is a critical need for their proper application (Estrin et al. 1999). This is possible since they are energy-intensive processes, and WSN nodes rely on tiny batteries for power (Pottie et al. 2000). As a result, another key need of WSN is the routing of messages between an undetermined number of neighbors (Camilli et al. 2007).

The sensors can convey its measurements through increasing their power usage by around 50% when transitioning to "processing and transmitting/receiving" when additional sensor messages are received (Camilli et al. 2007). This indicates that for all measurements in each device, around 1.2 seconds of activity time each minute must suffice. Because the measured data is not held in the sensors for lengthy periods of time due to memory and energy constraints, measurements should be delivered quickly, resulting in 48-byte messages (Madden et al. 2003).

Sensors are mainly classified according to their readiness in field deployment in terms of availability, scalability, and cost (Mahmoud and Fawzy 2021). In addition, they could be categorized based on their manufacturing materials, as indicated in Table 3.1. The wireless sensor network consists of data acquisition and data distribution networks, and these networks are supposed to be controlled by a central station (Ramson and Moni 2017). In addition, a smartphone-based irrigation system was used in 2015 (Jagüey et al. 2015) that was only based on soil moisture, with no consideration for other environmental factors.

Commercial sensors for agricultural and irrigation systems are prohibitively expensive for small farmers to deploy. Manufacturers, on the other hand, are presently offering low-cost sensors that may be linked to nodes to create low-cost irrigation management and farm monitoring systems (Veerachamy and Ramar 2021). Due to the demand for low-cost sensors for agriculture and water monitoring, new low-cost sensors are being investigated (Guruprasadh et al. 2017).

It was concluded from Figure 3.5a that the application of sensors, whether in agriculture or other fields (Figure 3.5b), is boosting, as the total number of publications increased dramatically from only 1 article in 1993 to 509 in 2020.

The implementer of a sensor network must be familiar with the network mechanics and have some understanding of the application domain. The developer must be able to properly code the application's inherent details in order for the entire set to perform properly and produce dependable results. This practitioner skills situation

TABLE 3.1
Most Common Sensors Utilized in Agricultural Activities and Their Manufacturing Materials

Sensor Name	Manufacturing Material	References
Multilevel soil moisture sensor	Copper rings placed along a PVC pipe	(Guruprasadh et al. 2017)
Water turbidity sensor	Tinted plastic	(Sendra et al. 2015)
Water salinity monitoring sensor	Copper coils	(Parra et al. 2013)
Water level measurement	–	(Annuar et al. 2015)
Ultrasonic level sensor	–	(Saraswati et al. 2012)
Capacitance level sensor	–	(Stacheder et al. 2009)
TAUPE-TDR sensor	Polyethylene	(Robinson et al. 2003)
State-of-ground (SOG) sensor	–	(Hübner 1999)

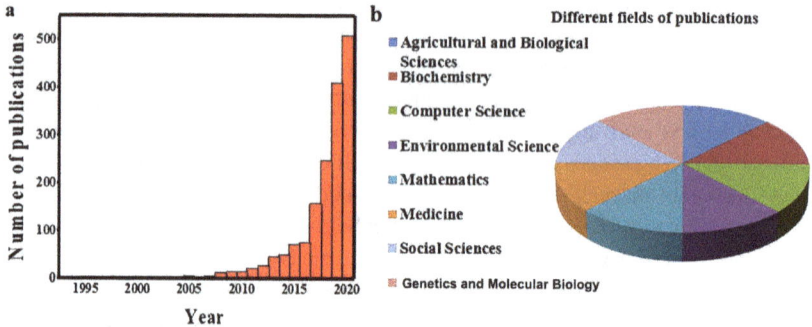

FIGURE 3.5 (a) Number of publications of sensor application. (b) Different fields of sensor applications.

Source: Data obtained from Scopus in June 2021.

harkens back to the early precision agriculture, when the technology was only available to specialists, posing a barrier to newcomers (Wang et al. 2006). Due to ongoing attempts to standardize programming interfaces and middleware infrastructure used by WSN, this difficult endeavor will most likely become considerably easier in the near future. Enhancements should make technology abilities, such as the ones used to create this program, available to the broader public as infrastructure (Lee 2000). The predicted reduction in sensor costs, as well as an improvement in its memory and processing at low energy cost, will most likely be used to enable this new functionality (Ye et al. 2004).

Improving water management may significantly enhance both food and water security in most countries, particularly in dry regions, where agriculture is the main income and economic activity (Kansara et al. 2015). Smart and electronic devices are now extremely common, and they can be used in a variety of fields of science and technology. Using smart and sensory devices via the IoT is one of the most effective ways to address the aforementioned concerns. Farmers can benefit from the system's protection features, which include power outage alerts, security alerts, water flow rate, and water level control.

The multi-intelligent control system (MICS) is a very dependable system that delivers a satisfactory solution for water management in different sectors. This system provides a unique water pumping and pumping station approach that is built, installed, and employed in the agricultural sector. MICS is composed of three control systems: an electro-pump controller, a reservoir water level controller, and an alarm control system. The system is controlled by AI and operated via ringtone or SMS from any location. To avoid electrical shocks and mechanical stresses, a soft starter mechanism was created and proposed for powering the electro-pump. A four-state switch was built and used, which allows the system to be run and operated manually, automatically, over the internet, and finally, in off mode. The result of the introduced model was that the water management system can be saved to 60% with increasing efficiency and productivity.

As shown in Figure 3.6, MICS develops five-layer IoT structures for suitable water management regulation. The five smart layouts are smart communication, smart

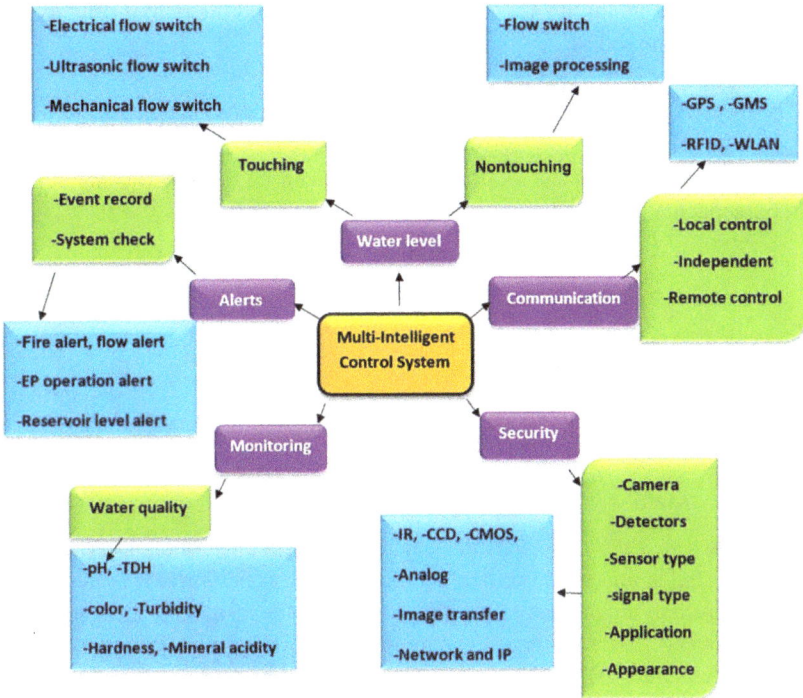

FIGURE 3.6 The framework of the multi-intelligent control system (MICS) of water grid.

security, smart monitoring, smart warnings, and smart water level monitoring. Each layer is made up of multiple sublayers. As illustrated in the diagram, each layout has a number of control systems. The quality of water, for example, may be easily assessed in terms of desirable parameters.

The mathematical model of MICS in determining the capacity of the electro-pump, as well as other electrical components, is presented in the following section.

The flow rate balance of the model is evaluated based on input and output of flow rate:

$$Q_{i(t)} - Q_o(t) = A dh / dt \tag{1}$$

where, A is the area of the tank; h the height of the water in the tank.

For a vented tank, the discharge velocity can be assessed using:

$$V_1 = A_2 V_2 / A_1 \tag{2}$$

Therefore, the velocity of the output valve can be stated as:

$$V_2^2 = (A_2 V_2 / A_1)^2 + 2gh. \tag{3}$$

As $A_1 \gg A_2$ the amount of $(A_2 V_2/A1)^2$ is around zero, and the output velocity of the streamline is expressed as:

$$V_2 = \sqrt{2gh} \tag{4}$$

Therefore, the output and the total inlet flow rate can be estimated using equations 5 and 6, respectively:

$$Q_o(t) = A_2 V_2 = \frac{\pi d^2}{4}\sqrt{2gh} \tag{5}$$

$$Q_i(t) = A^{dh/dt} + \frac{\pi d^2}{4}\sqrt{2gh} \tag{6}$$

3.5   CONCLUSIONS

It is necessary to have a system that can aid farmers and function as a guide to irrigate crop fields and to maximize use of water in agriculture. It is necessary to develop innovative techniques to properly utilize water resources for agriculture. It has to be noticed that a country's economy and growth are majorly dependent on agricultural revenues. Reliance on wireless and noninvasive setup of sensor technique will unequivocally result in increasing crop yield while also assisting farmers in making irrigation selections to boost agricultural cultivation. Crop production cost can be decreased by adopting cutting-edge agricultural technologies. By embedding wireless sensor networks into agricultural areas, the work is projected to get scaled up in the near future.

Multi-intelligent control system (MICS) was introduced and tested successfully in the agricultural sector with its three control systems: (1) the electro-pump control system, (2) the water reservoir control system, and (3) the pump station control system. The control system is based on IoT, and it uses electronic and intelligent equipment to monitor and control water consumption level.

3.6   ACKNOWLEDGMENTS

The authors acknowledge the support of Egypt's Academy of Scientific Research and Technology (ASRT) for the SA-Egypt Joint Project, as well as the Egyptian Science, Technology & Innovation Funding Authority (STDF), grant number 45888, under the umbrella of USAID/STDF collaborative project.

REFERENCES

Ali, A., Al Soud, M., Abdallah, E., Addallah, S., 2009. Water pumping system with PLC and frequency control. Jordan Journal of Mechanical and Industrial Engineering 3, 216–221.
Annuar, K.M., Ab Hadi, N., Saadon, I., Harun, M., 2015. Design and construction of liquid level measurement system. Journal of Advanced Research in Applied Mechanics 12, 8–15.

Awokuse, T.O., Xie, R., 2015. Does agriculture really matter for economic growth in developing countries? Canadian Journal of Agricultural Economics/Revue canadienne d'agroeconomie 63, 77–99.

Barkunan, S., Bhanumathi, V., Sethuram, J., 2019a. Smart sensor for automatic drip irrigation system for paddy cultivation. Computers & Electrical Engineering 73, 180–193.

Barkunan, S.R., Bhanumathi, V., Sethuram, J., 2019b. Smart sensor for automatic drip irrigation system for paddy cultivation. Computers & Electrical Engineering 73, 180–193.

Barman, A., Neogi, B., Pal, S., 2020. Solar-Powered Automated IoT-Based Drip Irrigation System. IoT and Analytics for Agriculture. Springer, Singapore, pp. 27–49.

Camilli, A., Cugnasca, C.E., Saraiva, A.M., Hirakawa, A.R., Corrêa, P.L., 2007. From wireless sensors to field mapping: Anatomy of an application for precision agriculture. Computers and Electronics in Agriculture 58, 25–36.

Chunduri, K., Menaka, R., 2019. Agricultural Monitoring and Controlling System Using Wireless Sensor Network. In: Wang, J., Reddy, G.R.M., Prasad, V.K., Reddy, V.S. (Eds.). Soft Computing and Signal Processing. Springer Singapore, Singapore, pp. 47–56.

Eli-Chukwu, C.N., 2019. Applications of artificial intelligence in agriculture: A review. Engineering, Technology Applied Science Research 9, 4377–4383.

Estrin, D., Govindan, R., Heidemann, J., Kumar, S., 1999. Next century Challenges: Scalable Coordination in Sensor Networks. In: Proceedings of the 5th annual ACM/IEEE International Conference on Mobile Computing and Networking, pp. 263–270.

García, L., Parra, L., Jimenez, J.M., Lloret, J., Lorenz, P., 2020. IoT-based smart irrigation systems: An overview on the recent trends on sensors and IoT systems for irrigation in precision agriculture. Sensors 20, 1042.

Gungor, V.C., Lambert, F.C., 2006. A survey on communication networks for electric system automation. Computer Networks 50, 877–897.

Guruprasadh, J., Harshananda, A., Keerthana, I., Krishnan, K.Y., Rangarajan, M., Sathyadevan, S., 2017. Intelligent Soil Quality Monitoring System for Judicious Irrigation. In: 2017 International Conference on Advances in Computing, Communications and Informatics (ICACCI). IEEE, pp. 443–448.

Gwenzi, W., Selvasembian, R., Offiong, N.-A.O., Mahmoud, A.E.D., Sanganyado, E., Mal, J., 2022. COVID-19 drugs in aquatic systems: A review. Environmental Chemistry Letters 20, 1275–1294.

Hempstead, M., Lyons, M.J., Brooks, D., Wei, G.-Y., 2008. Survey of hardware systems for wireless sensor networks. Journal of Low Power Electronics 4, 11–20.

Hosny, M., Fawzy, M., El-Borady, O.M., Mahmoud, A.E.D., 2021. Comparative study between Phragmites australis root and rhizome extracts for mediating gold nanoparticles synthesis and their medical and environmental applications. Advanced Powder Technology 32, 2268–2279.

Huang, L., Yan, T., Mahmoud, A.E.D., Li, S., Zhang, J., Shi, L., Zhang, D., 2021. Enhanced water purification via redox interfaces created by an atomic layer deposition strategy. Environmental Science: Nano 8, 950–959.

Hübner, C., 1999. Entwicklung hochfrequenter Messverfahren zur Boden-und Schneefeuchtebestimmung.

Iglesias, A., Garrote, L., Diz, A., Schlickenrieder, J., Martin-Carrasco, F., 2011. Re-thinking water policy priorities in the Mediterranean region in view of climate change. Environmental Science & Policy 14, 744–757.

Iglesias, A., Santillán, D., Garrote, L., 2018. On the barriers to adaption to less water under climate change: Policy choices in Mediterranean countries. Water Resources Management 32, 4819–4832.

Jagüey, J.G., Villa-Medina, J.F., López-Guzmán, A., Porta-Gándara, M.Á., 2015. Smartphone irrigation sensor. IEEE Sensors Journal 15, 5122–5127.

Jha, K., Doshi, A., Patel, P., Shah, M., 2019. A comprehensive review on automation in agriculture using artificial intelligence. Artificial Intelligence in Agriculture 2, 1–12.

Kamienski, C., Soininen, J.-P., Taumberger, M., Dantas, R., Toscano, A., Salmon Cinotti, T., Filev Maia, R., Torre Neto, A., 2019. Smart water management platform: Iot-based precision irrigation for agriculture. Sensors 19, 276.

Kamienski, C., Soininen, J.-P., Taumberger, M., Fernandes, S., Toscano, A., Cinotti, T.S., Maia, R.F., Neto, A.T., 2018. Swamp: An iot-Based Smart Water Management Platform for Precision Irrigation in Agriculture. In: 2018 Global Internet of Things Summit (GIoTS). IEEE, pp. 1–6.

Kansara, K., Zaveri, V., Shah, S., Delwadkar, S., Jani, K., 2015. Sensor based automated irrigation system with IOT: A technical review. International Journal of Computer Science and Information Technologies 6, 5331–5333.

Kodali, R.K., Sahu, A., 2016. An IoT based Soil Moisture Monitoring on Losant Platform. In: 2016 2nd International Conference on Contemporary Computing and Informatics (IC3I). IEEE, pp. 764–768.

Kumar, G., 2014. Research paper on water irrigation by using wireless sensor network. International Journal of Scientific Research Engineering & Technology (IJSRET), 3–4.

Kuyper, M., Balendonck, J., 1997. Application of Dielectric Soil Moisture Sensors for Real-Time Automated Irrigation Control. In: III International Symposium on Sensors in Horticulture, Volume 562. International Symposium on Sensors in Horticulture, Netherlands, pp. 71–79.

Lee, K., 2000. IEEE 1451: A Standard in Support Of Smart Transducer Networking. Proceedings of the 17th IEEE Instrumentation and Measurement Technology Conference [Cat. No. 00CH37066]. IEEE, pp. 525–528.

Liakos, K.G., Busato, P., Moshou, D., Pearson, S., Bochtis, D., 2018. Machine learning in agriculture: A review. Sensors 18, 2674.

Madden, S., Franklin, M.J., Hellerstein, J.M., Hong, W., 2003. The Design of an Acquisitional Query Processor for Sensor Networks. In: Proceedings of the 2003 ACM SIGMOD international conference on Management of Data, pp. 491–502. ACM digital library.

Mahmoud, A.E.D., 2020. Graphene-based nanomaterials for the removal of organic pollutants: Insights into linear versus nonlinear mathematical models. Journal of Environmental Management 270, 110911.

Mahmoud, A.E.D., Al-Qahtani, K.M., Alflaij, S.O., Al-Qahtani, S.F., Alsamhan, F.A., 2021a. Green copper oxide nanoparticles for lead, nickel, and cadmium removal from contaminated water. Scientific Reports 11, 12547.

Mahmoud, A.E.D., Fawzy, M., 2021. Nanosensors and Nanobiosensors for Monitoring the Environmental Pollutants. In: Makhlouf, A.S.H., Ali, G.A.M. (Eds.). Waste Recycling Technologies for Nanomaterials Manufacturing. Springer International Publishing, Cham, pp. 229–246.

Mahmoud, A.E.D., Fawzy, M., Abdel-Fatah, M.M.A., 2022a. Technical Aspects of Nanofiltration for Dyes Wastewater Treatment. In: Muthu, S.S., Khadir, A. (Eds.). Membrane Based Methods for Dye Containing Wastewater: Recent Advances. Springer Singapore, Singapore, pp. 23–35.

Mahmoud, A.E.D., Fawzy, M., Hosny, G., Obaid, A., 2020. Equilibrium, kinetic, and diffusion models of chromium(VI) removal using Phragmites australis and Ziziphus spina-christi biomass. International Journal of Environmental Science and Technology 18, 2125–2136.

Mahmoud, A.E.D., Fawzy, M., Radwan, A., 2016. Optimization of Cadmium (CD2+) removal from aqueous solutions by novel biosorbent. International Journal of Phytoremediation 18, 619–625.

Mahmoud, A.E.D., Hosny, M., El-Maghrabi, N., Fawzy, M., 2022b. Facile synthesis of reduced graphene oxide by Tecoma stans extracts for efficient removal of Ni (II) from water: batch experiments and response surface methodology. Sustainable Environment Research 32, 22.

Mahmoud, A.E.D., Umachandran, K., Sawicka, B., Mtewa, T.K., 2021b. 26 — Water resources security and management for sustainable communities. In: Mtewa, A.G., Egbuna, C. (Eds.). Phytochemistry, the Military and Health. Elsevier, pp. 509–522.

Mansoor, S., Farooq, I., Kachroo, M.M., Mahmoud, A.E.D., Fawzy, M., Popescu, S.M., Alyemeni, M.N., Sonne, C., Rinklebe, J., Ahmad, P., 2022. Elevation in wildfire frequencies with respect to the climate change. Journal of Environmental Management 301, 113769.

Mishra, B., Tiwari, A., Mahmoud, A.E.D., 2022. Microalgal potential for sustainable aquaculture applications: Bioremediation, biocontrol, aquafeed. Clean Technologies and Environmental Policy, 1–13.

Mogili, U.M.R., Deepak, B.B.V.L., 2018. Review on application of drone systems in precision agriculture. Procedia Computer Science 133, 502–509.

Monckton, D.C., 2019. Economic impacts of coal seam water for agricultural enterprises, lessons for efficient water management. Sustainable Water Resources Management 5, 333–346.

Muangprathub, J., Boonnam, N., Kajornkasirat, S., Lekbangpong, N., Wanichsombat, A., Nillaor, P., 2019. IoT and agriculture data analysis for smart farm. Computers and Electronics in Agriculture 156, 467–474.

Özerol, G., Dolman, N., Bormann, H., Bressers, H., Lulofs, K., Böge, M., 2020. Urban water management and climate change adaptation: A self-assessment study by seven midsize cities in the North Sea Region. Sustainable Cities and Society 55, 102066.

Papadopoulou, M.P., Charchousi, D., Spanoudaki, K., Karali, A., Varotsos, K.V., Giannakopoulos, C., Markou, M., Loizidou, M., 2020. Agricultural water vulnerability under climate change in cyprus. Atmosphere 11, 648.

Parekh, V., Shah, D., Shah, M., 2019. Fatigue Detection Using Artificial Intelligence Framework. Augmented Human Research 5, 5.

Parekh, V., Shah, D., Shah, M., 2020. Fatigue detection using artificial intelligence framework. Augmented Human Research 5, 1–17.

Parra, L., Ortuño, V., Sendra, S., Lloret, J., 2013. Low-Cost Conductivity Sensor based on Two Coils. In: Proceedings of the First International Conference on Computational Science and Engineering (CSE'13), Valencia, Spain, p. 107112.

Patel, D., Shah, D., Shah, M., 2020. The intertwine of brain and body: a quantitative analysis on how big data influences the system of sports. Annals of Data Science 7, 1–16.

Plessen, M.G., 2019. Freeform Path Fitting for the Minimisation of the Number of Transitions between Headland Path and Interior Lanes within Agricultural Fields. arXiv Preprint arXiv:1910.12034.

Pottie, G.J., Kaiser, W.J., 2000. Wireless integrated network sensors. Communications of the ACM 43, 51–58.

Ramson, S.J., Moni, D.J., 2017. Applications of wireless sensor networks—A survey. 2017 international conference on innovations in electrical, electronics, instrumentation and media technology (ICEEIMT). IEEE, pp. 325–329.

Robinson, D.A., Jones, S.B., Wraith, J.M., Or, D., Friedman, S.P., 2003. A review of advances in dielectric and electrical conductivity measurement in soils using time domain reflectometry. Vadose Zone Journal 2, 444–475.

Robles, T., Alcarria, R., de Andrés, D.M., de la Cruz, M.N., Calero, R., Iglesias, S., Lopez, M., 2015. An IoT based reference architecture for smart water management processes. Journal of Wireless Mobile. Networks Ubiquitous Computing and Dependable Applications 6, 4–23.

Sabitha, T., Santhiya, B., Saranya, T., Tamilmani, T., Krithika, R., 2017. Automatic irrigation and measuring soil strength. International Journal for Research in Applied Science & Engineering Technology 8.

Saraswati, M., Kuantama, E., Mardjoko, P., 2012. Design and Construction of Water Level Measurement System Accessible Through SMS. In: 2012 Sixth UKSim/AMSS European Symposium on Computer Modeling and Simulation. IEEE, pp. 48–53.

Savitha, M., UmaMaheswari, O., 2018. Smart crop field irrigation in IOT architecture using sensors. International Journal of Advanced Research in Computer Science 9, 302–306.

Sawicka, B., Umachandran, K., Fawzy, M., Mahmoud, A.E.D., 2021.27 — Impacts of Inorganic/Organic Pollutants on Agroecosystems and Eco-Friendly Solutions. In: Mtewa, A.G., Egbuna, C. (Eds.). Phytochemistry, the Military and Health. Elsevier, pp. 523–552.

Sendra, S., Parra, L., Lloret, J., Jiménez, J.M., 2015. Oceanographic multisensor buoy based on low cost sensors for Posidonia meadows monitoring in Mediterranean Sea. Journal of Sensors 2015.

Shekhar, Y., Dagur, E., Mishra, S., Sankaranarayanan, S., 2017. Intelligent IoT based automated irrigation system. International Journal of Applied Engineering Research 12, 7306–7320.

Stacheder, M., Koeniger, F., Schuhmann, R., 2009. New dielectric sensors and sensing techniques for soil and snow moisture measurements. Sensors 9, 2951–2967.

Talaviya, T., Shah, D., Patel, N., Yagnik, H., Shah, M., 2020. Implementation of artificial intelligence in agriculture for optimisation of irrigation and application of pesticides and herbicides. Artificial Intelligence in Agriculture 4, 58–73.

Tokarev, K., Rogachev, A., Protsyuk, M., Rudenko, A.Y., Chernyavsky, A., Tokareva, Y.M., 2020. Analysis of Promising Methods of Irrigation and Melioration Techniques of Crops in Arid Climate. In: IOP Conference Series: Earth and Environmental Science. IOP Publishing, p. 012047.

Tran, T., Ha, Q.P., 2015. Dependable control systems with internet of things. ISA Transactions 59, 303–313.

Vambol, S., Vambol, V., Mozaffari, N., Mahmoud, A.E.D., Ramsawak, N., Mozaffari, N., Ziarati, P., Khan, N.A., 2021. Comprehensive Insights into Sources of Pharmaceutical Wastewater in the Biotic systems. In: Khan, N.A., Ahmed, S., Vambol, V., Vambol, S. (Eds.). Pharmaceutical Wastewater Treatment Technologies: Concepts and Implementation Strategies. International Water Association Publishing.

Varatharajalu, K., Ramprabu, J., 2018. Wireless Irrigation System via Phone Call & SMS. International Journal of Engineering and Advanced Technology 8, 397–401.

Veerachamy, R., Ramar, R., 2021. Agricultural Irrigation Recommendation and Alert (AIRA) system using optimization and machine learning in Hadoop for sustainable agriculture. Environmental Science and Pollution Research, 1–20.

Venot, J.-P., Kuper, M., Zwarteveen, M., 2017. Drip Irrigation for Agriculture: Untold Stories of Efficiency, Innovation and Development. Taylor & Francis, UK.

Wang, N., Zhang, N., Wang, M., 2006. Wireless sensors in agriculture and food industry—Recent development and future perspective. Computers and Electronics in Agriculture 50, 1–14.

Wang, Y., Li, S., Qin, S., Guo, H., Yang, D., Lam, H.-M., 2020. How can drip irrigation save water and reduce evapotranspiration compared to border irrigation in arid regions in northwest China. Agricultural Water Management 239, 106256.

Ye, W., Heidemann, J., Estrin, D., 2004. Medium access control with coordinated adaptive sleeping for wireless sensor networks. IEEE/ACM Transactions on Networking 12, 493–506.

Yong, W., Shuaishuai, L., Li, L., Minzan, L., Ming, L., Arvanitis, K., Georgieva, C., Sigrimis, N., 2018. Smart sensors from ground to cloud and web intelligence. IFAC-PapersOnLine 51, 31–38.

Ziarati, P., El-Esawi, M., Sawicka, B., Umachandran, K., Mahmoud, A.E.D., Hochwimmer, B., Vambol, S., Vambol, V., 2019. Investigation of prospects for phytoremediation treatment of soils contaminated with heavy metals. Journal of Medical Discovery 4, 1–13.

Zou, H., Fan, J., Zhang, F., Xiang, Y., Wu, L., Yan, S., 2020. Optimization of drip irrigation and fertilization regimes for high grain yield, crop water productivity and economic benefits of spring maize in Northwest China. Agricultural Water Management 230, 105986.

4 Advances in Artificial Intelligence Applications in Sustainable Water Remediation

Alaa El Din Mahmoud, Nadeem Ahmad Khan, and Yung-Tse Hung

CONTENTS

4.1 INTRODUCTION

Computers may simulate the human brain using AI or machine intelligence, which is also known as artificial intelligence (AI) (Alam et al. 2022a). The fields of mathematics, information engineering, computer science, data science, logic, statistics, cognitive linguistics, and ontology all fall under the umbrella of artificial intelligence. We may expect a wide range of applications for AI in the future, from education to agriculture to metrology to transportation to health care to space exploration (Malviya and Jaspal 2021). Artificial intelligence (AI) relies heavily on models such as ant colonies and particle swarms as well as deep learning and genetic algorithms. This new technology uses all these concepts to decrease human effort while increasing process efficiency. Artificial intelligence (AI) models have been used in a number of studies to evaluate the efficacy of wastewater treatment processes. Diverse studies on the optimization and modeling of natural water systems have been conducted to determine the removal of specific contaminants or numerous pollutants (Khan et al. 2021; Mousazadeh et al. 2021; Mahmoud and Kathi 2022). AI models such as neural networks and hybrid models have been created to analyze the nonlinear pattern of pollutant removal in the treatment processes. It is estimated that the world's population nearly quadrupled over the twentieth century (Pouyanfar

DOI: 10.1201/9781003260455-4

53

et al. 2022). As the population has grown, so has the need for water, which has resulted in an increase in the amount of water use from natural sources (Dotaniya et al. 2022). Industrialization and urbanization have resulted in a rapid depletion of water resources (Mahmoud et al. 2020a; Huang et al. 2021). According to the 2019 World Water Development Report, the current water demand trend is predicted to continue until 2050. In order to conserve water, we must utilize it wisely and treat the wastewater we produce so it may be used for a variety of purposes. There are a number of sophisticated mathematical formulas and calculations involved in conventional techniques for wastewater treatment (Ding et al. 2021; Mahmoud et al. 2022c). These formulas and calculations include the design parameters, process parameters, and operational parameters (Mahmoud et al. 2016; 2018; Omran et al. 2022).

The removal efficiency of contaminants and uncertainty predictions were previously evaluated using statistical and multivariate data analysis approaches. There is no need for sophisticated mathematical formulae or extensive knowledge about how input and output variables are related to each other for AI technologies to accurately forecast R^2 (regression coefficients). Complex, interactive, and dynamic wastewater treatment challenges can be handled by AI technology (Deepnarain et al. 2020). The treatment processes' effluent quality may be predicted with their help, and quick action can be taken to keep treatment processes operating properly (Mahmoud 2022; Mahmoud et al. 2022a). Preventing a disaster can be achieved by taking prompt action with the use of automated sampling. Improved operational efficiency, reduced energy usage, and reduced costs are all made possible through sensor-based continuous monitoring (Matheri et al. 2021). Wastewater treatment will be transformed in the near future by artificial intelligence. There has been a lot of study in this area, and this chapter provides an overview of the many technologies that have been employed thus far (Misbah Biltayib et al. 2021). Figure 4.1 illustrates the annual publications of sustainable water remediation so far.

Figure 4.2 shows the top 25 countries that are responsible for publishing various publications in the field of sustainable water remediation. Various blue color

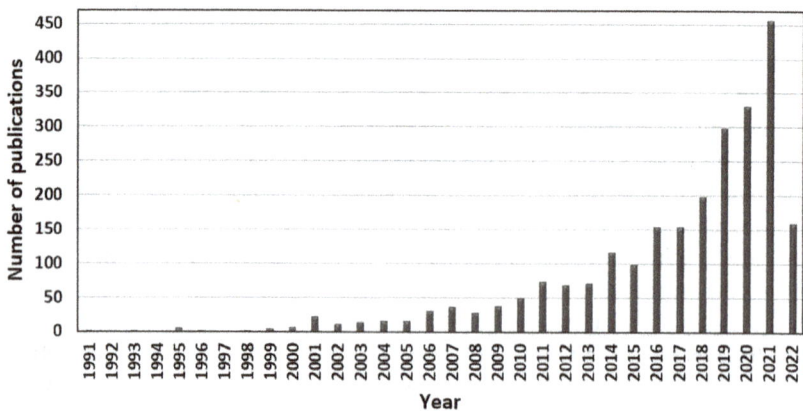

FIGURE 4.1 Annual publications production of sustainable water remediation indexed in Scopus (1991–2022).

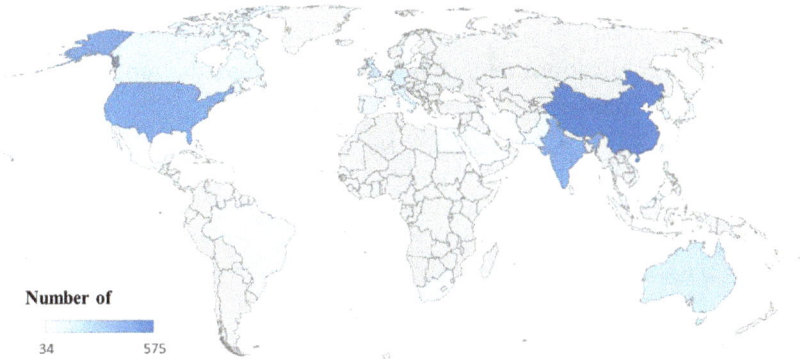

Number of

34   575

FIGURE 4.2 The number of publications based on geographical distribution.

variations indicate the level of publications productivity. Consequently, China, India, the United States, and the United Kingdom have the most published articles, reviews, book chapters, books, and conference papers in this this field. The gray color reflects the rest of the countries not counted in the calculations, as the top 25 countries are only mentioned. Hence, Table 4.1 illustrates most of the countries that have conducted research or literature survey on sustainable water remediation.

It is also interesting to illustrate the co-occurrence link between the terms of publications for sustainable water remediation (Figure 4.3). Sustainable development is the most relevant term and is most mentioned in the publications, especially in the articles.

UN Sustainable Development Goals (SDGs) include access to safe drinking water as a major component (Mahmoud et al. 2021b). New, efficient, and cost-effective methods for removing contaminants from wastewater have been developed during the past few decades (Hosny et al. 2021; Mahmoud et al. 2021a). Recent years have seen a rise in the use of optimization and modeling tools to measure performance and improve efficiency (Mahmoud 2020; Mahmoud et al. 2022b).

A major discipline of computer science known as artificial intelligence (AI) focuses on constructing smart systems and solving problems in a way similar to the human intelligence system. Human-relevant computer capabilities, including learning, problem-solving, reasoning, and observation, are at the heart of AI applications (Wen and Vassiliadis 1998). A wide range of industries, from health care to smart cities and transportation, e-commerce to banking and academics, is benefiting from AI. Machine learning, deep learning, and data analytics are all subcategories of artificial intelligence, or AI. For intelligent decision-making, blockchain, cloud computing, the Internet of Things (IoT), and the fourth industrial revolution (Industry 4.0), these strategies are most commonly utilized (Wang et al. 2022). For the most part, AI's growth is being fueled by its unique abilities to analyze past data, figure out patterns, and make decisions. Because AI-based systems include intelligence, flexibility, and intentionality into its suggested algorithms, AI's relevance continues to grow throughout time. All multidisciplinary domains can benefit from AI systems, and they've been used in a variety of ways to improve algorithms for optimization,

TABLE 4.1

The Number of Publications on Sustainable Water Remediation Based on the Most Productive Countries

No.	Country	Number of Publications	No.	Country	Number of Publications
1	China	575	28	Nigeria	28
2	India	439	29	Turkey	27
3	United States	439	30	Finland	25
4	United Kingdom	176	31	Denmark	24
5	Australia	168	32	Greece	24
6	Italy	134	33	Czech Republic	23
7	Germany	126	34	Singapore	22
8	South Korea	98	35	Switzerland	22
9	Canada	85	36	Vietnam	22
10	Spain	84	37	Argentina	18
11	Malaysia	74	38	New Zealand	18
12	Pakistan	70	39	Thailand	17
13	France	58	40	United Arab Emirates	16
14	Brazil	53	41	Chile	14
15	Saudi Arabia	52	42	Romania	14
16	Portugal	48	43	Russian Federation	14
17	Egypt	46	44	Iraq	13
18	South Africa	44	45	Indonesia	12
19	Mexico	41	46	Serbia	12
20	Hong Kong	40	47	Bangladesh	11
21	Japan	38	48	Sri Lanka	11
22	Sweden	38	49	Ireland	10
23	Iran	37	50	Morocco	10
24	Belgium	35	51	Tunisia	9
25	Taiwan	34	52	Algeria	8
26	Netherlands	31	53	Oman	8
27	Poland	29	54	Qatar	8

classification, regression, and forecasting. In order to improve the precision of optimum solution prediction, AI technologies are sometimes employed in conjunction with experimental design approaches, like response surface methodology (RSM). Artificial intelligence is being used in water treatment to alleviate some of the difficulties associated with traditional processes. The water industry is currently investing in artificial intelligence, and this investment is predicted to reach $6.3 billion by 2030, based on current market research (Asnake Metekia et al. 2022). Artificial intelligence (AI) has the potential to save water treatment companies 20–30% of their current operational costs by reducing the cost and optimizing the use of chemical compounds. The use of AI in water treatment has made the process simple because of its low implementation, adaptability, generalization, and design simplicity. Adaptive-network-based fuzzy inference system (ANBIS), feed-forward backpropagation neural network (FFBPNN), decision tree (DT), and recurrent neural network (RNN) are

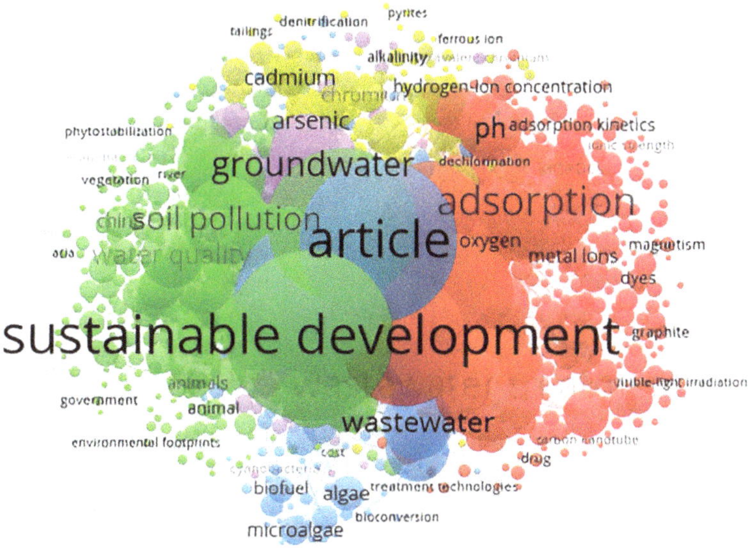

FIGURE 4.3 Co-occurrence network of publications for sustainable water remediation.

some of the most often utilized AI approaches in water treatment (ANFIS). Hybrid approaches such as MLP-ANN, ANN-PSO, PSO-GA, BP-ANN, feed-forward back-propagation (FFBP)–ANN, and support vector regression (SVR)–GA have also been investigated in water treatment (Alam et al. 2022b). The fundamental problem in AI applications is the availability of data, as AI requires enough previous data to fore-cast future events and provide system improvement. Water pollution remediation has been successfully modeled and optimized using various artificial intelligence tech-nologies, as shown in many research. The use of AI in water purification is currently hindered by a number of obstacles, as mentioned (Alam et al. 2022b). To evaluate the effectiveness of the water adsorption process for the removal of metals, dyes, organic chemicals, nutrients, medications, and pharmaceutical products (PCPs), this chapter critically assesses a variety of artificial intelligence (AI) techniques. The parameters used to evaluate the effectiveness of AI models are also examined, as are the input factors that influence process performance (Mahmoud et al. 2020). Finally, the considerable hurdles in the general implementation of AI in water treatment are addressed, and ideas for future study are also presented.

4.2   ADVANCES IN THE FIELD OF ARTIFICIAL INTELLIGENCE

Artificial intelligence (AI) is a useful technology for streamlining wastewater treat-ment processes and reducing their associated risks and costs. The use of AI models in wastewater treatment has grown exponentially over the last few decades (Niu et al. 2022). Water treatment performance is predicted, effluent quality is estimated, oper-ating parameters and treatment unit design are optimized, sensors for component estimation are designed, micropollutants and other emerging contaminants are optimized, maintenance strategies based on fault diagnosis are predicted, instructions for operation are optimized based on performance, and automation is implemented as part of treatment (He et al. 2021). If you want the finest outcomes, you must pick the optimal intelligent control approach for your treatment system based on the mechanisms it employs and the goal for which it is being employed. The use of the key terms "wastewater treatment," "artificial intelligence," and "model name" in the Scopus database has allowed researchers to investigate the applicability of various models in wastewater treatment (Gadkari 2022). Soft computing methods like artificial neural networks (ANNs) are increasingly being used to forecast water quality and other relevant aspects (Nasr et al. 2017). ANN is a system that processes data. Input and output data are both used in the computing process (Heydari et al. 2021). Neurons in the human brain are able to conduct sophisticated calculations in parallel. Parallel processing of nonlinear connections between inputs and outputs is mostly employed for pattern recognition challenges (Picos-Benítez et al. 2020). Nodes like sigmoid and hyperbolic in ANNs' activation functions are employed to execute nonlinear computations. A proper cost function may be used to adjust the weights between neurons (nodes) in an ANN to learn from observed data in order to improve the model. It is the hidden layers between the first and last ANN layers that are used to represent input and output, respectively, in ANNs composed of several layers (Wei et al. 2021). Complex models may be trained to improve ANN performance by increasing the number of hidden layers.

A machine learning approach known as k-NN is utilized for regression and classification. k-NN classifies new data based on similarities to previously classified data. There are two categories, W and Z, in a categorization issue. Whenever a new data point appears with a problem with the W or Z category, the new data point should be classified according to the Euclidean distance calculation. New points are added to category Z, which include points that have the greatest number of neighbors. As far as classification techniques go, k-NN is the most often used one (Alam et al. 2022b).

AI professionals employ the decision tree (DT) approach to solve classification and regression difficulties. DT's primary goal is to create a training model that can be used to predict class outcomes by learning "basic decision rules." There are nodes that represent the attributes or features of the data and edges that indicate the possible answers to a problem and the leaf nodes that reflect the actual output or class label. Because of its accuracy and ease of implementation, this method is widely used. The method can provide a variety of outcomes. From root to leaf node, all aspects of an issue are taken into account when using DT to find the best solution for a given set of circumstances (Alam et al. 2022b; Asnake Metekia et al. 2022).

It is possible to employ RF to solve classification- as well as regression-type issues. More decision trees in the RF suggest that it will be more resilient, much like in a forest. DTs are generated from data samples, predictions are made for each DT, and finally, the best solution is selected via a voting method. By averaging the results of the DTs, RF reduces the likelihood of overfitting. A decision tree is constructed for each of the randomly selected samples from a given dataset. After then, each decision tree's outcome is known. The next step is to vote on each of the anticipated outcomes and then choose the one with the most votes as the final outcome (Alam et al. 2022b; Asnake Metekia et al. 2022; Niu et al. 2022).

Fuzzy neural networks (FNNs) are an AI approach that combines fuzzy logic with neural networks. For example, fuzzy sets and fuzzy rules can be detected by modifying neural network approximation approaches. For pattern recognition, regression, and density estimation, FNN is commonly employed in the absence of a formal model (Alam et al. 2022b; Gadkari 2022; Niu et al. 2022).

There are many other types of neural networks, but the computerized neural networks (CNNs) are the most widely used and most usually referred to one that uses convolution as an alternative to matrix multiplication in at least one of its layers (FFNN). CNNs are mostly utilized for image and video identification as well as financial time series analysis. CNN uses "local sparse connections among subsequent layers, weight sharing, and pooling" as three fundamental ideas. When lowering feature size, pooling is utilized as an alternative to the previous two techniques. In a CNN, the hidden layers (convolution and pooling layers) are responsible for sophisticated feature extraction, while the classification layers (fully connected and output layers) are responsible for determining a judgment. High-dimensional data and sophisticated models benefit from the addition of extra layers and neurons. In order to use DNN, you'll need more computational power and face more demanding training requirements. DNN outperforms other ANNs when the datasets have a sufficient amount of data (Alam et al. 2022b; Wang and Man 2021; Wei et al. 2021).

Similar to regular ANNs, but with an extra memory state for the neurons, RNNs share the same parameters as their counterparts. An RNN is an FFNN in which data is sent from the input to the output layers during processing. In order to forecast the output, it records the output of a given layer and connects it back to the input. Variable length sequences are processed by RNNs using their internal state (memory). With three gates (the input, output, and forger gate) to determine the hidden state, the most common RNN is long short-term memory (LSTM). Here's a simple example of RNN: a neural network's nodes are compacted into a single layer to form RNN. X, Y, and Z are the parameters of the proposed RNN (Alam et al. 2022b; Heydari et al. 2021; Zhang et al. 2021).

Classification and regression issues may be solved using SVM, an AI-based approach. In order to determine the next step, it requires tagged training data for each category. SVM is based on the idea of mapping an input vector into a high-dimensional feature space. Different kernel functions, such as linear, polynomial, and radial basis functions, are used to produce the mapping, with datasets serving as the foundation for function selection. In order to enhance the margin between classes, SVM draws a hyperplane to distinguish between the two classes in feature space. When it comes to pattern recognition, SVM is a go-to algorithm. One support vector is provided for each class (Alam et al. 2022b; Mahapatra et al. 2021).

In the field of artificial intelligence, SOM is the most often utilized approach. The SOM consists of two layers: an input and an output. *Feature map* and *map layer* are other names for the final output layer. It is possible to directly assign weights to the output layer, and a weight vector with the same dimension as the input space is assigned to each SOM. In order to make predictive modeling easier, dimensionality reduction helps to limit the number of input variables in a dataset. More features make it more difficult (Khan et al. 2022; Shang et al. 2022).

An algorithm called GA uses heuristics to find the best potential answer from a large pool of candidates, much like population genetics and selection in nature. There are many possible answers, and a recursive approach is used to choose the best one out of them all. As a genetic algorithm (GA), all possible solutions are stored as strings of letters from a particular alphabet. New solutions are created by mutation and mating from the current population of members. For the most part, this method is used to identify the optimal answer by solving a problem (Khan et al. 2022; Shang et al. 2022).

As a result of its repetition mechanism, particle swarm optimization (PSO) has become a popular AI-based approach for solving optimization issues. In PSO, the search space is traversed by particles based on their velocity and location. As each particle moves towards the best known location, its position and velocity are updated over time in the search space. The velocity of each particle changes as it moves across the search space in accordance with a predetermined rule. More than one artificial intelligence (AI) approach is combined in AI hybrid techniques. Researchers have previously used a variety of hybrid approaches in a variety of sectors to combine the benefits of several methodologies. GA, PSO, RNN, and SVM are four of the most prevalent approaches used in conjunction with other techniques to improve accuracy. Some of the most commonly used hybrid techniques reported in the literature are GA-MLPANN, GA-RBANN, GA-FNN, GA-FL, SVM-SA, SVM-ASAGA, ANN–differential evolution (DE), ANN–genetic algorithm neural network (GANN), PSO–wavelet neural network (WNN), and PSO– ANN–genetic algorithm neural network (GANN) (ENN). Water treatment is another area where AI hybrid approaches have received a lot of interest (Khan et al. 2022; Shang et al. 2022).

The modeling and optimization of water treatment processes, such as the removal of contaminants from water (Gwenzi et al. 2022), have been documented in many research using AI approaches. These are some of the most often utilized AI approaches for the adsorption removal of metals and chemical compounds as well as nutrients and medicines from water. There is a correlation between the many factors in the water treatment process thanks to the use of AI. Batch adsorption of pollutants, for example, frequently uses pollutant concentration as an input variable, as well as adsorbent dose, duration, pH, agitation rate, and temperature as output factors. The removal efficiency (percent) or adsorption capacity is the primary output variable. The models' projected outcomes are checked for accuracy by calculating R^2, MSE, SSE, and RMSE (root-mean-square error), all of which are based on the coefficients of determination. Most of the time, the model's predictions matched exactly what was observed in the lab. AI has been anticipated in certain research to be able to remove many contaminants from water at the same time. According to these findings, artificial intelligence (AI) might be used to enhance the efficacy of real-world water treatment systems. Batch adsorption may also be used to forecast the removal performance of adsorbents in column experiments using AI approaches. Table 4.2 illustrates different types of models used, with their efficacy

TABLE 4.2

Types of Models Used with Their Efficacy

Wastewater/Water Characteristics	Type of System Used	Observations	Accuracy	References
Chemical oxygen demand	ANN model	Observed was a quite high correlation of predicted and actual values. It was also seen that hydrogen peroxide and ferrous ion were most influenced, and more than 70% removal efficacy was observed from the system.	98%	(Khan et al. 2022; Liu et al. 2021; Shang et al. 2022)
	ARMA-VAR model, BP-ANN, ANFLS	The main observations for these models were high accuracy in forecasting and ability to control performances.	86–99%	(Khan et al. 2022; Pouyanfar et al. 2022)
Biochemical oxygen demand	ANN	Quite robust, with average flow up to 1 million m3 day-1 reported.	96%	(Alam et al. 2022b; Misbah Biltayib et al. 2021)
	ARIMA	Mean absolute error reduced to 4% only.	99%	(Nourani et al. 2021)
Heavy metals	RBF-ANN	High perdition values.	99%	(Malviya and Jaspal 2021)
	ANFIS	High perdition values.	99%	(Shahsavar et al. 2021)
	GRNN	Bayesians algorithm gave high accuracy.	99%	(Malviya and Jaspal 2021)
Organic pollutants	ANN-GA	Color removal efficiency more than 90% observed.	98.9%	(Qi et al. 2020)
	ANN	RMSE and MAPE values were quite high for mode accuracy.	99%	(Matheri et al. 2021)

4.3 MONITORING AND PARAMETER OPTIMIZATION USING AI TECHNIQUES FOR WATER/WASTEWATER TREATMENT

Water treatment performance may be evaluated using ANN by looking at factors like COD, BOD, and TSS (total suspended solids). For wastewater biological treatment units (lagoons), Akratos and colleagues, as well as Hamed and other researchers, created a model based on artificial neural networking for the elimination of BOD in an independent examination (Nayak et al. 2018). Three alternative models were tested to evaluate the performance of a wastewater treatment plant for the variables COD, BOD, and TSS. They found that the ensemble neural network model was more accurate in predicting BOD (up 24%) and COD, as well as total nitrogen in the efflu-ent (up roughly 5% both), than any other model tested. The BOD concentrations in a particularly eutrophic river were also predicted using computer modeling. Using 16–4–1 modeling, phosphate was shown to be a key input variable that could be used to forecast the parameter. For the purpose of modeling the breakdown of penicil-lin-type drugs (amoxicillin, ampicillin, cloxacillin) in water by the Fenton technique interpreted as COD elimination, Emad and his colleagues used artificial intelligence (Nourani et al. 2021). The model's appropriateness was validated by a correlation coefficient of 0.997, which was derived using ANN. Time, pH, hydrogen peroxide concentration, and molar ratio, as well as ferrous ions, all had an impact on the elimination of COD. For sewage water, an ANN was used to predict COD removal efficiency (Xiang et al. 2021). AI was used to analyze the Khorramabad wastewater treatment plant in one of the case studies by Samaneh and colleagues.

Performances were evaluated in terms of BOD and COD and total solids in terms of the influent parameters, such as temperature, pH, and BOD and COD dissolved solids. With a maximum pollutant removal effectiveness of 87.68%, good correlation coefficients were found. Another study by Zaqoot et al. employed a linear regression model and an artificial neural network (ANN) to determine water quality (Zaqoot et al. 2016). ANN was found to provide a more accurate assessment of wastewater performance than linear regression when considering influent parameters such as pH, temperature, BOD, COD, and TSS, as evidenced by error analysis and regression coefficient values (Saheb et al. 2022). When it came to evaluating overall perfor-mance, regression analysis, connectionist systems, and goal programming were less accurate, although BOD concentrations were far more predictable. Olive mill effluent was also studied using the AI approach to examine the impact of operating parame-ters on the removal of chemical oxygen demand. Experiments were done to examine the effect of pH, electrical flow per unit cross-sectional area, and process duration on the elimination efficiency of chemical oxygen demand. Effectiveness of 42.7% was matched by an R^2 value of 0.92, which is consistent with the trial results. Table 4.3 illustrates the different types of networking methods employed for treatment units. Nitrogen (N_2) and phosphorous (P) removal was studied using mathematical models that took into consideration influent volume, biomass activity, and the quality of efflu-ent (Kamali et al. 2021). In order to compare ANN results with those of batch exper-iments, the adsorption properties of chemicals such as phenol and resorcinol were examined. Amounts of carbonaceous adsorbent, adsorbate concentrations, duration, and pH were the input variables (El Din Mahmoud and Fawzy 2016; Mahmoud et al. 2020b; Fungaro et al. 2021). The neural network's output confirmed the experimen-tal results on the removal efficiency of the pollutants. Clementine 11.1 software was used by A. Pinto et al. to forecast the presence of manganese and turbidity in a water sample. A year's worth of water samples were used as the starting point for this study. Cassia seed gum produced from Cassia species has been studied to determine the removal effectiveness of reactive dye. To figure out the extraction tendency, an error analysis and a correlation coefficient value were performed. For practical and projected values, a close connection was found (Malviya and Jaspal 2021).

Manu and Thalla used AI models like support vector machines (SVM) and neu-rofuzzy interference systems to assess the removal efficiency of nitrogen in real time

TABLE 4.3

Different Types of Networking Methods Employed for Treatment Units

S. No.	Networking System Used	Observation	References
1	Korhonen type of mapping system	Quite accurate for BOD and COD values	(Maleki et al. 2022; Wan et al. 2022)
2	Two-phase models system	Showed well-predicated values for treatment process	(Malviya and Jaspal 2021)
3	Hybrid AI system	Used for different parameters during wastewater treatment	(Alam et al. 2022a; Malviya and Jaspal 2021)
4	Multilayer network	Quite satisfactory results for physicochemical treatment	(Alam et al. 2022a; Malviya and Jaspal 2021)
5	Nonlinear models	Overfitting issues can be dealt	(Malviya and Jaspal 2021)
6	Integrated deep learning model	Eliminates abnormalities during treatment operations	(Malviya and Jaspal 2021)

from a wastewater treatment plant in Mangalore, India (ANFIS). Input variables used in the research were influent pH, total solids, COD, NH_3, and influent nitrogen. SVM was shown to be more efficient than ANFIS at predicting. Mustafa Yasmeen and coresearchers used AI to examine and simulate the oil-degrading capabilities of photo-process (Malviya and Jaspal 2021). To train the ANN, a backpropagation approach was utilized. Using this algorithm, it was discovered that factors such as hydrogen peroxide and ferrous ions had a substantial impact on oil breakdown. Zinc intake on defatted oil cake has been accurately predicted using artificial intelligence (AI). When it comes to both batch and continuous operations, parameters such as pH (starting concentration), temperature (flow rate), adsorbent bed height, and so on were examined. The correlation coefficient and error analysis showed that ANN can accurately anticipate the future. Rainwater and gray water systems have both been studied using AI-based modeling. As part of the development of this modeling system, fuzzy cognition mapping and MATLAB tools were used. Instrumentation, control, and automation in water treatment systems and units have been highlighted by Yuan et al. (2019). Rustum's Kohonen mapping ANN/fuzzy inference system approach to the modeling of the activated sludge process has also been thoroughly researched. AI, he said, was better than traditional approaches because it was more viable and adaptable. The Kohonen mapping technique was important in predicting the BOD in a timely manner (Alam et al. 2022a; Ding et al. 2021). Hybrid system performance was also improved. Two steps were presented in Srinivas's and other researchers' proposed AI paradigm, namely, an analytical and synthesis phase. Conventions were extracted from the input, and the impact of treatment procedures at mixed concentrations was recognized by an algorithm during the analysis phase. Based on the analytical phase, the optimal circumstances were determined in the second step. An integrated computational intelligence system with a multilayer design can monitor the functioning of a wastewater treatment plant (WWTP). Classical and knowledge-based methodologies have been surpassed in their effectiveness in identifying systemic abnormalities, even when applied to several operational units (Zeng et al. 2022).

A previous attempt to use artificial techniques in wastewater management resulted in a reduction of energy usage and an increase in the treatment plant's efficiency. Hybrid AI system based on an algorithm proved to be useful in managing and monitoring numerous efficiency-related factors in a Taiwanese wastewater treatment plant. AI has also been widely employed in the field of environmental engineering, which includes wastewater treatment, air quality monitoring, and the management of solid waste. To evaluate biological and chemical purification processes, a multilayer feed-forward network is the most appropriate ANN among those available. AI was used to assess water quality metrics and evaluate the models' performance. Data from a survey conducted over a period of around five years was used to estimate current conditions. In order to find the best algorithm, a computerized study was performed. The application of artificial intelligence was employed to anticipate the energy consumption of urban wastewater pumping. Software and statistical modeling were used to get the computational results. One study found a 16.7% reduction in power use over the course of 90 days, as well as an improvement in the development of positive public perception. Operation of changeable wastewater tanks and development of imaginary data for insight reinforcement benefitted from it. For monitoring, assessing, and automating treatment processes, Akron's treatment facility implemented machine intelligence—supervisory control and data acquisition (SCADA) and distributed control system (DCS). In China, an AI system was used to discover operating system errors and predict water quality in the wastewater management industry. Felix and colleagues also tried out a simulation method for managing the influent load and dissolved oxygen as one of the set point factors (Malviya and Jaspal 2021).

Bongard et al. used artificial intelligence (AI) to improve the efficiency of wastewater systems (Bongards et al. 2014). They determined that the use of AI and automation technologies was more cost-effective than the use of traditional engineering approaches. Set points in sewage treatment facilities may be automatically rescheduled using an AI-based nonlinear model that takes into account the fluctuation in the amount of flow in the system. Controller operations may be used to conveniently manage the filtration effectiveness of sewage water, as well as to forecast the process's past and future behavior. AI models were used to test the efficacy of Tabriz, Iran's wastewater treatment facilities. Network models for household wastewater were evaluated by Guo Y. M. Using the harmony search algorithm introduced by Geem and colleagues, they were able to improve the usual optimization procedure for wastewater treatment and demonstrate its effectiveness. It includes a watering area with better water quality, as well as cost-effective construction and an optimal filtering procedure. It was possible to eliminate anomalous characteristics while also increasing predictability thanks to the combination of deep neural networks and support vector machines, which were used at a treatment center in the United States (Malviya and Jaspal 2021). Machine learning and statistics were used to solve complex environmental concerns, like wastewater management. AI (artificial intelligence) approaches were used to quickly build a variety of knowledge patterns which were then compared to a standard dataset.

4.4 CONCLUSION AND RECOMMENDATIONS

Artificial intelligence (AI) has the potential to revolutionize the wastewater treatment process. Artificial intelligence (AI) is a powerful technique in water treatment for removing a variety of impurities from the water supply and wastewater. There have been a number of AI models (single and hybrid) that have accurately predicted the performance of different adsorbent materials for the removal of a variety of pollutants from water and wastewater. The full potential of AI technologies in water and wastewater treatment applications must be realized despite the numerous challenges. These issues may be addressed by the use of adequate data, hybrid AI techniques, and further research at the pilot plant level. Despite these challenges, recent research indicates that AI technologies have a promising future in water and wastewater treatment applications. AI systems are widely used because of their high levels of precision and accuracy, as well as their ability to function without interruption. The application of artificial intelligence (AI) in wastewater treatment monitoring has proved effective. Researchers are still interested in exploring the following areas: contamination detection and removal, detection of levels for the start of the process, release of chemicals in wastewater treatment plant effluents, contamination identification, and subsequent selection of alternative treatment techniques. Instrumentation-based automation is widely used in the water and wastewater treatment industry; however, there is a dearth of collaboration between practitioners and academics. In addition, ecologists, engineers, IT professionals, and computer specialists must participate in the exchange of cutting-edge water technology concepts. There is still a long way to go in the digital transformation process. There is a problem with the existing use of software because of the inability to integrate information and varied methodologies together with the cost–benefit assessment of data-driven technology. There are several downsides to the science of AI, including high expenses, a lack of experience advancement, and the generation of jobs. Hybrid models have been able to handle unpredictable water input, but they haven't been tested in bioinspired methods. The internal networking and feasibility of on-field engagement require further investigation into AI. Research is needed to bridge the gap between model developers, specialists, and water and wastewater treatment engineers and to broaden the boundaries of current knowledge. Using cutting-edge technology to develop highly effective and economically efficient treatment systems is a must in order to solve this issue. Intelligent hybrid systems may hold the key to the solution in the future.

4.5 ACKNOWLEDGMENTS

Alaa El Din Mahmoud acknowledges the support of Egypt's Academy of Scientific Research and Technology (ASRT) for the SA-Egypt Joint Project.

REFERENCES

Alam, G., Ihsanullah, I., Naushad, M., Sillanpää, M., 2022a. Applications of artificial intelligence in water treatment for optimization and automation of adsorption processes: Recent advances and prospects. Chem. Eng. J. 427, 130011. https://doi.org/10.1016/j.cej.2021.130011

Alam, G., Ihsanullah, I., Naushad, M., Sillanpää, M., 2022b. Applications of artificial intelligence in water treatment for optimization and automation of adsorption processes: Recent advances and prospects. Chem. Eng. J. 427, 130011. https://doi.org/https://doi.org/10.1016/j.cej.2021.130011

Asnake Metekia, W., Garba Usman, A., Hatice Ulusoy, B., Isah Abba, S., Chirkena Bali, K., 2022. Artificial intelligence-based approaches for modeling the effects of spirulina growth mediums on total phenolic compounds. Saudi J. Biol. Sci. 29, 1111–1117. https://doi.org/https://doi.org/10.1016/j.sjbs.2021.09.055.

Bongards, M., Gaida, D., Trauer, O., Wolf, C., 2014. Intelligent automation and IT for the optimization of renewable energy and wastewater treatment processes. Energy Sustain. Soc 4, 19.

Deepnarain, N., Nasr, M., Kumari, S., Stenström, T.A., Reddy, P., Pillay, K., Bux, F., 2020. Artificial intelligence and multivariate statistics for comprehensive assessment of filamentous bacteria in wastewater treatment plants experiencing sludge bulking. Environ. Technol. Innov. 19, 100853. https://doi.org/https://doi.org/10.1016/j.eti.2020.100853.

Ding, Y., Jin, Y., Yao, B., Khan, A., 2021. Artificial intelligence based simulation of Cd(II) adsorption separation from aqueous media using a nanocomposite structure. J. Mol. Liq. 344, 117772. https://doi.org/https://doi.org/10.1016/j.molliq.2021.117772.

Dotaniya, M.L., Meena, V.D., Saha, J.K., Dotaniya, C.K., Mahmoud, A.E.D., Meena, B.L., Meena, M.D., Sanwal, R.C., Meena, R.S., Doutaniya, R.K., Solanki, P., Lata, M., Rai, P.K., 2022. Reuse of poor-quality water for sustainable crop production in the changing scenario of climate. Environ. Dev. Sustain. 1–32.

El Din Mahmoud, A., Fawzy, M., 2016. Bio-based methods for wastewater treatment: Green sorbents. In: Phytoremediation. Springer, pp. 209–238.

Fungaro, D.A., Silva, K.C., Mahmoud, A.E.D., 2021. Aluminium tertiary industry waste and ashes samples for development of zeolitic material synthesis. J. Appl. Mater. Technol. 2, 2, 66–73.

Gadkari, S., 2022. Chapter 23 —Application of artificial intelligence methods for the optimization and control of bioelectrochemical systems, In: Jadhav, D.A., Pandit, S., Gajalakshmi, S., Shah, M.P. (Eds.), Scaling Up of Microbial Electrochemical Systems. Advances in Green and Sustainable Chemistry. Elsevier, US, pp. 437–455. https://doi.org/https://doi.org/10.1016/B978-0-323-90765-1.00023-X

Gwenzi, W., Selvasembian, R., Offiong, N.-A.O., Mahmoud, A.E.D., Sanganyado, E., Mal, J., 2022. COVID-19 drugs in aquatic systems: A review. Environ. Chem. Lett. 20, 1275–1294.

He, L., Bai, L., Dionysiou, D.D., Wei, Z., Spinney, R., Chu, C., Lin, Z., Xiao, R., 2021. Applications of computational chemistry, artificial intelligence, and machine learning in aquatic chemistry research. Chem. Eng. J. 426, 131810. https://doi.org/https://doi.org/10.1016/j.cej.2021.131810.

Heydari, B., Abdollahzadeh Sharghi, E., Rafiee, S., Mohtasebi, S.S., 2021. Use of artificial neural network and adaptive neuro-fuzzy inference system for prediction of biogas production from spearmint essential oil wastewater treatment in up-flow anaerobic sludge blanket reactor. Fuel 306, 121734. https://doi.org/https://doi.org/10.1016/j.fuel.2021.121734.

Hosny, M., Fawzy, M., El-Borady, O.M., Mahmoud, A.E.D., 2021. Comparative study between Phragmites australis root and rhizome extracts for mediating gold nanoparticles synthesis and their medical and environmental applications. Adv. Powder Technol Volume 32, 2268–2279.

Huang, L., Yan, T., Mahmoud, A.E.D., Li, S., Zhang, J., Shi, L., Zhang, D., 2021. Enhanced water purification via redox interfaces created by an atomic layer deposition strategy. Environ. Sci.: Nano 8, 950–959.

Kamali, M., Appels, L., Yu, X., Aminabhavi, T.M., Dewil, R., 2021. Artificial intelligence as a sustainable tool in wastewater treatment using membrane bioreactors. Chem. Eng. J. 417, 128070. https://doi.org/https://doi.org/10.1016/j.cej.2020.128070.

Khan, A.H., Abutaleb, A., Khan, N.A., El Din Mahmoud, A., Khursheed, A., Kumar, M., 2021. Co-occurring indicator pathogens for SARS-CoV-2: A review with emphasis on exposure rates and treatment technologies. Case Studies Chem. Environm. Eng. 4, 100113.

Khan, H., Khan, S.U., Hussain, S., Ullah, A., 2022. Modelling of transmembrane pressure using slot/pore blocking model, response surface and artificial intelligence approach. Chemosphere 290, 133313. https://doi.org/https://doi.org/10.1016/j.chemosphere.2021.133313.

Liu, W., Xu, Y., Fan, D., Li, Y., Shao, X.-F., Zheng, J., 2021. Alleviating corporate environmental pollution threats toward public health and safety: The role of smart city and artificial intelligence. Saf. Sci. 143, 105433. https://doi.org/https://doi.org/10.1016/j.ssci.2021.105433.

Mahapatra, D.M., Mishra, P., Thakur, S., Singh, L., 2021. Leveraging artificial intelligence in bioelectrochemical systems. Trends Biotechnol. 40, 535–538. https://doi.org/https://doi.org/10.1016/j.tibtech.2021.11.005.

Mahmoud, A.E.D. 2020. Graphene-based nanomaterials for the removal of organic pollutants: Insights into linear versus nonlinear mathematical models. J. Environ. Manage. 270, 110911.

Mahmoud, A.E.D., 2022. Recent Advances of TiO2 Nanocomposites for Photocatalytic Degradation of Water Contaminants and Rechargeable Sodium Ion Batteries. In: Shalan, A.E., Hamdy Makhlouf, A.S., Lanceros-Méndez, S. (Eds.). Advances in Nanocomposite Materials for Environmental and Energy Harvesting Applications. Springer International Publishing, Cham, pp. 757–770.

Mahmoud, A.E.D., Al-Qahtani, K.M., Alflaij, S.O., Al-Qahtani, S.F., Alsamhan, F.A. 2021a. Green copper oxide nanoparticles for lead, nickel, and cadmium removal from contaminated water. Sci. Rep. 11, 12547.

Mahmoud, A.E.D., Fawzy, M., Hosny, G., Obaid, A., 2020a. Equilibrium, kinetic, and diffusion models of chromium(VI) removal using Phragmites australis and Ziziphus spina-christi biomass. Int. J. Environ. Sci. Technol. 18, 2125–2136.

Mahmoud, A.E.D., Fawzy, M., Radwan, A., 2016. Optimization of Cadmium (CD2+) removal from aqueous solutions by novel biosorbent. Int. J. Phytoremed. 18, 619–625.

Mahmoud, A.E.D., Franke, M., Braeutigam, P., 2022c. Experimental and modeling of fixed-bed column study for phenolic compounds removal by graphite oxide. J. Water Proc. Eng. 49, 103085.

Mahmoud, A.E.D., Franke, M., Stelter, M., Braeutigam, P., 2020b. Mechanochemical versus chemical routes for graphitic precursors and their performance in micropollutants removal in water. Powder Technol. 366, 629–640.

Mahmoud, A.S., Farag, R.S., Elshfai, M.M., 2020. Reduction of organic matter from municipal wastewater at low cost using green synthesis nano iron extracted from black tea: Artificial intelligence with regression analysis. Egypt. J. Pet. 29, 9–20. https://doi.org/https://doi.org/10.1016/j.ejpe.2019.09.001

Mahmoud, A.E.D., Hosny, M., El-Maghrabi, N., Fawzy, M. 2022a. Facile synthesis of reduced graphene oxide by Tecoma stans extracts for efficient removal of Ni (II) from water: batch experiments and response surface methodology. Sustain. Environ. Res. 32, 18, 22, 2125–2136.

Mahmoud, A.E.D., Hosny, M., El-Maghrabi, N., Fawzy, M. 2022b. Facile synthesis of reduced graphene oxide by Tecoma stans extracts for efficient removal of Ni (II) from water: batch experiments and response surface methodology. Sustainable Environment Research 32, 22

Mahmoud, A.E.D., Kathi, S., 2022. Chapter 7 — Assessment of biochar application in decontamination of water and wastewater. In: Kathi, S., Devipriya, S., Thamaraiselvi, K. (Eds.). Cost Effective Technologies for Solid Waste and Wastewater Treatment. Elsevier, pp. 69–74.

Mahmoud, A.E.D., Stolle, A., Stelter, M., Braeutigam, P., 2018. Adsorption technique for organic pollutants using different carbon materials. Abstracts of Papers of the American Chemical Society. Washington, DC 20036 USA.

Mahmoud, A.E.D., Umachandran, K., Sawicka, B., Mtewa, T.K. 2021b. 26 — Water resources security and management for sustainable communities. In: Mtewa, A.G., Egbuna, C. (Eds.), Phytochemistry, the Military and Health. Elsevier, pp. 509–522

Maleki, R., Jahromi, A.M., Ghasemy, E., Khedri, M., 2022. Chapter 1 — Smart sensing technologies for wastewater treatment plants, In: Asadnia, M., Razmjou, A., Beheshti, A.B.T.-A.I. and D.S. In E.S. (Eds.), Cognitive Data Science in Sustainable Computing. Academic Press, pp. 1–17. https://doi.org/https://doi.org/10.1016/B978-0-323-90508-4.00003-4

Malviya, A., Jaspal, D., 2021. Artificial intelligence as an upcoming technology in wastewater treatment: A comprehensive review. Environ. Technol. Rev. 10, 177–187. https://doi.org/10.1080/21622515.2021.1913242

Matheri, A.N., Ntuli, F., Ngila, J.C., Seodigeng, T., Zvinowanda, C., 2021. Performance prediction of trace metals and cod in wastewater treatment using artificial neural network. Comput. Chem. Eng. 149, 107308. https://doi.org/https://doi.org/10.1016/j.compchemeng.2021.107308

Misbah Biltayib, B., Bonyani, M., Khan, A., Su, C.-H., Yu, Y.-Y., 2021. Predictive modeling and simulation of wastewater treatment process using nano-based materials: Effect of pH and adsorbent dosage. J. Mol. Liq. 343, 117611. https://doi.org/https://doi.org/10.1016/j.molliq.2021.117611

Mousazadeh, M., Naghdali, Z., Rahimian, N., Hashemi, M., Paital, B., Al-Qodah, Z., Mukhtar, A., Karri, R.R., Mahmoud, A.E.D., Sillanpää, M., Dehghani, M.H., Emamjomeh, M.M., 2021. Chapter 9 — Management of environmental health to prevent an outbreak of COVID-19: A review. In: Hadi Dehghani, M., Karri, R.R., Roy, S. (Eds.). Environmental and Health Management of Novel Coronavirus Disease (COVID-19). Academic Press, pp. 235–267.

Nasr, M., Mahmoud, A.E.D., Fawzy, M., Radwan, A., 2017. Artificial intelligence modeling of cadmium(II) biosorption using rice straw. Appl. Water Sci. 7, 823–831.

Nayak, M., Dhanarajan, G., Dineshkumar, R., Sen, R., 2018. Artificial intelligence driven process optimization for cleaner production of biomass with co-valorization of wastewater and flue gas in an algal biorefinery. J. Clean. Prod. 201, 1092–1100. https://doi.org/https://doi.org/10.1016/j.jclepro.2018.08.048

Niu, C., Li, X., Dai, R., Wang, Z., 2022. Artificial intelligence-incorporated membrane fouling prediction for membrane-based processes in the past 20 years: A critical review. Water Res. 216, 118299. https://doi.org/https://doi.org/10.1016/j.watres.2022.118299

Nourani, V., Asghari, P., Sharghi, E., 2021. Artificial intelligence based ensemble modeling of wastewater treatment plant using jittered data. J. Clean. Prod. 291, 125772. https://doi.org/https://doi.org/10.1016/j.jclepro.2020.125772

Omran, M.A., Fawzy, M., Mahmoud, A.E.D., Abdullatef, O.A., 2022. Optimization of mild steel corrosion inhibition by water hyacinth and common reed extracts in acid media using factorial experimental design. Green Chem. Lett. Rev. 15, 216–232.

Picos-Benítez, A.R., Martínez-Vargas, B.L., Duron-Torres, S.M., Brillas, E., Peralta-Hernández, J.M., 2020. The use of artificial intelligence models in the prediction of optimum operational conditions for the treatment of dye wastewaters with similar structural characteristics. Process Saf. Environ. Prot. 143, 36–44. https://doi.org/https://doi.org/10.1016/j.psep.2020.06.020.

Pouyanfar, N., Harofte, S.Z., Soltani, M., Siavashy, S., Asadian, E., Ghorbani-Bidkorbeh, F., Keçili, R., Hussain, C.M., 2022. Artificial intelligence-based microfluidic platforms for the sensitive detection of environmental pollutants: Recent advances and prospects. Trends Environ. Anal. Chem. 34, e00160. https://doi.org/https://doi.org/10.1016/j.teac.2022.e00160.

Qi, J., Hou, Y., Hu, J., Ruan, W., Xiang, Y., Wei, X., 2020. Decontamination of methylene Blue from simulated wastewater by the mesoporous rGO/Fe/Co nanohybrids: Artificial intelligence modeling and optimization. Mater. Today Commun. 24, 100709. https://doi.org/ https://doi.org/10.1016/j.mtcomm.2019.100709

Saheb, Tahereh, Dehghani, M., Saheb, Tayebeh, 2022. Artificial intelligence for sustainable energy: A contextual topic modeling and content analysis. Sustain. Comput. Informatics Syst. 35, 100699. https://doi.org/https://doi.org/10.1016/j.suscom.2022.100699

Shahsavar, M.M., Akrami, M., Gheibi, M., Kavianpour, B., Fathollahi-Fard, A.M., Behzadian, K., 2021. Constructing a smart framework for supplying the biogas energy in green buildings using an integration of response surface methodology, artificial intelligence and petri net modelling. Energy Convers. Manag. 248, 114794. https://doi.org/https:// doi.org/10.1016/j.enconman.2021.114794.

Shang, Q., Chi, W., Zhang, P., Ling, Y., Liu, X., Cui, G., Liu, W., Shi, X., Tang, B., 2022. Optimization of Bi2O3/TS-1 preparation and photocatalytic reaction conditions for low concentration Erythromycin wastewater treatment based on artificial neural network. Process Saf. Environ. Prot. 157, 297–305. https://doi.org/https://doi.org/10.1016/j. psep.2021.11.031.

Wan, X., Li, X., Wang, X., Yi, X., Zhao, Y., He, X., Wu, R., Huang, M., 2022. Water quality prediction model using Gaussian process regression based on deep learning for carbon neutrality in papermaking wastewater treatment system. Environ. Res. 211, 112942. https://doi.org/https://doi.org/10.1016/j.envres.2022.112942.

Wang, Z., Man, Y., 2021. Chapter 7 — Artificial intelligence algorithm application in wastewater treatment plants: Case study for COD load prediction. In: Ren, J., Shen, W., Man, Y., Dong, L. (Eds.), Applications of Artificial Intelligence in Process Systems Engineering. Elsevier, pp. 143–164. https://doi.org/https://doi.org/10.1016/ B978-0-12-821092-5.00009-7.

Wang, J.-H., Zhao, X.-L., Guo, Z.-W., Yan, P., Gao, X., Shen, Y., Chen, Y.-P., 2022. A full-view management method based on artificial neural networks for energy and material-savings in wastewater treatment plants. Environ. Res. 211, 113054. https://doi.org/https://doi. org/10.1016/j.envres.2022.113054.

Wei, Y., Yu, J., Du, Y., Li, H., Su, C.-H., 2021. Artificial intelligence simulation of Pb(II) and Cd(II) adsorption using a novel metal organic framework-based nanocomposite adsorbent. J. Mol. Liq. 343, 117681. https://doi.org/https://doi.org/10.1016/j. molliq.2021.117681.

Wen, C.-H., Vassiliadis, C.A., 1998. Applying hybrid artificial intelligence techniques in wastewater treatment. Eng. Appl. Artif. Intell. 11, 685–705. https://doi.org/https://doi. org/10.1016/S0952-1976(98)00036-0.

Xiang, X., Li, Q., Khan, S., Khalaf, O.I., 2021. Urban water resource management for sustainable environment planning using artificial intelligence techniques. Environ. Impact Assess. Rev. 86, 106515. https://doi.org/https://doi.org/10.1016/j.eiar.2020.106515.

Yuan, Z., Olsson, G., Cardell-Oliver, R., van Schagen, K., Marchi, A., Deletic, A., Urich, C., Rauch, W., Liu, Y., Jiang, G., 2019. Sweating the assets—the role of instrumentation, control and automation in urban water systems. Water Res. 155, 381–402.

Zaqoot, H.A., Hamada, M., El-Tabash, M., 2016. Investigation of drinking water quality in the kindergartens of Gaza Strip Governorates. J. Tethys 4, 088–099.

Zeng, K., Hachem, K., Kuznetsova, M., Chupradit, S., Su, C.-H., Nguyen, H.C., El-Shafay, A.S., 2022. Molecular dynamic simulation and artificial intelligence of lead ions removal from aqueous solution using magnetic-ash-graphene oxide nanocomposite. J. Mol. Liq. 347, 118290. https://doi.org/https://doi.org/10.1016/j.molliq.2021.118290.

Zhang, C., Sun, W., Wei, H., Sun, C., 2021. Application of artificial intelligence for predicting reaction results in advanced oxidation processes. Environ. Technol. Innov. 23, 101550. https://doi.org/https://doi.org/10.1016/j.eti.2021.101550.

5 Water Quality and Water Pollution Monitoring Using Artificial Intelligence

Ghadir A. El-Chaghaby, Sayed Rashad,
Murthy Chavali, El-Shimaa A. Rawash,
Muhammad Abdul Moneem, and Soliman R. Radwan

CONTENTS

5.1 INTRODUCTION

Water bodies are crucial in reducing the effects of climate change. Additionally, they stand for a vital resource that billions of people can use in a variety of ways. Understanding their quality is also very important. The accurate forecasting of the water quality time series may offer guidance for early warning of water

DOI: 10.1201/9781003260455-5

contamination and assist policy makers in the more efficient management of water resources (Ablieieva et al. 2022). Water is an essential element of the ecosystem, but because of both natural and human-caused activities, the quality of surface water and groundwater has been declining for a long time. Hydrological, atmospheric, climatic, topographical, and lithological elements are examples of natural factors that affect water quality (Oyedotun and Ally 2021). Mining, livestock production, trash generation, and waste disposal (municipal, industrial, and agricultural); increased sediment runoff or soil erosion as a result of land-use change; and heavy metal contamination are a few examples of anthropogenic activities that negatively influence water quality (Uddin et al. 2021).

In the areas of reducing pollution, improving water quality, and addressing the issue of water scarcity, the sixth sustainable development goal highlights the necessity of having access to clean water (Ighalo and Adeniyi 2020). With the primary goal of identifying the issue and addressing it holistically depending on the intended use, water quality illustrates the physical, chemical, and biological status of water bodies. Depending on the intended use, different water quality standards apply, such as for public water supply, fish and wildlife habitat, recreation, agriculture, and industry (Giri 2021).

Water and wastewater treatment facilities, as well as several other natural and industrial systems that depend on the resource, must have access to sustainable and clean water (Panagopoulos 2022). Treatment facilities must deal with complicated regulatory procedures to fulfill rising standards of quality, in addition to catering to customer wants and enhancing infrastructure for quality of life (Faherty 2021). This is only complicated by the fact that nations continue to have severely contaminated waterways, which damage aquatic and terrestrial life in addition to human life. These problems are noticeably getting worse as nations continue to industrialize and modernize (Ebenstein 2012). Researchers from all over the world have looked into ways to improve our water-related applications (Vu and Wu 2022). In many research groups, the focus on developing and modeling optimal, economical, and intelligent models has been adequate.

It thus is essential to have precise and effective technologies for monitoring water quality and tracing pollution sources, figuring out when and where pollutants are discharged into rivers and figuring out how much of each is released. Assessing the quality of the water and identifying the causes of water pollution are necessary for achieving these objectives (Hojjati-Najafabadi et al. 2022). Water quality features that result from pollutant inputs at various point sources can be observed by measurements taken at monitoring stations. Therefore, identifying point sources is necessary to create policies to stop pollution from waste discharges. It is thus obvious that to safeguard aquatic systems, it is crucial to evaluate the features of water quality and identify the point sources of contaminants (Wang et al. 2019).

In this respect, water quality monitoring programmers provide a reliable assessment of water quality, which allows decision makers to understand, interpret, and use the data in support of management activities aimed at protecting the resource (Huang et al. 2022).

The relationship between water quality and quantity as well as watershed preservation is well-known and understood in the field of watershed management. However, because of the poor water quality brought on by industrialization, declining water quality in rivers, lakes, and groundwater has become a global issue. As a result, many nations have decided to reform their water governance systems in order to achieve sustainable development using an integrated strategy, as is advised. In many aspects, society is becoming more digital, and it is anticipated that this trend will continue (Syafri et al. 2020).

While digitalization takes many forms, artificial intelligence (AI), including machine learning, has exploded in popularity in recent years. Transportation, energy, health care, and manufacturing are all notable application domains. In comparison to these other fields, AI use in the water domain is still quite limited (Sreedevi et al. 2022). Artificial intelligence techniques have become increasingly popular in recent years (AI). As a result, initiatives for guidance on how to develop "responsible AI" that is aligned with human and ethical values have sprung up. The use of AI-based techniques in the water domain is relatively low when compared to that in other sectors, such as energy, health care, and transportation (Alam et al. 2022).

In this chapter, we are discussing water quality and water pollution by emphasizing emerging water contaminants. We are also highlighting the use of artificial intelligence for water quality monitoring, underlining its benefits and various techniques. The chapter also discusses, in brief, the main water quality parameters and the emerging water contaminants. The application of different artificial intelligence techniques for water quality and water pollution monitoring is also discussed.

5.1.1 WATER QUALITY AND WATER POLLUTION

Water quality illustrates the physical, chemical, and biological status of water bodies, with the primary goal of identifying the issue and addressing it holistically, depending on the intended use. Depending on the intended use, different water quality standards apply, such as for public water supply, fish and wildlife habitat, recreation, agriculture, and industry (Giri 2021).

The quality of most ambient water bodies, such as rivers, lakes, and streams, is determined by precise quality standards. Water specifications for various applications/use also have their own set of standards. Irrigation water, for example, must not be overly saline, nor must it contain hazardous compounds that can be passed to plants or soil, harming ecosystems. Water quality for industrial application necessitates a variety of qualities depending on the unique industrial activities (Conine et al. 2021).

Water is crucial to all facets of our lives, but as a result of urbanization, industrialization, and population growth, pollution is constantly worsening and causing a decline in water quality. Finding the pollutants contaminating the water is essential for maintaining the quality of life. In most cases, determining the quality of water is a laborious task that takes time and requires manual laboratory analysis and statistical inferences. Systems that track and identify water pollution in real time have been developed all over the world (Chen et al. 2022). To measure water quality, various metrics are used. The most typical variables are depicted in Table 5.1.

Pollution in aquatic bodies is becoming more prevalent as the world's population and industries grow. As a result, the quality of natural-source water in general, and

TABLE 5.1

Water Quality Parameters

Parameter	What It Defines	Reference
pH	Acidity or alkalinity of water.	(Li et al. 2022)
Turbidity	Amount of dissolved, nonfilterable solids present in the water.	(Ajala et al. 2022)
Temperature	Has a significant impact on aquatic systems; influences the amount of dissolved oxygen and gas transfer rates.	(Ajala et al. 2022)
Electrical conductivity	Measure of how well water can conduct electricity beneficial in terms of the ionic content of the water.	(Baxa et al. 2021)
Chloride	Mineral that is naturally present in water. While excessive amounts are generally not harmful to humans, exceeding 250 mg/l causes the water to taste saltier and may have negative effects on agricultural activities.	(Akintan et al. 2022)
Dissolved oxygen	Demonstrates how easily oxygen dissolves in water.	(Dehghani et al. 2022)
Total hardness	Amount of magnesium and calcium concentrations in the water determines whether water is suitable for domestic and commercial use.	(Ram et al. 2021)
Total solids	Water's suspended and dissolved solids. It denotes the presence of impurities in the water, such as amounts of calcium, phosphorus, and sulfur.	(Patale and Tank 2022)
Total dissolved solids (TDS)	Quantity of inorganic and organic soluble solids left in the water.	(Wilson et al. 2021)
Biological oxygen demand (BOD)	Measures how much oxygen is used by aquatic life, particularly bacteria and protozoa.	(Sobczyk et al. 2021)
Chemical oxygen demand (COD)	Quantity of oxygen used during the oxidation of existing inorganic material as well as during the breakdown of organic material.	(Ghorbani and Salem 2021)
Fecal coliforms (FC)	Bacteria that are primarily found in the intestines of warm-blooded animals and are present in both human and animal waste.	(Olalemi and Akinwumi 2022)

surface water in particular, must be continuously monitored (Sharma et al. 2021). The determination of certain characteristics allows for monitoring and assessment of water quality (which is also useful for creating quality indices). A biological index or physicochemical parameters can also be used to track water quality (Fatima et al. 2022).

River contamination has risen alarmingly as a result of climate change and anthropogenic activities. In order to develop more effective management policies and advanced early warning systems, river basin water quality (WQ) prediction, risk assessment, and pollutant classification techniques have been heavily researched in recent decades (Paepae et al. 2021). The next problem is dealing with water-related data, which is difficult to manage due to nonlinearity, nonstationarity, and ambiguous properties caused by unpredictable natural changes, interdependent relationships, human interference, and complexity (Tiyasha et al. 2020).

5.1.2 Water Quality Monitoring

To assess the quality of the marine environment, it is widely employed to measure the causes (pollutants) and effects (ecosystem impact) in the water. Using a successful experimental design as the foundation, monitoring is a strategy that lasts for a number of years. The data gathered by a monitoring program is used for a number of objectives. They can be employed for model validation and verification to see if potential effects on the marine environment fall within acceptable bounds established at the start of the activity, or for compliance monitoring, to make sure that both pollutants and ecosystem impacts from specific activities do not exceed standard values set by the authorities or relevant legislation. They can also be used to track trends and spot long-term environmental changes (Karydis and Kitsiou 2013).

To evaluate the quality of water for ecosystem health and hygiene, industrial usage, agricultural use, and home use, regular water quality monitoring of the resources is very important. The examination of water quality may involve complex compound factors, which can lead to several concerns about the water's overall quality. For large samples with concentrations of several components, determining the quality of the water is difficult. Conventional approaches for assessing water quality rely on comparing experimentally derived parameter values with currently accepted standards (Tirkey et al. 2013).

Data on the specific chemical components of each water sample may be obtained using conventional procedures for water quality analysis. Numerous valuable research and compliance monitoring needs are now met by this analyte-by-analyte method. However, these methods have basic and significant limitations in accounting for the "universe" of millions of compounds that may be present, since they require a priori targeting of a certain analyte or analytes (Doyle et al. 2015). Measuring causes (pollutants) and consequences (ecosystem impact) in the water is frequently used to evaluate the quality of the marine environment. *Monitoring* is a term used to describe a technique that is based on an effective experimental design and lasts for many years. A monitoring program's data collection serves a variety of purposes.

They can be used for compliance monitoring to ensure that both pollutants and ecosystem impacts from specific activities do not exceed standard values set by the authorities or relevant legislation or for model validation and verification to determine whether potential effects on the marine environment are within acceptable limits set at the beginning of the activity. They can also be used to monitor patterns and identify long-term environmental changes (Karydis and Kitsiou 2013).

5.2 EMERGING CONTAMINANTS IN THE WATER ENVIRONMENT: SOURCES, RISKS, AND DETECTION METHODS

"Emerging" refers to a pollutant that has a novel origin, an alternative human exposure pathway, or novel treatment methods (Gogoi et al. 2018). "Emerging contaminants" (ECs) are substances, whether man-made or naturally occurring, or microbes that are not regularly observed in the environment but have the potential to infiltrate it and have known or suspected negative ecological and/or health consequences on humans, and they consist of different types of compounds (Figure 5.1) (Rosenfeld and Feng 2011).

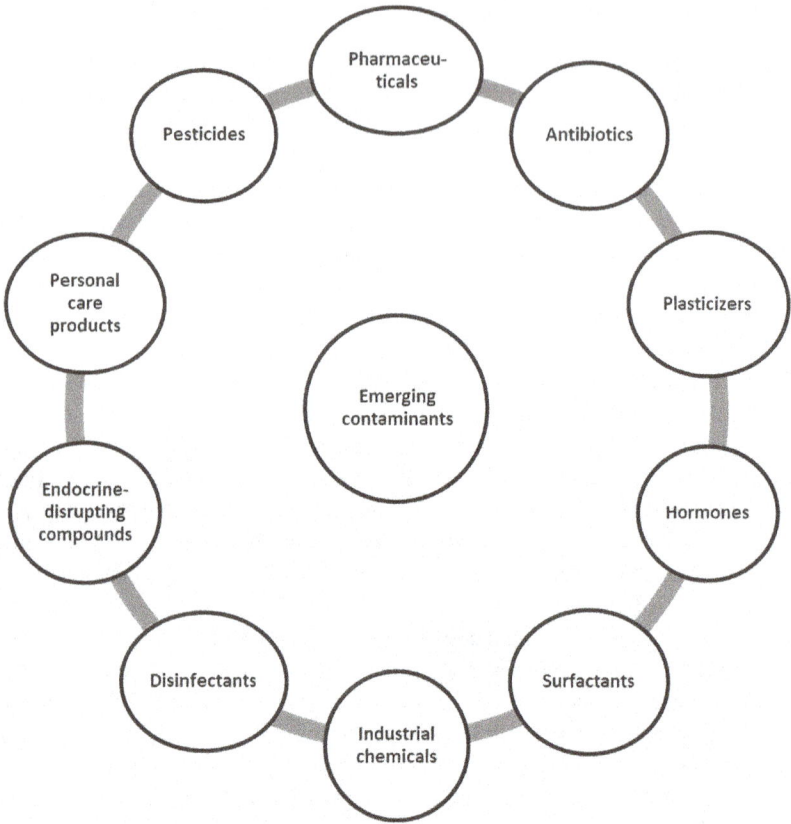

FIGURE 5.1   Emerging water contaminants.

Emerging pollutants are closely related to environmental quality standards. Data on the environmental chemistry, human toxicity, ecotoxicology, and epidemiology of a novel substance increase when it starts to raise concerns. In the end, this leads to government intervention to set environmental standards or criteria to assure proper protection. Compounds that are previously controlled are frequently re-evaluated with the inclusion of fresh data in a similar order (Sauvé and Desrosiers 2014).

The key issue with ECs is that the amount of data that is now available for the majority of these pollutants is sparse and very small, and the corresponding detection techniques and apparatus either do not yet exist or are in the early stages of development. This is why finding these microcontaminants in the environment only became feasible with the slow advancement of analytical detecting and measuring techniques (Stefanakis and Becker 2016).

According to the literature, two types of EC sources in the water environment can be distinguished: (1) *point sources*, where pollutants are produced from many sources and enter the environment in a manner that is geographically constrained and can be statistically described, and (2) *nonpoint sources*, where most pollution

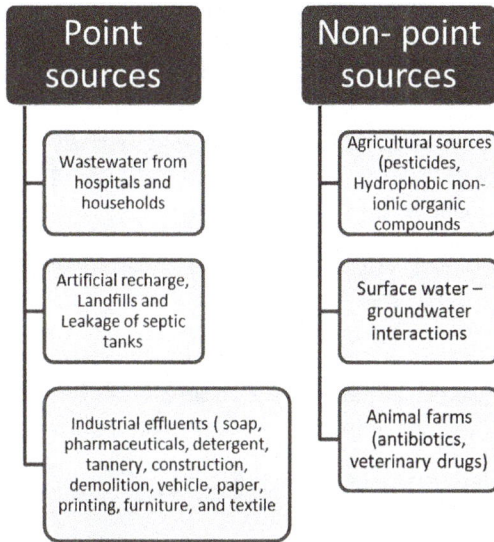

FIGURE 5.2 Sources of water pollution.

sources' origins are unknown and difficult to identify (Mukhopadhyay et al. 2022). Figure 5.2 summarizes the main sources of ECs (Kumar et al. 2022; Mukhopadhyay et al. 2022).

Because of the chemical nature of some compounds of the ECs group, they are easily absorbed in water and carried through the water cycle, posing a substantial risk to aquatic creatures and people (Rodriguez-Narvaez et al. 2017). Concerns about the existence of these developing pollutants in the environment include aberrant physiological processes and reproductive damage, higher cancer occurrences, the development of antibiotic-resistant microorganisms, and the possible enhanced toxicity of chemical combinations (Gogoi et al. 2018).

EC can have a negative influence on aquatic and terrestrial species, as well as human societies. Antibiotics may create bacterial pathogen resistance, resulting in a changed microbial community structure in nature and affecting the upper food chain. Endocrine-disruptive compounds have been shown to produce estrogenic effects in rats and hormonal effects that enhance the risk of breast cancer in humans. It is said to work as an antiandrogen, causing feminizing adverse effects in men. Antimicrobiological preservatives used in cosmetics, toiletries, and even food may have modest estrogenic action. Disinfectants and antiseptics are poisonous or biocidic agents that contribute to microbial resistance (Gogoi et al. 2018).

Identification and measurement of ECs in water or wastewater have emerged as a key scientific problem requiring very sophisticated analytical methods that can detect in nanograms per liter. The main analytical technique accessible for these sorts of chemicals is mass spectrometry (MS), in combination with chromatography, either with gas or liquid. Even in very complex matrices like surface water or wastewater, individual contaminants may now be correctly identified and quantified thanks to recent major breakthroughs, especially in terms of selectivity, sensitivity, and specificity. Due to this, the usage of MS technology for the detection and quantification of ECs has increased (Rodriguez-Narvaez et al. 2017).

5.3 ARTIFICIAL INTELLIGENCE TOOLS FOR WATER QUALITY AND POLLUTION MONITORING

Artificial intelligence (AI) originally achieved its name, repetition capability, and early successes in 1956. The AI discipline, which focuses on computers with a humanlike consciousness, is presently playing a key role across a variety of fields (Dwivedi et al. 2021). AI techniques allowed academics to imitate human knowledge in precisely specified fields by combining expressive understanding, procedural knowledge, and reasoning. Additionally, improvements in AI methods have made it possible to build intelligent management systems by utilizing shells inside well-known platforms like MathLab, Visual Basic, and C++ (Agrawal et al. 2021).

The field of computer science known as artificial intelligence (AI) focuses on creating intelligent machines that solve issues in a manner akin to that of a human intelligence system. Artificial intelligence (AI) models have demonstrated remarkable success and superiority in dealing with nonlinear data due to their higher accuracy, reliability, robustness, cost-effectiveness, problem-solving capability, decision-making capability, efficiency, and effectiveness (Pothe 2022).

The use of artificial intelligence (AI) to govern an environmental process entails several subprocesses that must be controlled or automated. In a pollution-control facility, for example, there are numerous levels at which to approach the problem of management and control. Instruments that monitor, detect, and manipulate process variables are available at the most basic level (Rahman et al. 2021). The instruments are frequently linked to a control structure capable of enforcing control legislation. The supervisory host computer is the next level, which is generally linked to some control hardware through network connections. The supervisory host computer runs programs one step above the fundamental control operations, such as the database. In turn, the supervisory host computer may be linked to a plant-wide and, ultimately, a corporate-wide computer system (Chan and Huang 2003). There are several types of artificial intelligence (AI) tools (Ighalo et al. 2021) (Figure 5.3).

5.3.1 ARTIFICIAL NEURAL NETWORKS

To address problems with water quality, modeling of water quality has been created utilizing cutting-edge computer and artificial intelligence (AI) approaches. By foreseeing changes in water quality, artificial neural networks (ANNs) have benefited in the monitoring of water quality systems (Al-Adhaileh and Alsaade 2021). **Artificial neural networks (ANN)** have attracted a lot more attention due to their success in prediction, grouping, and classification. This strategy is quickly replacing traditional ones as a viable and well-liked one. ANN refers to intelligent bionic models as well as

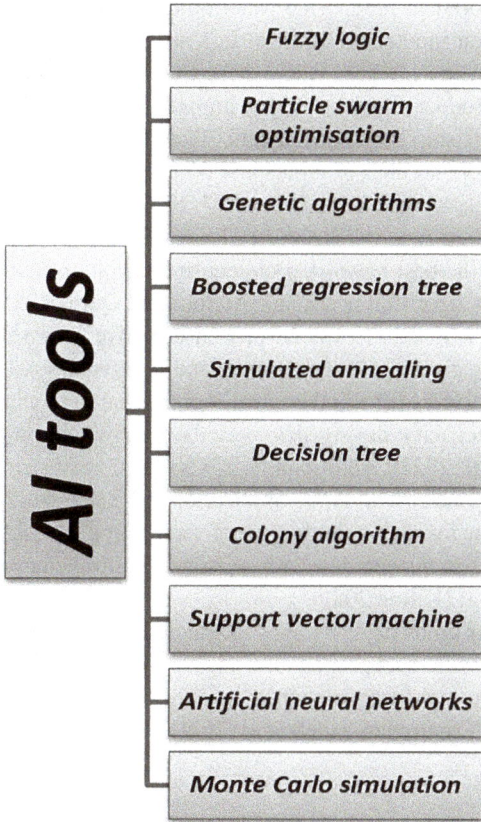

FIGURE 5.3 Artificial intelligence tools.

nonlinear, adaptive dynamics, large-scale systems made up of many linked neurons. ANN models have been frequently used to solve water quality issues (Chu et al. 2013).

The input layer, output layer, and hidden layer—which might be one or more—make up the fundamental structure of an ANN. Numerous units that were coupled to one another made up each layer; these elements were known as nodes or neurons. A neuron was a component that carried out a variety of operations, including ingesting incoming data, giving it weights, adding the weight and its corresponding input data, and adding a bias. The activation function, a function, was then used to alter the outcome. Each layer's individual neurons were uniquely connected to one another (Ighalo et al. 2021).

To evaluate the quality of the water, a lot of data must be analyzed. Artificial neural network (ANN) models come to the fore when traditional approaches (such as linear and nonlinear regression) fall short of researchers' expectations. A group of models known as ANNs has an architecture inspired by biological neural networks. ANNs are viewed by scientists as a group of synthetic neurons that have been organized into an interconnected network (Tan et al. 2022).

The neural network can predict the water quality index and can identify implicit relationships between inputs and outputs. It is sufficient to train the network, after which it will be able to forecast values based on past performance. Additionally, ANN models can produce forecasts with a high degree of accuracy even when there is a nonlinear relationship between the data (Trach et al. 2022). An appropriate network structure, a large number of inputs and outputs, and a large number of simulation epochs are necessary when building an ANN. Experience and trial and error are used to choose the best network structure. To approximate complex nonlinear relationships, selecting the best network architecture, activation function, loss function, and an optimization algorithm is crucial (Alaloul and Qureshi 2020).

The use of ANN models has been proven effective in a variety of processes, including estimating river flows (flow, level, and flow volume), issuing flood warnings, managing reservoirs for flood control, figuring out a stream's water potential, producing hydroelectricity during dry periods, and organizing transportation in streams (Chebii et al. 2022). Additionally, ANNs have been successfully applied in hydrology. The most successful ANN applications in recent studies have been in the hydrology and water resources fields (runoff forecasting, rainfall-runoff modeling, incoming runoff, reservoir operation, dispersion in natural channels, and suspended sediment forecasting) (Kilinc 2022).

The most popular kinds of ANN used in water quality systems are the multilayer perceptron (MLP) with backpropagation algorithm, radial basis neural networks (RBNN), adaptive neural-fuzzy inference system (ANFIS), and deep learning neural networks (DLNN) (Emamgholizadeh et al. 2014).

5.3.1.1 Multilayer Perceptron (MLP)

MLP networks are common types of ANN widely used in many complex mathematical problems that lead to solving complex nonlinear equations. MLP consists of several steps, including (i) data preparation, including training, validation, and testing; (ii) designation of the structure of the model, including several hidden layers and neurons; (iii) choosing the type of activation functions; (iv) training algorithm; and in the end, (v) evaluating coupled with validating the performance of the developed model (Qishlaqi et al. 2017).

The drawbacks of MLP are the unclear direct relation between independent and dependent variables and the suitable functioning of the model being dependent on training quality (Alam et al. 2022).

5.3.1.2 Radial Basis Function Neural Network (RBNN)

An ANN type that might be used for a variety of issues is called RBNN. The central placement of the hidden node is determined during a grouping step of training for other forms of ANN; however, the RBNN model is more favorable. RBNN was created as one of the most popular three-layer neural feed-forward networks for calculating water quality metrics. The design of this model was reduced into two weight levels, with fundamental function parameters in the first layer, and linear combinations of those basic functions for processing the output as well as water quality parameters in the second layer (Astuti et al. 2020).

RBNN is similar in topology to the MLP network, but RBNN has the advantage of nonsuffering from local minima in the same way as the multilayer perceptron. RBNN networks are also good at modeling nonlinear data and could be trained in one stage rather than using an iterative process as in MLP and also learn the given application.

Although, when the dataset from practical water quality assessment is complicated with more data samples, RBNN needs more hidden neurons to enhance its learning capability, which results to complicating structural size coupled with high computational complexity (Wang et al. 2022).

5.3.1.3 Adaptive Neural-Fuzzy Inference System (ANFIS)

In 1993, Jang debuted ANFIS for the first time (Shah et al. 2021). Fuzzy logic and ANN were combined to create ANFIS. It successfully incorporated a fuzzy inference system's learning capabilities into neural networks. Any real continuous function on a small set can be accurately approximated using this method. The member function (MF) of the ANFIS model could be tuned, and fuzzy rules could be produced using two different optimization techniques (hybrid and backpropagation) (Azad et al. 2019). The best ANFIS model was selected for reducing the modeling complexity that yielded the lowest modeling error with a minimum rules number, and then cross-validation was employed to evaluate the final outcome from the ANFIS model (Tiri et al. 2018). The main disadvantages of ANFIS were the probability to be trapped in local optima, causing poor performance (Azad et al. 2019).

5.3.1.4 Deep Learning Neural Networks (DLNN)

To overcome the disadvantages of different ANNs, especially in modeling prediction and high computational complexity in the absence of effective features of datasets, much attention was paid towards a new era of ANN technique called deep learning neural networks (DLNN) to enhance the capability of feature extraction and data representation extraction (Alam et al. 2022).

5.3.2 Fuzzy Logic

Fuzzy logic, a generalization of formal logic and classical set theory, is one of the techniques that reduce uncertainty and make numerical problem-solving easier. By extending classical logic, fuzzy logic can be used to address issues with a high degree of subjectivity. Scientist Lotfi Zadeh made the initial suggestion for the application of fuzzy logic in 1965.

Fuzzy logic is able to handle multilingual, hazy, and ambiguous data and is characterized as a logical, reliable, and transparent process of acquiring and using data that generates opportunities for decision-making in the environment (Charisis et al. 2022). Fuzzy logic is unique in its capacity to resolve complicated environmental problems with numerous input variables and intricate interdependencies. To assess water quality, fuzzy logic tools and capabilities are employed to calculate the WQI.

Modeling ecological systems is a difficult scientific task because input and output statements made by researchers are frequently inaccurate. To address this issue, fuzzy logic can be used to create environmental monitoring indicators (Uddin et al. 2021).

Modeling ecological systems is a difficult scientific task because input and output statements made by researchers are frequently inaccurate. This problem can be solved by developing environmental monitoring indicators using fuzzy logic. Due to its simplicity, fuzzy logic is successfully utilized to simulate natural-language-based water quality evaluation. The assessment of the water quality index using an algebraic expression yields less-precise results than fuzzy inference methods that use linguistic calculations. Fuzzy inference systems have been used to generate water quality indices because these methods can provide alternative approximations when aims and boundaries are vague or unclear (Trach et al. 2022).

5.3.3 PARTICLE SWARM OPTIMIZATION

A stochastic population-based optimization method impacted by social interactions is called particle swarm optimization (PSO). This approach is commonly employed in numerous contexts, including water resources. The PSO evolutionary method was first proposed by Kennedy and Eberhart in 1995. The simplified social system of a flock of birds flying in search of food resource locations towards an unknown goal (fitness function) served as the model for the particle swarm idea. Birds are used in the PSO algorithm, and they plan and coordinate their motions to fly directly to their destination (Jahandideh-Tehrani et al. 2020).

Using evolutionary computing, the particle swarm optimization (PSO) algorithm is used. This method is based on a flock of birds (which are thought of as particles), with each bird looking for food and attempting to travel the shortest distance possible in the search space. The N-dimensional problem's search space is first filled with particles (Gad 2022). These particles could estimate the objective function in the current position of each particle, showing a possible solution. Then, the swarm of particles flies repeatedly in search of the best answer. The best positions obtained by the particle and swarm, respectively, are designated as p-best and g-best. G-best and p-best update each particle's position to achieve the best performance. Due to its excellent performance and simplicity, this algorithm is frequently used in optimization problems (Freitas et al. 2020).

5.3.4 MONTE CARLO SIMULATION

Monte Carlo simulation is one of the available water quality evaluation and prediction methodologies commonly used to anticipate the evolution of water quality. Monte Carlo simulation is a simulation approach that employs a large number of repeated random tests to generate numerical answers to intriguing issues (Bazeli et al. 2022). It forecasts the needed indicators in a computer model by modeling the real process and random behavior of the system. This strategy approaches the problem as a series of simulated real-world experiments. It calculates probability and other indicators by counting the number of occurrences of an event in simulated time. Monte Carlo simulation is a helpful tool for solving actual engineering issues in a variety of domains (Wang et al. 2019).

Each water quality indicator's value varies in the Monte Carlo simulation for each recursion cycle and expected time index. These values are both predictable and arbitrary. The consistency demonstrates that a specific water quality mechanism model and a starting state are the basis for the expected values of water quality indicators. Each water quality indicator is affected by noise at each time index, and the projected values in each recursive cycle vary, demonstrating unpredictability. When there are enough trials, it is possible to approximate the probability distribution functions of water quality indicators at a given time index using the law of large numbers (Wang et al. 2019).

5.3.5 Basic Genetic Algorithm

The basic genetic algorithm (BGA) is a kind of intelligent optimization method that imitates the processes of natural selection, crossover, and mutation in genetics. BGA is a robust and effective stochastic search method which is based on a large number of optimization experiments using numerical functions and Holland's theory. Engineering jobs such as reservoir operation optimization, numerical model parameter optimization, inverse problem research, and others have seen a significant increase in the application of BGA (Parouha and Verma 2021).

BGA begins with an initial population of a number of individuals and uses random selection, crossover, and mutation to create new individuals that are better adapted to the environment. Numerous evolutionary steps eventually lead to the best individual (generations). Every person in a population is a potential solution to an optimization problem, and the ideal person is the solution to the problem at hand (Zhao et al. 2022). The following are some benefits of BGA over other optimization algorithms: (1) the optimization objective function may be continuous or discrete; (2) it has the property of global search and automatic convergence to the optimal solution; (3) it is robust in handling complex nonlinear problems; and (4) the principle is simple, easy to understand, versatile, and highly manipulable.

Given the BGA's extensive use, numerous advancements have been made, such as an adaptive algorithm of crossover and mutation probability and a specialized technology of crossover operation (Parouha and Verma 2021).

5.3.6 Support Vector Machines (SVM)

SVM was a supervised machine learning algorithm that Boser et al. first introduced in 1992 (Ighalo et al. 2021). Both classification and regression issues were solved using SVM. The best "off-the-shelf" supervised method was determined to be SVM. SVM was a two-class linear classifier. The goal of the linear classifier was to identify a hyperplane that correctly categorized data points. Since the underlying system processes were nonstationary, nonlinear, and poorly defined beforehand, SVM had the benefit of properly forecasting time series data (Sun et al. 2022).

Support vector machines (SVMs), which are essentially kernel-based procedures and are based on Vapnik–Chervonenkis (VC) theory, are a relatively new machine learning methodology that has lately become one of the most effective methods for classifying patterns and approximating functions. SVM may concurrently reduce model dimensions and estimate errors. It is less prone to overfitting and has strong generalizability. SVMs use kernel functions to map high-dimensional feature space to make the original inputs linearly separable. SVMs have been effectively used in a variety of study domains to solve the classification and regression challenges (Jahandideh-Tehrani et al. 2020).

Similar to ANN, SVM is capable of capturing the complex, nonlinear functional relationships within data without necessitating a thorough comprehension of the underlying physical processes or the requirement to prespecify a functional form of the model, giving them an edge over many of the models typically employed for modeling water resource variables. In contrast to the empirical risk minimization (ERM) concept used by the majority of conventional ANN models, the support vector machine employs the structural risk minimization (SRM) approach. Instead of focusing on minimizing the training error, SRM emphasizes minimizing an upper bound on the generalization error. Based on this idea, SVM produces the ideal network structure. Furthermore, SVM is analogous to solving a linear restricted quadratic programming problem, resulting in a unique and globally optimum solution for each SVM issue (Liu and Lu 2014).

Table 5.2 summarizes some studies highlighting the use of AI techniques for water quality monitoring, and some case studies for the application of different artificial intelligence techniques for water quality assessment are summarized in Table 5.3.

5.4 AI-ASSISTED TECHNIQUES FOR WATER POLLUTANTS SENSING AND REMOVAL

Several process engineering domains, including design, plant planning and optimization, configuration, training, computer-aided teaching, operator decision support, process control, simulation, alarm management, troubleshooting, and scheduling, can benefit from the application of artificial intelligence–assisted methodologies (Ahmad et al. 2022). Water process engineering is witnessing an unanticipated shift from conventional to AI-assisted techniques since molecular sciences and AI are progressing at the same rate and, more importantly, there is a large convergence between these two domains. Artificial intelligence (AI)–assisted removal and detecting of many sorts of contaminants is being used on the spot.

Drinking water disinfection can prevent waterborne infections. However, chemical disinfectants will react with organic and inorganic substances in source water, resulting in the formation of disinfection by-products (DBPs). More than 600 DBPs have been found in drinking water to date. Multiple linear/log-linear regression (MLR) models are the most commonly used DBP prediction models. Furthermore, the multiple linear regression model's prediction accuracies are limited due to the intricate nonlinear interaction between DBP production and other inputs (Jeong et al. 2016). Artificial neural networks (ANNs), a modeling technique inspired by the human nervous system, were the second most common DBP prediction models. It is distinguished by its ability to build nonlinear correlations between independent factors and dependent variables, as well as to extract nuanced information and complicated knowledge from representative datasets. Several researchers have shown that ANNs can predict DBPs (Deng et al. 2021).

In drinking water treatment, fuzzy logic–based algorithms have been used to forecast coagulant dose from raw water properties or to evaluate plant design and

TABLE 5.2

Water Quality Checking Using Artificial Intelligence

Type of Approach	Method	Water Quality Parameter	Reference
Artificial Neural Network (ANN)	Backpropagation neural networks	COD, NH3, DO, sediment	(Al-yaari et al. 2022; Ghose and Samantaray 2018; Zhao et al. 2007)
	Generalized regression NN	COD	(Ay and Kişi 2017)
	Wavelet neural network	DO	(Xu and Liu 2013)
	Feed-forward neural network	BOD, DO, temperature, total nitrogen, WQI	(Dogan et al. 2009; Elkiran et al. 2018)
	Radial basic neural networks	NH3, COD, DO, TDS, Turbidity, suspended sediment	(Ay and Kişi 2017; Ahmed 2017)
	Adaptive network-based fuzzy inference system (ANFIS)	BOD5	(Kumar et al. 2020; Khaled et al. 2018)
	Multilayer feed-forward neural networks	pH, BOD, DO, temperature, turbidity, TP, TN, boron	(Kim and Seo 2015)
	Multilayer perceptron neural networks	BOD, WQI, DO, COD, TDS, turbidity, electrical conductivity	(Najah et al. 2011; Najah et al. 2013; (Raheli et al. 2017)
ANN-ARIMA		DO, NH3-N, temperature, boron	(Faruk et al. 2010)
Wavelet-ANN		EC	(Barzegar et al. 2016)
Wavelet-ANFIS		EC, TDS, turbidity	(Najah et al. 2012)
k-nearest neighbors		TDS, EC	(Sattari et al. 2016)
Support vector machine (SVM)		COD, DO, BOD, CODMn, NH3-N, BOD5	(Fijani et al. 2019; Kisi and Parmar 2016)
FFNN and ANFIS	DO RMSE, MSE, and DC	BOD, DO, pH, and T	(Kumar et al. 2020; Karaboga and Kaya 2018)
FFNN and RBFNN	R, MSE, and coefficient of efficiency (E)	BOD and COD	(Ahmed 2017)
MLPNN	RMSE, MAE, and R2	pH, T, EC, river discharge (Q), DO	(Ay and Kişi 2017)
RBFNN	R, MAP, and RMSE	DO and NH3-N	(Kumar et al. 2016)

operations at DWTPs. Fuzzy logic was utilized to create a raw water quality index and evaluate its impact on water treatment (Palanichamy et al. 2022). Other water-related uses of FIS include optimizing methane generation in fixed-bed reactors, controlling nitrogen removal processes in wastewater, and predicting the total suspended

TABLE 5.3

Some Case Studies for Water Quality Checks Using AI

Technique	Water Quality Parameter	Country	Water Source	Reference
MLP model	pH, HCO_3, Cl, Na, SO_4^{-2}, Mg, and Ca	Iran	Tireh River	(Qishlaqi et al. 2017)
MLP networks	Total salt concentration, TDS, residual sodium carbonates, and boron	India	Parakai Lake	(Sahaya Vasant and Adish Kum 2019)
ANNs models, including MLP, RBNN, and ANFS	EC, pH, calcium, magnesium, sodium, turbidity, phosphate, nitrate and nitrite DO, BOD, and COD	Egypt	Karoon Lake	(Emamgholizadeh et al. 2014)
MLP and RBNN	Nitrite	Iran	Groundwater	(Ehteshami et al. 2016)
SVM	EC, pH, Ca, Mg, Na, phosphate, nitrate and sulfate, TDS (total dissolved solids), bicarbonates, chromium, fluoride, and chloride	India	Groundwater	(Agrawal et al. 2021)
SVM, MLP, and DLNN	DO, pH, BOD, nitrates, conductivity, facial coliform, and total coliform	India	Drinking water	(Aldhyani et al. 2020)
MLP and RBNN	TDS, EC, and turbidity	Malaysia	Johor River Basin	(Najah et al. 2013)
ANN	pH, DO BOD, COD, ammonia nitrogen, and total suspended solids (TSS)	Malaysia	Pontian River	(Sulaiman et al. 2019)

solids of a clariflocculator effluent in an industrial wastewater treatment facility. Using data-driven models like artificial neural networks, several ways of predicting chemical dosing rates at preoxidation and coagulation stages in full-scale DWTP have been investigated (Godo-Pla et al. 2021).

For forecasting the removal percentage of lead ions from aqueous solution utilizing magnetic graphene oxide supported on nylon 6, an artificial neural network with three algorithms, fuzzy inference system, and adaptive neuro-fuzzy inference system was utilized (Ali 2013). A multivariate postchlorination dosage control system based on an artificial neural network was utilized to control chlorine dosing in a water treatment plant's postchlorination stage (Felipe Henriques Librantz et al. 2018). Linear/log-linear regression models (LRM) and radial basis function artificial neural networks (RBF ANN) were used to predict the presence of trihalomethanes (THMs) in tap water (Xu et al. 2022). Several examples of water pollution monitoring using AI are given in Table 5.4.

TABLE 5.4

Water Pollution Monitoring Using Artificial Intelligence

Technique	Approach	Water Applications	Reference
Fuzzy Inference System (FIS)	Artificial intelligence Decision-making, system control	Chlorine dosage, set point control, environmental pollution control	(Cherkassky 1998; Afroozeh et al. 2018)
Support Vector Machines/ Regressions (SVM/ SVR)	Supervised machine learning Regression, classification/ pattern analysis	Membrane-process parameter modeling Chemical oxygen demand (COD) modeling Biological oxygen demand (BOD) Dissolved oxygen modeling	(Cortes and Vapnik 1995; Chua 2003; Goodfellow et al. 2016)
Artificial Neural Network (ANN)– General	Supervised machine learning Regression, classification	Chlorine dosage/set point DBP formation modeling Adsorption process parameter modeling Membrane-process parameter modeling Dissolved-oxygen concentration modeling	(Yegnanarayana 2006; Uhrig 1995)
Recurrent Neural Network (RNN)/ Long Short-Term Memory (LSTM)	Supervised machine learning Regression, classification	Dissolve oxygen concentration modeling	(Zhang et al. 2020; Smagulova and James 2019)
Neural Network (CNN)	Supervised machine learning Regression, classification, segmentation	DBP formation modeling	(Kim 2017; Gu et al. 2018)
Extreme Learning Machine (ELM)	Supervised machine learning Regression, classification	Dissolved-oxygen concentration	(Zhu et al. 2005; Huang et al. 2004)
Fuzzy Inference System (FIS)	Artificial intelligence Decision-making, system control	Chlorine dosage, set point control Environmental pollution control	(Moraga et al. 2004; Kosko and Isaka 1993)
Genetic Algorithm/ Genetic Programming (GA/GEP)	An evolutionary and stochastic algorithm Regression, classification	DBP formation	(Katoch et al. 2020; Ling 2007)

5.5 CONCLUSION

Existing models used to estimate water quality are not user-friendly and sufficient, and their application is typically subject to significant limitations. The choice of a good numerical model can be difficult for inexperienced program users; therefore, it is useful to incorporate a modern heuristic understanding of model manipulation as well as conceptual manipulation of calibration parameters. The most recent advancements in AI technology offer a means of bridging the gap between model professionals and designers.

This chapter looked at the state-of-the-art models for predicting water quality and the development of AI integration into those models. The embedded model can benefit significantly from the ANN technique. This method can assist inexperienced users in determining whether function modeling–generated digital models are genuine occurrences. Several plans were investigated, evaluated, and offered for prospective development. With the ever-increasing potential of AI technologies for more development in this regard, it is anticipated to be promising.

REFERENCES

Ablieieva, I., Artyukhova, N., Krmela, J., Malovanyy, M., & Berezhnyi, D. 2022. Fluidized bed dryers in terms of minimizing environmental impact and achieving sustainable development goals. *Drying Technology* 0(0): pp. 1–11.

Afroozeh, M., Reza S. M., Davallo, M., Mirnezami, S.Y., Motiee, F., & Khosravi, M. 2018. Application of artificial neural network, fuzzy inference system and adaptive neuro-fuzzy inference system to predict the removal of Pb(II) ions from the aqueous solution by using magnetic graphene/nylon-6. *Chemical Sciences Journal* 9(2), 1000185: pp. 1–7

Agrawal, P., et al. 2021. Exploring artificial intelligence techniques for groundwater quality assessment. *Water* 13(9): pp. 1172.

Ahmad, T., et al. 2022. Energetics Systems and artificial intelligence: Applications of industry 4.0. *Energy Reports* 8: pp. 334–361.

Ahmed, A.A.M. 2017. Prediction of dissolved oxygen in Surma River by biochemical oxygen demand and chemical oxygen demand using the artificial neural networks (ANNs). *Journal of King Saud University—Engineering Sciences* 29(2): pp. 151–158.

Ajala, L.O., et al. 2022. Insights into purification of contaminated water with activated charcoal derived from hamburger seed coat. *International Journal of Environmental Science and Technology* 19(7): pp. 6541–6554.

Akintan, O., Olusola, J., Falade, J., & Adeyeye, J. 2022. Physicochemical, bacteriological and water quality index assessment of hand dug well (HDW) water suitability for drinking. *International Journal of Environmental Analytical Chemistry* 13(9): pp. 1–22.

Al-Adhaileh, M.H., & Alsaade, F.W. 2021. Modelling and prediction of water quality by using artificial intelligence. *Sustainability (Switzerland)* 13(8): pp. 1–18.

Alaloul, W.S., & Qureshi, A.H. 2020. Data Processing Using Artificial Neural Networks. In D. G. Harkut (ed) *Dynamic Data Assimilation*, IntechOpen: Rijeka.

Alam, G., Ihsanullah, I., Naushad, M., & Sillanpää, M. 2022. Applications of artificial intelligence in water treatment for optimization and automation of adsorption processes: Recent advances and prospects. *Chemical Engineering Journal* 427: p. 130011.

Aldhyani, T.H.H., Al-Yaari, M., Alkahtani, H., & Maashi, M. 2020. Water Quality Prediction Using Artificial Intelligence Algorithms. *Applied Bionics and Biomechanics* 2020: p. 12.

Ali, E.H. 2013. Antifungal act ivities of methanolic extracts of some marine algae (Ulvaceae and Dictyotaceae) of Benghazi Coasts, *Libya. The Egyptian Society of Experimental Biology* 9(1): pp. 75–79.

Al-Yaari, M., Aldhyani, T.H., & Rushd, S. 2022. Prediction of arsenic removal from contaminated water using artificial neural network model. *Applied Sciences* 12(3), 999: pp. 1–13.

Astuti, A.D., et al. 2020. Artificial intelligence approach to predicting river water quality: A review. *Journal of Environmental Treatment Techniques* 8(3): pp. 1093–1100.

Ay M., & Kişi Ö. 2017. Estimation of dissolved oxygen by using neural networks and neuro-fuzzy computing techniques. *KSCE Journal of Civil Engineering* 21(5): pp. 1631–1639.

Azad, A., Karami, H., Farzin, S., Mousavi, S.F., & Kisi, O. 2019. Modeling river water quality parameters using modified adaptive neuro fuzzy inference system. *Water Science and Engineering* 12(1): pp. 45–54.

Barzegar, R., Adamowski, J., & Moghaddam, A.A. 2016. Application of wavelet artificial intelligence hybrid models for water quality prediction: A case study in Aji-Chay River, Iran. *Stochastic Environmental Research and Risk Assessment* 30(7): pp. 1797–1819.

Baxa, M., et al. 2021. Dissolved oxygen deficits in a shallow eutrophic aquatic ecosystem (fishpond)—Sediment oxygen demand and water column respiration alternately drive the oxygen regime. *Science of The Total Environment* 766: p. 142647.

Bazeli, J., et al. 2022. Health risk assessment techniques to evaluate non-carcinogenic human health risk due to fluoride, nitrite and nitrate using Monte Carlo simulation and sensitivity analysis in Groundwater of Khaf County, Iran. *International Journal of Environmental Analytical Chemistry* 102(8): pp. 1793–1813.

Chan, C.W., & Huang, G.H. 2003. Artificial intelligence for management and control of pollution minimization and mitigation processes. *Engineering Applications of Artificial Intelligence* 16(2): pp. 75–90.

Charisis, V., Hadjidimitriou, S., & Hadjileontiadis, L.J. 2022. Fiseval-A novel project evaluation approach using fuzzy logic: The paradigm of the i-Treasures project. *Expert Systems with Applications* 202: p. 117260.

Chebii, S.J., Mukolwe, M.M., & Ong'or, B.I. 2022. River flow modelling for flood prediction using artificial neural network in ungauged Perkerra catchment, Baringo County, Kenya. *Water Practice and Technology* 17(4): pp. 914–929.

Chen, S.S., Kimirei, I.A., Yu, C., Shen, Q., & Gao, Q. 2022. Assessment of urban river water pollution with urbanization in East Africa. *Environmental Science and Pollution Research* 29(27): pp. 40812–40825.

Cherkassky, V., 1998. Fuzzy Inference Systems: A critical review. In Computational Intelligence: Soft Computing and Fuzzy-Neuro Integration with Applications, Springer: Berlin/Heidelberg, Germany, pp. 177–197.

Chu, H.B., Lu, W.X., & Zhang, L. 2013. Application of artificial neural network in environmental water quality assessment. *Journal of Agricultural Science and Technology* 15: pp. 343–356.

Chua, K.S. 2003. Efficient computations for large least square support vector machine classifiers. *Pattern Recognition Letters* 24: pp. 75–80.

Conine, A.L., Rickard, S.E., Onion, A.M., Wiegert, E.J., & Smith, A.J. 2021. Use of power analysis to determine the number of samples needed to assess water quality in lakes and flowing waters. *Integrated Environmental Assessment and Management* 10.1002/ieam.4571 18: 1621–1628.

Cortes, C., & Vapnik, V. 1995. Support-vector networks. *Machine Learning* 20: pp. 273–297.

Dehghani, R., Torabi Poudeh, H., & Izadi, Z. 2022. Dissolved oxygen concentration predictions for running waters with using hybrid machine learning techniques. *Modeling Earth Systems and Environment* 8(2): pp. 2599–2613.

Deng, Y., et al. 2021. New methods based on back propagation (BP) and radial basis function (RBF) artificial neural networks (ANNs) for predicting the occurrence of haloketones in tap water. *Science of The Total Environment* 772: p. 145534.

Dogan, E., Sengorur, B., & Koklu, R. 2009. Modeling biological oxygen demand of the Melen River in Turkey using an artificial neural network technique. *Journal of Environmental Management* 90(2): pp. 1229–35.

Doyle, E., et al. 2015. Effect-based screening methods for water quality characterization will augment conventional analyte-by-analyte chemical methods in research as well as regulatory monitoring. *Environmental Science and Technology* 49(24): pp. 13906–13907.

Available at: https://pubs.acs.org/doi/full/10.1021/es5053254 [Accessed June 24, 2022].

Dwivedi, Y.K., et al. 2021. Artificial Intelligence (AI): Multidisciplinary perspectives on emerging challenges, opportunities, and agenda for research, practice and policy. *International Journal of Information Management* 57: p. 101994.

Ebenstein, A. 2012. The consequences of industrialization: Evidence from water pollution and digestive cancers in China. *Review of Economics and Statistics* 94(1): pp. 186–201.

Ehteshami, M., Farahani, N.D., & Tavassoli, S. 2016. Simulation of nitrate contamination in groundwater using artificial neural networks. *Modeling Earth Systems and Environment* 2(1): pp. 1–10.

Elkiran, G., Nourani, V., Abba, S.I., & Abdullahi, J. 2018. Artificial intelligence-based approaches for multi-station modelling of dissolved oxygen in the river. *Global Journal of Environmental Science and Management* 4(4): pp. 439–450.

Emamgholizadeh, S., Kashi, H., Marofpoor, I., & Zalaghi, E. 2014. Prediction of water quality parameters of Karoon River (Iran) by artificial intelligence-based models. *International Journal of Environmental Science and Technology* 11(3): pp. 645–656.

Faherty, A. 2021. Tapped out: How Newark, New Jersey's lead drinking water crisis illuminates the inadequacy of the federal drinking water regulatory scheme and fuels environmental injustice throughout the nation. *Environmental Claims Journal* 33(4): pp. 304–327.

Faruk Ömer D. 2010. A hybrid neural network and ARIMA model for water quality time series prediction. *Engineering Applications of Artificial Intelligence* 23(4): pp. 586–594.

Fatima, S.U., et al. 2022. Geospatial assessment of water quality using principal components analysis (PCA) and water quality index (WQI) in Basho Valley, Gilgit Baltistan (Northern Areas of Pakistan). *Environmental Monitoring and Assessment* 194(3): p. 151.

Felipe Henriques Librantz, A., Cosme Rodrigues dos Santos, F., & Gustavo Dias, C. 2018. Artificialneural networks to control chlorine dosing in a water treatment plant. Available at: https://doi.org/10.4025/actascitechnol.v40i1.37275 [Accessed June 28, 2022].

Fijani, E., Barzegar, R., Deo, R., Tziritis, E., & Konstantinos, S. 2019. Design and implementation of a hybrid model based on a two-layer decomposition method coupled with extreme learning machines to support real-time environmental monitoring of water quality parameters. *Science of the Total Environment* 648: pp. 839–853.

Freitas, D., Lopes, L.G., & Morgado-Dias, F. 2020. Particle swarm optimisation: A historical review up to the current developments. *Entropy* 22(3): p. 362.

Gad, A.G. 2022. Particle Swarm optimization algorithm and its applications: A systematic review. *Archives of Computational Methods in Engineering* 29: pp. 2531–2561.

Ghorbani, M., & Salem, S. 2021. Solar treatment of sewage discharged from industrial estate for reduction of chemical oxygen demand over Degussa P-25 titania. *Chemosphere* 265: p. 129123.

Ghose, D.K., & Samantaray, S. 2018. Modelling sediment concentration using back propagation neural network and regression coupled with genetic algorithm. *Procedia Computer Science* 125: pp. 85–92.

Giri, S. 2021. Water quality prospective in twenty first century: Status of water quality in major river basins, contemporary strategies and impediments: A review. *Environmental Pollution* 271: p. 116332.

Godo-Pla, L., et al. 2021. Control of primary disinfection in a drinking water treatment plant based on a fuzzy inference system. *Process Safety and Environmental Protection* 145: pp. 63–70.

Gogoi, A., et al. 2018. Occurrence and fate of emerging contaminants in water environment: A review. *Groundwater for Sustainable Development* 6: pp. 169–180.

Goodfellow, I., Bengio, Y., & Courville, A., 2016. Deep Learning, MITP: Frechen, Germany, The MIT Press, ISBN: 978-0262035613.

Gu, J., Wang, Z., Kuen, J., Ma, L., Shahroudy, A., Shuai, B., Liu, T., Wang, X., Wang, G., & Cai, J., 2018. Recent advances in convolutional neural networks. *Pattern Recognition* 77: pp. 354–377.

Hojjati-Najafabadi, A., Mansoorianfar, M., Liang, T., Shahin, K., & Karimi-Maleh, H. 2022. A review on magnetic sensors for monitoring of hazardous pollutants in water resources. *Science of The Total Environment* 824: p. 153844.

Huang, G., Zhu, Q., Siew, C., 2004. Extreme Learning Machine: A New Learning Scheme of Feedforward Neural Networks. In Proceedings of the 2004 IEEE International Joint Conference on Neural Networks (IEEE Cat. No.04CH37541) *Budapest, Hungary*, pp. 25–29 July.

Huang, Y., et al. 2022. Forward-looking roadmaps for long-term continuous water quality monitoring: bottlenecks, innovations, and prospects in a critical review. *Environmental Science and Technology* 56(9): pp. 5334–5354.

Ighalo, J.O., & Adeniyi, A.G. 2020. A comprehensive review of water quality monitoring and assessment in Nigeria. *Chemosphere* 260: p. 127569.

Ighalo, J.O., Adeniyi, A.G., & Marques, G. 2021. Artificial intelligence for surface water quality monitoring and assessment: A systematic literature analysis. *Modeling Earth Systems and Environment* 7(2): pp. 669–681. Available at: https://link.springer.com/article/10.1007/s40808-020-01041-z [Accessed June 24, 2022].

Jahandideh-Tehrani, M., Bozorg-Haddad, O., & Loáiciga, H.A. 2020. Application of particle swarm optimization to water management: an introduction and overview. *Environmental Monitoring and Assessment* 192(5): pp. 1–18.

Jeong, C.H., et al. 2016. Monohaloacetic acid drinking water disinfection by-products inhibit follicle growth and steroidogenesis in mouse ovarian antral follicles in vitro. *Reproductive Toxicology* (*Elmsford, N.Y.*) 62: pp. 71–76.

Karaboga, D., & Kaya, E. 2018. Adaptive network-based Fuzzy Inference System (ANFIS) training approaches: A comprehensive survey. *Artificial Intelligence Review* 52: pp. 2263–2293.

Karydis, M., & Kitsiou, D. 2013. Marine water quality monitoring: A review. *Marine Pollution Bulletin* 77(1–2): pp. 23–36.

Katoch, S., Chauhan, S.S., & Kumar, V. 2020. A review on genetic algorithm: Past, present, and future. *Multimedia Tools and Application* 80: pp. 8091–8126.

Khaled, B., Abdellah, A., Noureddine, D., Haddam, S., & Sabeha, A. 2018. Modelling of biochemical oxygen demand from limited water quality variable by ANFIS using two partition methods. *Water Quality Research Journal of Canada* 53(1): pp. 24–40.

Kilinc, H.C. 2022. Daily streamflow forecasting based on the hybrid particle swarm optimization and long short-term memory model in the orontes basin. *Water* 14(3): p. 490.

Kim, P. 2017. Convolutional Neural Network. In Matlab Deep Learning. Apress: Berkeley, CA, pp. 121–147.

Kim, S.E., & Seo, I.W. 2015. Artificial neural network ensemble modeling with conjunctive data clustering for water quality prediction in rivers. *Journal of Hydro-Environment Research* 9(3): pp. 325–339.

Kisi, O., & Parmar, K.S. 2016. Application of least square support vector machine and multivariate adaptive regression spline models in long-term prediction of river water pollution. *Journal of Hydrology* 534: pp. 104–112.

Kosko, B., & Isaka, S., 1993. Fuzzy logic. *Scientific American* 269: pp. 76–81.

Kumar, D., Pandey, A., Sharma, N., & Flügel, W.A. 2016. Daily suspended-sediment simulation using a machine learning approach. *Catena* 138: pp. 77–90.

Kumar, K., Davim, J.P., Adedeji, P.A., Masebinu, S.O., Akinlabi, S.A., & Madushele, N. 2020. Adaptive Neuro-Fuzzy Inference System (ANFIS) Modelling in Energy System and WATER RESOURCES. In Optimization Using Evolutionary Algorithms and Metaheuristics: Applications in Engineering, Essay, CRC Press, Taylor & Francis Group: Abingdon, UK.

Kumar, R., et al. 2022. A review on emerging water contaminants and the application of sustainable removal technologies. *Case Studies in Chemical and Environmental Engineering* 6: p. 100219. Available at: https://linkinghub.elsevier.com/retrieve/pii/S266601642200041X [Accessed June 24, 2022].

Li, W., et al. 2022. Influence of the solution pH on the design of a hydro-mechanical magne-to-hydraulic sealing device. *Engineering Failure Analysis* 135: p. 106091.

Ling, W.-K. 2007. Nonlinear Digital Filters, Academic Press: Cambridge, MA, Part of the Book Series: The Springer International Series in Engineering and Computer Science.

Liu, M., & Lu, J. 2014. Support vector machine—an alternative to artificial neuron network for water quality forecasting in an agricultural nonpoint source polluted river? *Environmental Science and Pollution Research* 21(18): pp. 11036–11053.

Moraga, C., Trillas, E., & Guadarrama, S. 2004. Multiple-Valued Logic and Artificial Intelligence Fundamentals of Fuzzy Control Revisited. In Artificial Intelligence in Logic Design, Springer: Dordrecht, The Netherlands, pp. 9–37.

Mukhopadhyay, A., Duttagupta, S., & Mukherjee, A. 2022. Emerging organic contaminants in global community drinking water sources and supply: A review of occurrence, processes and remediation. *Journal of Environmental Chemical Engineering* 10(3): p. 107560.

Najah, A., El-Shafie, A., Karim, O.A., & El-Shafie, A.H. 2013. Application of artificial neural networks for water quality prediction. *Neural Computing and Applications* 22(Suppl.1): pp. 187–201.

Najah, A., El-Shafie, A., Karim, O.A., & El-Shafie, A.H. 2013. Application of artificial neural networks for water quality prediction. *Neural Computing and Applications* 22(Suppl.1): pp. 187–201.

Najah, A.A., El-Shafie, A., Karim, O.A., & Jaafar, O. 2012. Water quality prediction model utilizing integrated wavelet-ANFIS model with cross-validation. *Neural Computing and Applications* 21(5): pp. 833–841.

Najah, A., Karim, O.A., Jaafar, O., & El-shafie, A.H. 2011. An application of different artificial intelligence techniques for water quality prediction. *International Journal of the Physical Sciences* 6(22): pp. 5298–5308.

Olalemi, A.O., & Akinwumi, I.M. 2022. Microbial health risks associated with rotavirus and enteric bacteria in River Ala in Akure, Nigeria. *Journal of Applied Microbiology* 132(5): pp. 3995–4006.

Oyedotun, T.D.T., & Ally, N. 2021. Environmental issues and challenges confronting surface waters in South America: A review. *Environmental Challenges* 3: p. 100049.

Paepae, T., Bokoro, P.N., & Kyamakya, K. 2021. From fully physical to virtual sensing for water quality assessment: A comprehensive review of the relevant state-of-the-art. *Sensors* 21(21): p. 6971.

Palanichamy, J., et al. 2022. Simulation and prediction of groundwater quality of a semi-arid region using fuzzy inference system and neural network techniques. *Journal of Soft Computing in Civil Engineering* 6(1): pp. 110–126.

Panagopoulos, A. 2022. Brine management (saline water & wastewater effluents): Sustainable utilization and resource recovery strategy through Minimal and Zero Liquid Discharge (MLD & ZLD) desalination systems. *Chemical Engineering and Processing—Process Intensification* 176: p. 108944.

Parouha, R.P., & Verma, P. 2021. State-of-the-Art reviews of meta-heuristic algorithms with their novel proposal for unconstrained optimization and applications. *Archives of Computational Methods in Engineering* 28(5): p. 4049–4115.

Patale, V.V., & Tank, J.G. 2022. Evaluation of the edaphic and water properties of Diu coast (Saurashtra, Gujarat, India) in relation to the population density of Avicennia marina. *Applied Water Science* 12(4): p. 78.

Pothe, A.S. 2022. Artificial Intelligence and its Increasing Importance. In J. Karthikeyan, T. S. Hie, & N. Y. Jin içinde (eds). Book Learning Outcomes Of Classroom Researchs: 74–81.

Qishlaqi, A., Kordian, S., & Parsaie, A. 2017. Field measurements and neural network modeling of water quality parameters. *Applied Water Science* 7(1): pp. 523–523.

Raheli, B., Aalami, M.T., El-Shafie, A., Ghorbani, M.A., & Deo, R.C. 2017. Uncertainty assessment of the multilayer perceptron (MLP) neural network model with implementation of the novel hybrid MLP-FFA method for prediction of biochemical oxygen demand and dissolved oxygen: A case study of Langat River. *Environmental Earth Sciences* 76(14): p. 503.

Rahman, M., et al. 2021. A framework for Learning System for Complex Industrial Processes. In K. Kyprianidis & E. Dahlquist (eds). AI and Learning Systems, Rijeka: IntechOpen, p. en.92899.

Ram, A., et al. 2021. Groundwater quality assessment using water quality index (WQI) under GIS framework. *Applied Water Science* 11(2): p. 46.

Rodriguez-Narvaez, O.M., Peralta-Hernandez, J.M., Goonetilleke, A., & Bandala, E.R. 2017. Treatment technologies for emerging contaminants in water: A review. *Chemical Engineering Journal* 323: pp. 361–380.

Rosenfeld, P.E., & Feng, L.G.H. 2011. Emerging contaminants. *In Risks of Hazardous Wastes*, William Andrew Publishing, pp. 215–222, DOI: 10.1016/b978-1-4377-7842-7.00016-7.

Sahaya Vasant, S., & Adish Kum, S. 2019. Application of artificial neural network techniques for predicting the water quality index in the Parakai Lake, Tamil Nadu, India. *Applied Ecology and Environmental Research* 17(2): pp. 1947–1958.

Sattari, M.T., Joudi, A.R., & Kusiak, A. 2016. Estimation of water quality parameters with data-driven model. *Journal—American Water Works Association* 108(4): pp. E232—E239.

Sauvé, S., & Desrosiers, M. 2014. A review of what is an emerging contaminant. *Chemistry Central Journal* 8(1): p. 15. Available at: /pmc/articles/PMC3938815/ [Accessed June 23, 2022].

Shah, M.I., et al. 2021. Modeling surface water quality using the adaptive neuro-fuzzy inference system aided by input optimization. *Sustainability* 13(8): p. 4576.

Sharma, P., Kumar, S., & Pandey, A. 2021. Bioremediated techniques for remediation of metal pollutants using metagenomics approaches: A review. *Journal of Environmental Chemical Engineering* 9(4): p. 105684.

Smagulova, K., & James, A.P., 2019. Overview of Long Short-Term Memory Neural Networks. In Deep Learning Classifiers with Memristive Networks, Springer: Cham, Switzerland, pp. 139–153.

Sobczyk, M., Pajdak-Stós, A., Fiałkowska, E., Sobczyk, Ł., & Fyda, J. 2021. Multivariate analysis of activated sludge community in full-scale wastewater treatment plants. *Environmental Science and Pollution Research* 28(3): pp. 3579–3589.

Sreedevi, A.G., Nitya Harshitha, T., Sugumaran, V., & Shankar, P. 2022. Application of cognitive computing in healthcare, cybersecurity, big data and IoT: A literature review. *Information Processing & Management* 59(2): p. 102888.

Stefanakis, A.I., & Becker, J.A. 2016. A Review of Emerging Contaminants in Water: Classification, Sources, and Potential Risks. In Impact of Water Pollution on Human Health and Environmental Sustainability, IGI Global, pp. 55–80. Available at: www.igi-global.com/chapter/a-review-of-emerging-contaminants-in-water/140170 [Accessed June 23, 2022].

Sulaiman, K., Hakim Ismail, L., Adib Mohammad Razi, M., Shalahuddin Adnan, M., & Ghazali, R. 2019. Water quality classification using an Artificial Neural Network (ANN). *IOP Conference Series: Materials Science and Engineering* 601(1): p. 012005.

Sun, Z., et al. 2022. A review of earth artificial intelligence. *Computers and Geosciences* 159: p. 105034.

Syafri, S., et al. 2020. Water quality pollution control and watershed management based on community participation in Maros City, South Sulawesi, Indonesia. *Sustainability* 12(24):10260. https://doi.org/10.3390/su122410260.

Tan, W.Y., Lai, S.H., Teo, F.Y., & El-Shafie, A. 2022. State-of-the-art development of two-waves artificial intelligence modeling techniques for river streamflow forecasting. *Archives of Computational Methods in Engineering.* https://doi.org/10.1007/s11831-022-09763-2 15(1), 1585–1612.

Tiri, A., Belkhiri, L., & Mouni, L. 2018. Evaluation of surface water quality for drinking purposes using fuzzy inference system. *Groundwater for Sustainable Development* 6(January): pp. 235–244.

Tirkey, P., et al. 2013. Water quality indices-important tools for water quality assessment: A review environmental consequence and reclamation option for coal mining activities in Charhi & Kuju of Jharkhand, India view project thesis view project water quality indices-important tools for water quality assessment: A review. *International Journal of Advances in Chemistry (IJAC)* 1(1) 15–28. Available at: www.researchgate.net/publication/262730848 [Accessed June 24, 2022].

Tiyasha, Tung, T.M., & Yaseen, Z.M. 2020. A survey on river water quality modelling using artificial intelligence models: 2000–2020. *Journal of Hydrology* 585: p. 124670.

Trach, R., et al. 2022. A Study of Assessment and prediction of water quality index using fuzzy logic and ANN models. *Sustainability (Switzerland)* 14(9): p. 5656.

Uddin, M.G., Nash, S., & Olbert, A.I. 2021. A review of water quality index models and their use for assessing surface water quality. *Ecological Indicators* 122: p. 107218.

Uhrig, R.E. 1995. Introduction to Artificial Neural Networks. In Proceedings of the IECON'95, 21st Annual Conference on IEEE Industrial Electronics, Orlando, FL, 6–10 November.

Vu, C.T., & Wu, T. 2022. Recent progress in adsorptive removal of per- and poly-fluoroalkyl substances (PFAS) from water/wastewater. *Critical Reviews in Environmental Science and Technology* 52(1): pp. 90–129.

Wang, G., et al. 2022. Artificial neural networks for water quality soft-sensing in wastewater treatment: A review. *Artificial Intelligence Review* 55(1): pp. 565–587.

Wang, P., et al. 2019. Exploring the application of artificial intelligence technology for identification of water pollution characteristics and tracing the source of water quality pollutants. *Science of The Total Environment* 693: p. 133440.

Wang, X., et al. 2019. A novel water quality mechanism modeling and eutrophication risk assessment method of lakes and reservoirs. *Nonlinear Dynamics* 96(2): pp. 1037–1053. Available at: https://doi.org/10.1007/s11071-019-04837-6.

Wilson, M.M., Michieka, R.W., & Mwendwa, S.M. 2021. Assessing the influence of horticultural farming on selected water quality parameters in Maumau stream, a tributary of Nairobi River, Kenya. *Heliyon* 7(12): p.e08593.

Xu, L., & Liu, S. 2013. Study of short-term water quality prediction model based on wavelet neural network. *Mathematical and Computer Modelling* 58(3–4): pp. 801–807.

Xu, Z., et al. 2022. Using simple and easy water quality parameters to predict trihalomethane occurrence in tap water. *Chemosphere* 286(Pt 1). Available at: https://pubmed.ncbi.nlm.nih.gov/34303907/ [Accessed June 28, 2022] 286(Pt 1), 131586.

Yegnanarayana, B. 2006. Artificial Neural Networks, Prentice-Hall of India: Delhi, India, ISBN 9788120312531.

Zhang, H., Shinn, M., Gupta, A., Gurfinkel, A., Le, N., & Narodytska, N. 2020. Cover Verification of recurrent neural networks for cognitive tasks via reachability analysis. *Frontiers Applied Artificial Intelligence* 325: pp. 1690–1697.

Zhao, W., Wang, L., & Mirjalili, S. 2022. Artificial hummingbird algorithm: A new bio-inspired optimizer with its engineering applications. *Computer Methods in Applied Mechanics and Engineering* 388: p. 114194.

Zhao, Y., Nan, J., Cui, F., & Guo, L. 2007. Water quality forecast through the application of BP neural network at Yuqiao reservoir. *Journal of Zhejiang University-Science A* 8(9): pp. 1482–1487.

Zhu, Q.-Y., Qin, A.K., Suganthan, P.N., & Huang, G.-B. 2005. Evolutionary extreme learning machine. *Pattern Recognition* 38: pp. 1759–1763.

6 Water Quality Monitoring Using Sensors and Models

*Mudiaga Chukunedum Onojake, Chidi Obi,
and Alaa El Din Mahmoud*

CONTENTS

DOI: 10.1201/9781003260455-6

6.1 INTRODUCTION AND BACKGROUND INFORMATION

Water pollution is a serious global issue that both industrialized and developing countries must pay serious attention to (Umair et al. 2020). The quality and quantity of water domain are considered pertinent for the safety of both humans and aquatic species. Water is crucial in many aspects of human life, and its quality is diminishing as a result of anthropogenic contamination produced by activities such as urbanization, industrialization, municipal activities, population growth, excessive fertilizer use in agriculture, gaseous emissions, solid and liquid wastes deposited in the environment, and antibiotics, heavy metals, etc. disposed into water bodies (Onojake et al. 2011; Mahmoud et al. 2016; Gwenzi et al. 2022). In recent years, the growing environmental degradation caused by development, population increase, and global warming has demanded an investigation of its negative impact on the environment, particularly on water supplies, and its implications (Mahmoud and Fawzy 2016; Mahmoud et al. 2021; Mansoor et al. 2022). Water pollution in the seas, lakes, and rivers has generated a worldwide demand for more comprehensive environmental monitoring systems, particularly in the field of water quality monitoring. It is essential to detect pollutants that contaminate water in order to maintain a quality of life. Water quality detection is a time-consuming and difficult operation that necessitates manual laboratory analysis and statistical conclusions (Gazzaz et al. 2012).

Water pollution monitoring is the comprehensive process of evaluating the physicochemical and microbiological loads of water. It also includes the compilation of data from many forms of water bodies (Chapman 1996; Geissen et al. 2015). Monitoring water pollution gives data on the health of surface waters. It reveals the safety condition of the waters, if they are safe to drink, fish, or swim in. A variety of techniques is used to determine water quality. Monitoring experts obtain water, sediment, and fish tissue samples to assess the concentrations of parameters such as nutrients, metals, oils, pesticides, and so on. Temperature, flow, sediments, and the potential for erosion of stream banks and lakeshores may all be monitored. The richness and diversity of aquatic plant and animal life, as well as the capacity of test organisms to thrive in sampled water, are also often used to assess water quality. Pollution monitoring can be carried out continuously at fixed locations (stations) or on a need-to basis at selected locations (Mahmoud and Fawzy 2021). Additionally, it can be undertaken randomly at various locations around a region or on an emergency basis (Latiu et al. 2013). One of the major reasons that water pollution has become a concern is the fact that it affects not just point sources but also diffuse sources. Agricultural runoff, groundwater pollution, and hydrocarbons in urban stormwater runoff are examples of diffuse sources. Furthermore, upland runoff transports pollutants and excess sediments from upstream, as well as air pollution, such as sulfur and nitrogen deposition. Water's qualitative and quantitative aspects must be managed together since they are intricately connected and interdependent. Water scarcity, for example, affects the ability of industrial and urban land runoff and combined sewage overflow into rivers to be diluted, but poor water quality reduces the quantity of potable water, exacerbating the water shortage crisis (Bartram and Ballance 1996). Due to the increasing anthropogenic activities and concerns about water contamination, there is a compelling need for integrated and routine water monitoring to serve as a standard for accomplishing systematic water management. Assessing water quality in situ is labor-intensive and exorbitant (Arun et al. 2021). But due to the alarming rate of water contamination and degradation, precise detection is now required.

The fundamental variable used to express water portability is pH, which is a measurement of the acidity or basicity of a water sample. A sensitive sensor device is necessary for precisely monitoring pH values, which deal with a very small quantity of ionic concentration (Kumar and Puri 2012; Qin et al. 2015). Temperature is one of the frequently used physical measures, as it can affect other characteristics, such as photosynthesis of aquatic plants, their sensitivity to pollutants, parasites, and illnesses. The evaluation of oxygen content is imperative because of its impact on aquatic species. The presences of low or high concentrations of DO negatively impact the quality of water (Rasin and Abdullah 2009; Mulholland et al. 2005). Interface circuits are used to link sensors (electrodes) like temperature, pH, and oxygen partial pressure to the microcontroller. These sensors are useful for monitoring physical and chemical water parameters at remote locations due to their mobility and near real-time data collecting and logging capabilities (Gholizadeh et al. 2016; Akhter et al. 2021).

6.2 WATER QUALITY MONITORING

Water as an aqueous material has the affinity to attract numerous chemicals and particulates in dissolution and suspension. The assessment of water indicators is essentially imperative in the management of water pollution. The progress in water quality restoration was accomplished by controlling the relatively easily identified and regulated point sources of pollution since 1972. Water quality monitoring (WQM) systems' primary goals are data accuracy, data correctness, timely reporting, data accessibility, and completeness. Conventional monitoring systems are unable to match the high-accuracy requirements of WQM systems when employed to identify toxins or pollutants in real time (Adu-Manu et al. 2020). Figure 6.1 is a model representation of water quality monitoring process.

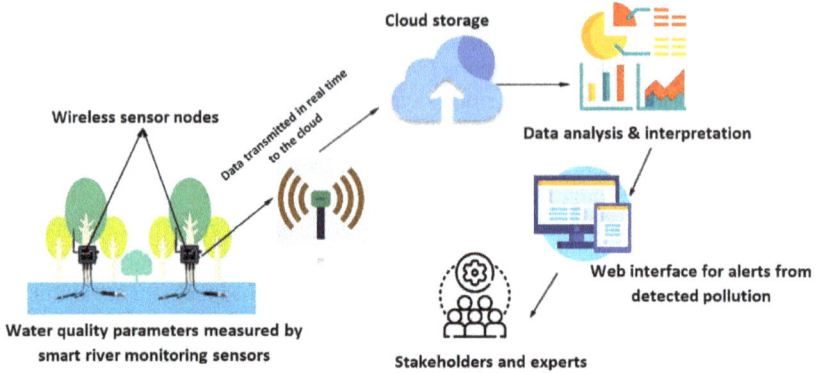

FIGURE 6.1 Model representation of water quality monitoring process.

Source: Adu-Manu et al. (2020).

However, it has been reported that surface water is linked to more than half of all diseases worldwide. Water quality monitoring studies are centered mainly on physicochemical and bacteriological evaluation of water samples from various sources. According to Jamie and Richard (1996), water quality monitoring includes watching, observing, checking, and keeping track of regulating and controlling the different sectors of the economy, not only for safety purposes, but also to ascertain suitability for use.

6.3 ANTHROPOGENIC ACTIVITIES AND WATER POLLUTION

A wide range of anthropogenic (agricultural, industrial, and urban) activities contaminate surface water and groundwater in ways that are primarily due to human activities (Akhtar et al. 2021; Mahmoud and Kathi 2022; Mahmoud et al. 2022). A few of the organic pollutants that end up in the aquatic environment are pesticides, solvents, halogenated chemicals, monocyclic and polycyclic hydrocarbons, etc. All these things are the result of anthropogenic and natural processes. Erosion, terrestrial runoff, irregular landfills, and other pollutants from industry, domestic use, and agriculture are also major contributors to water pollution and biodiversity loss. Pollutants such as sediment from erosion, agricultural and animal waste, heat created by asphalt and industrial operations, microorganisms from sewage, pets and farm animals, mining and smelting from industrial sources, household cleaners, oil and gasoline, etc. are all examples of pollutants that alter water quality (Dotaniya et al. 2022). Ecological mechanisms that naturally filter water domain are distorted by human interference. Nature has a way of removing pollutants from the water in healthy aquatic environments. Microorganisms, for example, break down organic waste into nutrients that plants can use. Sediment is trapped in wetlands, preventing it from flowing into rivers and lakes. Decomposition of hazardous waste and poisonous compounds is aided by wetland habitats. Biological deterioration and dilution may occur naturally in rivers if it is not inundated with waste. Even if these natural mechanisms were able to break down contaminants such as mercury, polychlorinated benzenes (PCBs), and certain pesticides, they could not completely remove them.

6.4 OBJECTIVES AND SIGNIFICANCE OF WATER QUALITY MONITORING

One essential component of the 2030 Sustainable Development Goals (SDGs) is improving water quality, which aims to reduce water pollution, eradicate dumping, minimize the release of hazardous chemicals, and reduce untreated wastewater by half (United Nation 2015).

Data on the physicochemical and microbiological inputs can be obtained using statistical sampling in water quality monitoring (Sanders and Nee 1987). Detection of drinking water standard compliance, environmental status assessment, and trend analysis of temporal water quality changes are just a few of the various aims and objectives that may be accomplished. Well-organized and periodic special investigations to determine the level of water contamination can all be used to meet the objectives.

Because water resources are typically used for a variety of conflicting objectives, any monitoring used to collect the data should consider the data needs of the stakeholders involved (Helmer 1994). As a result, depending on how many evaluation goals must be satisfied, there are two categories of monitoring programs:

> Single-objective or specialized monitoring can be set up to focus on a single field of investigation, such as pH, alkalinity, specific cations, nutrients, chlorophyll pigments, sodium, calcium, chloride, and so forth. Multi-objective monitoring, which may cover a variety of water uses, can provide and establish baseline data for many assessment programs, such as drinking water supply, industrial manufacturing, fisheries, or aquatic life, by measuring the concentration of critical pollutants in water data, thereby relating a large set of complex variables into usable information for the public (Adelagun et al. 2021).

6.5 WATER POLLUTANTS AND THEIR IMPACTS

Pollutants in water may be categorized into the following groups:

6.5.1 THERMAL POLLUTION

Rainwater is heated by paved surfaces and roofs, and this heat is then transferred to rivers and streams. Water used to lower the thermal heat of industrial and power plants also absorbs heat. The DO content of warm water is higher than that of cold water. This results in a reduction of the level of dissolved oxygen (DO) in water bodies due to heat. Aquatic species that depend on dissolved oxygen may die as a result, and their reproductive processes may be interrupted as these aquatic species are more susceptible to deterioration.

6.5.2 SEDIMENT OR SUSPENDED MATTER

Soil eroding from the land is the primary source of pollution. Suspension of soil and other materials in water that are not soluble clogs gills and ruins aquatic species' feeding and breeding grounds. Sediment can also cloud water, limit photosynthesis, and disrupt aquatic food webs. Plant nutrients are water-soluble nitrates and phosphates. Sewage treatment plants, fertilizer runoff, vehicle exhaust, air deposition, feed, farm manure, pesticides, and degraded soil are examples of human-caused discharges. Soil, wild animal excrement, decomposing organic material, the atmosphere, and groundwater discharge are examples of natural sources.

6.5.3 PROFUSE GROWTH OF AQUATIC PLANTS (ALGAE BLOOMS)

When algae die, bacteria decompose it, which depletes the water's dissolved oxygen. If the dissolved oxygen levels in the water are too low, aquatic creatures might die or get agitated. Premature births and the "blue baby syndrome" in infants younger than a year old are both linked to high nitrate levels in drinking water.

Organic chemicals and water-soluble inorganic chemicals—oil and gasoline, cleansing chemicals, detergents, toxic metal compounds, such as those containing mercury and lead—are a wide range of different substances that form this category. The health of humans and aquatic life can be jeopardized by exposure to toxic quantities of certain substances.

6.5.4 LITTER AND DEBRIS

Intentional and accidental littering, whether on purpose or by mistake, fast food, smoking-related trash, unlawful dumping (tires, refrigerators), fishing, and other activities contribute to the contamination of our oceans and beaches.

6.5.5 GENETIC POLLUTION

The introduction of nonnative plants (such as milfoil and hydrilla) into aquatic systems can compete with the native species, decrease biodiversity, and result in financial losses. Ballast from ships is one way of introducing contaminants into maritime environments. In addition to microscopic creatures, ballast water may also contain the first stages of larger plants and animals' lives.

6.6 WATER SAMPLING TECHNIQUES

Strict scientific procedures are frequently involved in the water sampling technique. Streamflow and physical water measurements are nearly always obtained, but water samples are frequently required for chemical investigations as well. A representative sample of the river or stream must always be used in research. Water samples from a stream can be taken in so many ways using different types of tools. The stream's size, depth, and flow rate should be taken into account. The manner of sampling is also dependent on the type of water domain. If the water in a tiny stream is well-mixed, the adoption of grab sampling technique may be an option. Sediment can influence the chemical analysis of a water sample if it is disturbed. Taking frequent water samples that are used to represent the whole stream or river is essential. Because the water quality and properties may vary so much from one river to another, collecting a sample from a bigger one necessitates more time and effort. All the water in a huge river takes a long time to mix together. Individual samples across the river must be taken at regular intervals in order to provide a complete picture of the river's water quality. Collecting samples from such water source is possible with the aid of a boat trip. In water sampling techniques, theories on field instrumentation and operation of equipment used for water sampling, quality assurance, and quality control (QA/QC), including how the water samples are prepared and analyzed in a laboratory, are properly explained (Wilson 2020).

6.7 WATER QUALITY MONITORING PARAMETERS

Various criteria/indicators are used to determine water quality, and some of these qualities include conductivity, total hardness, density, salinity, alkalinity, acidity, etc.

6.7.1 PH

This is measurement of degree of acidity or alkalinity of a given solution. The pH of natural water is usually between 6.5 and 8.5. The acidity or basicity of natural water determines the solubility and also the chemical states of substances in water. Every aspect of water supply and treatment is pH-dependent. Also, the metabolism of many aquatic species is catalyzed by acid or base, so it depends on the degree of acidity or alkalinity of the water.

Mathematically, the degree of acidity and alkalinity is expressed as:

$$pH = -\log_{10} [H^+] \qquad (6.1)$$

The range of pH starts from 1.0 to 14.0, with pH values below 7 being acidic and above 7 being alkaline.

6.7.2 CONDUCTIVITY

This describes the ability of ions to move through a solution and conduct electricity. Conductivity measurements are extremely useful for monitoring water quality supplies and in chemical oceanography for salinity determinations. The proper unit to represent the mathematical expression of conductivity is the siemens. Sometimes, micro Siemens (μs) are used for better representation. Conductivity increases with temperature, thereby increasing the extent of conducting electricity (conductance) and also with the formula of the salt (electrolyte) and their concentration in the sample (weak and strong electrolytes). The measurement of conductivity is usually carried out using a conductivity meter (cell) at either 20°C or 30°C, and values for natural waters generally fall within 5 to 120 s. It has been established that distilled water has low specific conductance (the reason deionized is preferred) and organic materials dissolve in water without producing ions. This is the reason that a salt solution has a high electrical conductivity (Kohlrausch law of independent migration of ions), while high concentrations of sugar would go undetected by this method (Ibezim-Ezeani and Obi 2017).

6.7.3 TOTAL HARDNESS

This is referred to as the composition of calcium and magnesium carbonates in water. Hardness in water is obtainable by the dissolution of the carbonates of calcium and magnesium. The degree of hardness depends on the concentrations of these components in water (Table 6.1). The presence of hydrogen carbonates and sulfates in water gives rise to two classes of hardness, namely, temporary and permanent hardness. Hard water has a negative impact on both domestic and industrial processes. Water hardness is usually determined using spectrometric techniques or by direct titration with ethylenediaminetetraacetic acid (EDTA) using Eriochrome Black T as an indicator.

6.7.4 SALINITY

This term is used to clarify the dissolved mineral (salt) content of water. Chlorine salinity indicates the sum of chloride ion (Cl^-) concentrations present. The salinity value is high in strong water or brines. Sodium chloride (NaCl) salinity is similar except that the chloride content determined by analysis is expressed as NaCl.

Salinity extensively affects farms in local communities and urban areas. High salinity is usually caused by large presence of salt and the concomitant weathering of rocks in water domains. Salinity is a difficult problem to control, but governments and individuals can reduce its impact by planting salt-tolerant plants, practicing crop rotation, and planting both annual and perennial plants. However, the key factor to salinity control is the maintenance of equilibrium between evaporation and precipitation.

TABLE 6.1

Classification of Water into Various Categories (USGS 2019)

Group	Unit (mg/L)
Soft	0–60
Moderately hard	61–120
Hard	121–180
Very hard	> 180

6.7.5   TURBIDITY

This means the formation of haze due to the suspension of impurities in water. The variation of turbidity of water bodies from river to river is a function of the quantity of impurities present. For instance, a muddied river has higher turbidity than natural water. *Turbidity* also refers to the optical formation of water through scattering of electromagnetic radiation (light) by the impurities in water. Experiments reveal that turbidity is highly dependent on the intensity of radiant energy. The standard unit of turbidity of water as proposed by United States Environmental Protection Agency (USEPA) is not more than 0.3 NTUs.

6.7.6   CHEMICAL OXYGEN DEMAND (COD)

This describes the degree of pollution in water domain. It represents the oxidation of organic matter by a strong chemical oxidant.

6.7.7   TOTAL DISSOLVED SOLID (TDS)

This is defined as the measure of the dissolved combined components of all inorganic and organic particles in their ionic, molecular, and microgranular forms present in a solution. It is a quantitative measurement that is reported in parts per million (ppm).

6.7.8   BIOCHEMICAL OXYGEN DEMAND (BOD)

This analysis is the most frequently used for determining the organic matter load of water and is reported in milligram per liter (mg/L). The direct discharge of effluent, sewage, agricultural, etc. has a serious effect on the aquatic habitat. However, these effluents are oxidized by microorganisms present in the water bodies. Under this condition, the Standard Methods for the Examination of Water and Wastewater (1998) noted that the depleted oxygen cannot be replaced by introducing oxygen back into the water domain for larger organisms such as fish and macroinvertebrates to thrive.

BOD is mostly carried out in 5 days, and sometimes 20 days, when the ultimate rate is required in the dark at 20°C. The cardinal point of this indicator is to assess water domain quality needed by village dwellers for domestic and commercial activities (Najafzadeh and Ghaemi 2019). However, the continued presence of components of organic origin in water bodies makes it unsafe for drinking, and the proportional application of chemicals for disinfection produces by-products such as trihalomethanes and other toxic compounds that are injurious or even lethal to humans.

BOD measurement provides empirical information and not absolute results. However, the measurement of BOD_5 is first carried out by making sure that the sample bottle is airtight and filled with the water sample to be tested. It is then incubated in the dark at 20°C for a five-day period. Hence, DO is evaluated at the start and end of the incubation. Then, BOD_5 is quantified from the variation of the initial and final results of the dissolved oxygen.

Secondly, BOD_5 is also tested by filling the sample bottle with incremental amounts of the sample water to be tested and then making it up with deionized water.

The sample bottles are left standing in the dark for five days at 20°C. During this period, the microorganisms present degrade the organic component using the DO present. Then, DO is recorded at the fifth day. Finally, interplay between the concentration of oxygen consumed during the five-day exercise and the corresponding increase in the water sample is used to determine the BOD_5.

The quantity of oxygen depleted by microorganisms followed pseudo–first order reaction. However, without reaeration, the rate of oxygen depletion is usually described by equation 6.2:

$$L = L_o e^{-kt} \tag{6.2}$$

Where, L represents the concentration of O_2 at time t, L_o is the initial concentration of oxygen in a water sample, and k is the rate constant (for sewage water, $k = 0.17$ per day) (Snoeyink and Jenkins 1980).

Similarly, equation 6.3 is usually applied to describe the oxidation of BOD in the water sample as it is the inverse of the oxygen consumption.

$$L = L_0 - L_o e^{-kt} \tag{6.3}$$

Where, L is the concentration of organic matter at time t, and other variables remain the same as in the preceding equation.

However, an alternative method according to Snoeyink and Jenkins (1980) in determining L_o is based on the determination of the BOD over a five-day period and fit the data generated in equation 5.3 using k as given previously and solving for L_o.

The ultimate BOD, L_o, can be evaluated by the application of Thomas slope method, as represented in equation 6.4:

$$\left(\frac{1}{y}\right)^{\frac{1}{3}} = \left(L_o k\right)^{-\frac{1}{3}} + \left(\frac{k^{\frac{2}{3}}}{6L_o^{\frac{1}{3}}}\right) t \tag{6.4}$$

Where t = time, and y = the BOD in mg/L at time t. A plot of $\left(\frac{1}{y}\right)^{\frac{1}{3}}$ against t gives a linear plot with $\left(L_o k\right)^{-\frac{1}{3}}$ as intercept and $\left(\frac{k^{\frac{2}{3}}}{6L_o^{\frac{1}{3}}}\right)$ as slope. From this, the variables L_o and k can be determined.

6.7.9 TOTAL SUSPENDED SOLID (TSS)

This is referred to as the weight of insoluble, suspended matter filtered from a measured volume of water, preferably expressed as milligrams per liter (mg/L). Plastic membrane test filters retaining solids larger than 0.45 μm are commonly used to determine TSS.

6.7.10 WATER TEMPERATURE

Water temperature influences DO in water, which invariably influences the survival of aquatic resources. However, increasing the temperature also increases the rate of water reactions (Mariola 2020). In addition, climate change in the Arctic may affect water temperatures and, by extension, affect the distribution and survival of many fish. Increase in temperature can also indicate the effects of human activities. Water temperature is best measured on-site.

6.7.11 NUTRIENT POLLUTION

Nutrient shortages are identified as a global problem contributing to water pollution. The main components of nutrients are phosphorus and nitrogen. However, the nutrient inputs include nitrates, phosphate, sulfates, chlorides, etc. The nutrients from different points entering the lakes, streams, rivers, estuaries, etc. necessitate the proliferation of aquatic species (most especially algae blooms) (Zorigto et al. 2020). It is evident that water bodies require some nutrients for aquatic species to thrive, but excessive introduction could be unpalatable.

The United States Environmental Protection Agency (USEPA) (2012) has revealed that the nutrient loads of various water sources are massive and must be monitored holistically.

6.7.11.1 Nitrate Pollution

This pollution initiates excessive quantities of nitrate in various water domains. Nitrate in ground and surface waters originates primarily from point and nonpoint sources. Ahmed et al. (2017) assert that excessive introduction of fertilizers containing nitrogen in agriculture has been one of the primary sources of high nitrate in groundwater.

In addition, livestock feeding, barnyards, and human activities are the important sources of nitrate in groundwater (Groen et al. 1988). Gupta (1981) noted that commercial activities may contribute to a high level of nitrate in groundwater.

6.7.11.2 Sources of Nutrients

The main causes of the river's poor water quality and aquatic habitat loss are elevated levels of two nutrients, namely, nitrogen and phosphorus. These nutrients occur naturally in soil, animal waste, plant materials, etc. The excess nutrients are supplied to the river through two sources, namely, point and nonpoint sources. Point source refers to discrete and identifiable source of pollution that is not driven by runoff, while nonpoint source broadly refers to diffuse origins of contaminations, such as nutrients from agricultural runoff, groundwater, septic systems, etc.

6.7.12 Heavy Metal Pollution and Effects

Heavy metals are mostly referred to as transition metals with high atomic number and with density five times greater than water. Some of these include transition (D-block) and inner transition metals, such as mercury (Hg), cadmium (Cd), arsenic, chromium (Cr), lead (Pb), etc.

The fragile ecosystem is being threatened by the introduction of heavy metals into the food chain. It also affects the biodegradability of organic pollutants, thereby making them less degradable and thus doubling pollution effects to the environment. Heavy metals also deter metabolism and result in malfunction or death of affected cells.

The adverse impact of heavy metal pollution/toxicity has proven to be the foundation of global health crisis (Duru et al. 2008). There are copious metals that attack the ecosystem, and thus its effects are enormous. Antimony, when one is exposed to high concentrations within a short time frame, causes instant throwing up and diarrhea. The excessive intake of cadmium causes bore defects in humans and animals. Mercury compounds cause both congenital and spontaneous malformations in humans. Long exposure to chromium causes skin irritation and ulcer in humans. It also causes kidney and liver damage. The intake of high concentrations of lead will result in direct tissue interaction, which includes tissue desiccation and gastrointestinal (GI) track distortion. In addition, a large intake of lead causes convulsions in humans.

6.7.13 Effluent Standards and Charges

The term "effluent" has been defined as liquid, solid, or gaseous products discharged by a process, treated or untreated, which usually may not be toxic.

The adverse effect of these industrial effluents has made both governmental and nongovernmental agencies set up standards for water quality monitoring. An effluent standard is the maximum number of specified pollutants allowed in discharged sewage as established by regulating agencies to achieve a desirable stream standard. An environmental quality standard is the maximum limit permitted by pollutants in a specific medium (e.g., air or water).

In the United States of America (USA), standards are put in force with respect to the permissible limits. Regulating bodies impose effluent charges, which are fixed fees levied by a regulating body against polluters for each quantity of waste discharged into public water. In Nigeria, Decree Number 58 of 1988, as amended by Decree Number 59 of 1992, set up the Federal Environmental Protection Agency (FEPA) to enforce and arraign defaulters to the relevant authorities.

Other includes:

(1) National effluent limitation regulation; its main objective is to ensure the installation treatment plant for pollutants.
(2) Pollution abatement in waste-generating industries and facilities; its main goals are the imposition of restrictions on the release of toxic substances and the placement of pollution-monitoring measures to ensure that possible limits are not exceeded.

The effluent limitation guidelines and guidelines for drinking water quality according to WHO (1984, 2004) are presented in Tables 6.2 and 6.3.

6.7.13.1 Effluent Limitation Guidelines

TABLE 6.2

Effluent Limitation Guidelines for Specific Industries (WHO 1984)

Industry	Parameter	Effluent (mg/1) Allowable Limit for Discharge into Waters
Agrochemicals	TSS	15
	Phosphate	13
	Fluoride	1
	pH	9–9
	Free ammonia	0–1
	Arsenic	0.1
Automobile	pH	6–9
	Iron	0.20
	Cadmium	0.10
	Nickel	0.05
	Copper	0.06
	Lead	0.01
	Arsenic	0.1
Brewery	pH	6–9
	COD	80
	BOD	30
Petrochemical	Temperature (°C)	30
	pH	6.5–8.5
	Phenol	0.5
	Ammonia	0.2
	Lead	0.05
Petroleum	Temperature (°C)	30
	pH	6.5–8.5
	Phenol	0.05
	Ammonia	0–05
	Sulfides	0.20
	BOD	10
	COD	40
	Chromium	0.3
	Lead	0.05
	Cadmium	0.01
	Cyanide	0.01
Plastics and synthetics	BOD	10
	COD	40
	Phenolics	0.50
	Oil and grease	10.0
	TSS	30
	Chromium	0.10

(Continued)

TABLE 6.2

Continued

Industry	Parameter	Effluent (mg/1) Allowable Limit for Discharge into Waters
Soap and detergents	COD	40
	BOD	15
	Oil and grease	10
	pH	6.0–9.0
Sugar processing	pH	6–9
	BOD	20
	COD	8
	TSS	30
	Chromium	0.10
	Phenols	0.10
	Sulfides	0.20
	Chloroform	400 mm/100 ml
Sugar processing	Color	None
	Odor	None

TABLE 6.3

Guidelines for Drinking Water Quality (WHO 2004)

S/N	Element/Compound	Acceptable Limit (mg/l)	MCL (mg/L)
1	Total dissolved solid (TDS)	500	1500
2	Total hardness (TH ($CaCO_3$))	100	500
3	Detergents (ABS)	0.5	1
4	Aluminum (Al)	0.2	0.3
5	Iron (Fe)	0.3	1
6	Manganese (Mn)	0.1	0.2
7	Copper (Cu)	1	1.5
8	Zinc (Zn)	5	15
9	Sodium (Na)	200	400
10	Nickel (Ni)	0.05	0.1
11	Chloride (Cl^-)	200	400
12	Fluoride (F^-)	1	1.5
13	Sulfate (SO_4^{2-})	200	500
14	Nitrate (NO_3^-)	25	70
15	Silver (Ag)	0.01	0.05
16	Magnesium (Mg)	50	120
17	Calcium (Ca)	100	200
18	Potassium (K)	10	12
19	Color units (platinum cobalt)	10	15
20	Turbidity (NTU)	1	5
21	pH	6.5–8.5	9.5
22	Temperature (°C)	8–25	

Source: *MCL represents maximum contaminant limit.

6.7.14 Water Quality Monitoring Models

Water quality monitoring model is a driving force that evaluates the pollution loads on both the entrance and mixing of the discharges in the relevant water bodies as well as the upliftment of the existing water sources (Madhusmita et al. 2020).

According to Refsguard and Henriksen (2004), these models are mathematical equations/expressions and computer-based software representing the pollution fate, transport, and degradation of a water body.

The World Bank Group (1999) noted that efficient and effective modeling program typically focuses on a few monitoring parameters of immediate attention, such as BOD, nutrient loads, DO, microbiological inputs, etc.

6.7.14.1 Water Quality Monitoring Model Categorization

The World Bank Group (1999) also noted that water quality monitoring models are classified based on the following factors:

  (i) The complexity of models.
 (ii) The nature and type of receiving water bodies.
(iii) The type of water quality parameter loadings.

However, the more complex the models, the more exorbitant and cumbersome their applications for a given case study. The complexities of these models are dependent on the following:

  (i) Factors of quality indicators. This factor seriously affects the adoption of any model to remedy any situation. The greater the number of parameters involved, the more complex the model will be in the prediction.
 (ii) The level and number of pollution sites. The variations in pollution data affect the complex nature of the model adopted.
(iii) The level of temporal case studies. The water quality monitoring models for predicting long-term stationary water points are easier and simpler than the short-term dynamic variations.
(iv) Complex nature of the water bodies under study. The prediction models for lakes that are homogeneous are simpler than those for moderate and large-sized rivers.
 (v) The of type of adsorbent and heavy metals involved. The adoption of adsorption models for the prediction of heavy metal loads is dependent on the native groups present at the surface and the type of heavy metals (adsorbate) present. Long-term stationary water domain containing heavy metals is easier to predict than short-term equilibrium water bodies.

There are numerous water quality monitoring models, but in this chapter, we are going to focus on a few essential ones, and they include the following:

6.7.14.2 Water Quality Assessment Methodology (WQAM)

This is an easy method or mathematical tool (not computer based) used for the fundamental prediction of changes in parameters due to variations in pollution load.

A typical illustration is observed when WQAM is applied in the analysis of moderately polluted tropical estuaries sited near a nearby vegetable canning plant (World Bank Group 1999). The report showed significant impact on the estuary's dissolved oxygen (DO) and nutrient. The researchers postulate that if the plant were brought online, DO would reduce from 4.5 ppm to 3.0 ppm, which could be problematic for water species in the estuary. In contrast, phosphorus concentrations were observed to rise 0.5 ppm to 2.0 ppm, and this could lead to the formation of algae blooms and could invariably affect fishing activities in that location.

6.7.14.2.1 QUAL2E

This refers to the upgraded stream water quality models. It is a computer-based model that makes use of software in its analysis and is a variant of the QUAL model series by USEPA (1994). It is commonly applied in the assessment of the resultant effects of variations in identifiable pollution (Praveen et al. 2020). It is also essentially effective and efficient for the analysis of algal concentration and dissolved oxygen. This model is used mostly in the United States and other parts of the world.

6.7.14.2.2 WASP

This is referred to as a stimulation analysis program for water domains. It is a flexible, compartmental modeling technique for stimulating very large pollutants like nitrogen, phosphorus, dissolved oxygen, BOD, algae, heavy metals, temperature, etc. in almost any type of water body. It is regarded as the most powerful and complex model involving resource and expertise for successful application. It is extensively applied to water quality assessments in water bodies. Currently, WASP has eight versions (WASP 1–8). It is software that can be freely downloaded from www.epa.gov (USEPA) (1991). The manipulation of the WASP model is based on the conservation of mass and momentum equation, as expressed here:

$$\frac{\partial Q}{\partial t} + \frac{1}{B}\left(\frac{\partial Q}{\partial x}\right) = q_s \tag{6.5}$$

$$\frac{\partial Q}{\partial t} + \frac{\partial}{\partial x}\left(\frac{\partial W}{A}\right) + gA\left(\frac{\partial Z}{\partial x} + \frac{Q[Q]}{K^2}\right) = 0 \tag{6.6}$$

Where Z = water surface elevation; Q = flow rate; B = wetted width; A = wetted cross-sectional area; t = time; x = the distance along the channel; K = conveyance of the channel; g = gravitational acceleration; q = the discharge per unit channel length (Madhusmita et al. 2020).

6.7.14.2.3 CE-QUAL-RIV 1

This is referred to as a dynamic, one-dimensional (longitudinal) model for turbulent movement of water bodies. It is used to analyze variations in essential measurement parameters during flood situations.

6.7.14.2.4   HEC-5Q

This is referred to as a computer model used to analyze variations in flood-prone areas. This is essentially used for analyzing the flow of water in reservoirs and the corresponding downstream river basin.

6.7.14.2.5   Adsorption Isotherm and Kinetic Models

There are numerous adsorption isotherm and kinetic models for the prediction of heavy metal loads in water bodies, but the basic models, such as Freundlich, Langmuir, BET, pseudo-first order, and pseudo-second order, will be evaluated.

6.7.14.2.5.1   Freundlich Adsorption Isotherm   Freundlich, in 1909, proposed an experimental equation expressing relationship between metal uptake and the mass of the adsorbent at constant temperature (Obi and Eigbiremonlen 2016; Obi et al. 2020). This equation, referred to as the Freundlich adsorption isotherm, is expressed nonlinearly as (Mahmoud et al. 2018; Mahmoud 2020):

$$q_e = K_L C_e^{\,n} \tag{6.7}$$

The linear equation is given as:

$$\text{In } q_e = \text{In } K_F + n \text{ In } C_e \tag{6.8}$$

Where q_e = equilibrium metal uptake; C_e = the concentration at equilibrium; K_F and n = constants. Though, Freundlich isotherm failed to predict value of metal uptake at high concentrations or pressures.

6.7.14.2.5.2   Langmuir Adsorption Isotherm   Langmuir, in 1916, put forward adsorption isotherm based on some hypotheses, one of which is that there is a balance between heavy metal ions and unadsorbed heavy metal molecules.

$$A(g) + B(S) \underset{\text{desorption}}{\overset{\text{Adsorption}}{\rightleftharpoons}} AB \tag{6.9}$$

Where, $A\ (g)$ = unadsorbed heavy metal molecule; $B\ (s)$ = unoccupied metal surface; and AB = adsorbed metal ions.

Accordingly, he proposed a relationship between the surface coverage and the concentration of heavy metal ions.

$$\theta = \frac{KC}{1 + KC} \tag{6.10}$$

Where θ = the surface coverage by the metal ions; C = concentration; K = the equilibrium constant. The basic limitation of Langmuir adsorption equation is its validity at low concentrations.

At lower concentrations, the surface coverage reduces to:

$$\theta = KC \tag{6.11}$$

At high concentrations, the surface coverage reduces to unity as expressed:

$$\theta = \frac{KC}{KC} = 1 \tag{6.12}$$

Langmuir linearized equation with respect to heavy metal ions removal is given as:

$$\frac{C_e}{q_e} = \frac{1}{K_L q_e} + \frac{C_e}{q_e} \tag{6.13}$$

Where K_L = the Langmuir adsorption constant related to the intensity of adsorption.

6.7.14.2.5.3 BET Adsorption Isotherm   BET simply means Brunauer, Emmett, and Teller, referring to the scientists whose theory was anchored on multilayer formation, and the mathematical representation is expressed as:

$$\frac{1}{V\left[\left(\frac{P_o}{P}\right) - 1\right]} = \frac{C-1}{V_m C}\left(\frac{P}{P_o}\right) + \frac{1}{V_m C} \tag{6.14}$$

Where C = BET constant, which is equal to $\exp\left(\frac{E_1 - E_L}{RT}\right)$; E_1 and E_L = the heats of adsorption and vaporization for the first and last layers; Vm = the monolayer volume; P = the equilibrium pressure; P_o = the saturated pressure; and V = the volume of adsorbed gas quantity.

A plot of $\dfrac{1}{V\left[\left(\frac{P_o}{P}\right) - 1\right]}$ against $\left(\dfrac{P}{P_o}\right)$ will be a linear plot from which V_m and C

will be determined from the slope and intercept. From the value of V_m obtained, the total and specific surface areas of the sample material can be evaluated using the following equations:

$$S_t = \frac{V_m NS}{V} \tag{6.15}$$

$$S_{BET} = \frac{S_t}{a} \tag{6.16}$$

Where S_t = the total surface area of sample material; S_{BET} = BET-specific surface area; N = Avogadro's number; S = the area of gas molecule adsorbed; and a = the mass of the sample.

6.7.14.2.5.4 Adsorption Kinetics Adsorption kinetics is defined as the rate at which these heavy metal ions or other particles are immobilized at the surface. The phenomenon of adsorption involves both physical and chemical impacts. Physical adsorption occurs due to van der Waals forces, while chemical adsorption emanates owing to true chemical bond formation between heavy metal ions and the adsorbent. For the purpose of this chapter, we shall look at two models, which follow.

6.7.14.2.5.4.1 Pseudo-First Order Model This is also referred to as the Lagergren model. The mathematical equation is written as:

$$\frac{d_{qt}}{dt} = k_1 \left(q_e - qt \right) \tag{6.17}$$

Where q_t = the metal uptake at time t; q_e = equilibrium metal uptake; k_1 = rate contant per min. The integral of equation 5.17 from t = 0 to t = 1 and q_t = q_t yields a linear expression in equation 6.18:

$$\ln \left(q_e - q_t \right) = \ln q_e - k_1 t \tag{6.18}$$

The value of k_1 is determined by plotting $\ln(q_e - q_t)$ against t.

6.7.14.2.5.4.2 Pseudo-Second Order Model The model equation is written nonlinearly as:

$$\frac{d_{qt}}{dt} = k_2 \left(q_e - q_t \right)^2 \tag{6.19}$$

Where k_2 = pseudo-second order rate constant and applying the integral limits for t (0, t) and qt (0, qt), and the linearized form (Tan and Hameed 2017) is expressed as:

$$\frac{t}{q_t} = \frac{1}{k_2 q_e^2} + \frac{t}{q_e} \tag{6.20}$$

6.7.15 RESEARCH CONCERNING THE INTERNET OF THINGS (IoT)

The use of sensors may provide innovative and cost-effective approaches to the problems of contamination detection and water quality analysis. Sensors for parameter measurements were proposed to perform in reality. Wireless communication devices coupled to the sensors transfer those readings to a controller (Ferguson and Redish

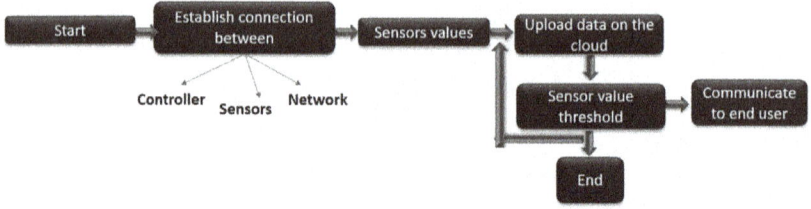

FIGURE 6.2 Operating flow diagram of a smart water quality monitoring system.

2011). These sensor readings are afterwards saved to data storage by the controllers via wireless communication technology and reflected in a tailored application (Lakshmikantha et al. 2021). Figure 6.2 shows the operating flow chart of a smart water quality monitoring system.

This would be followed by the design of a generic Internet of Things system instance. Encinas et al. (2017) utilized pond water quality monitoring sensors for monitoring temperature, pH, nitrates, carbonates, and dissolved oxygen in their study. The Arduino module and ZigBee transmitters and receivers were utilized in the project. The computer receives readings from a ZigBee transmitter coupled to a sensor and returns back to the ZigBee receiver. The readings are then kept locally and transferred to the web cloud, where they may be seen in the Android application.

In order to keep track of a polluted body of water, some farmers utilize a solar panel and a sensor node consisting of a variety of sensors, including those for dissolved oxygen (DO), nitrate (NO_3), and carbonates (CO_3). In the event that a sensor reading falls outside the acceptable range, an alarm is raised and a warning is sent to the farmer. The farmer can monitor sensor data in real time and retrieve past data through a mobile application. Vijai and Sivakumar (2016) employed an IoT framework in reality to monitor turbidity, chlorine, ORP, nitrates, and other contaminants. A system for monitoring most water quality indicators, for instance pH, conductivity, temperature, and turbidity, using sensors to take the measurements was developed. These indicators are crucial in determining whether or not water is suitable for aquatic life (Birje et al. 2016). Cloete et al. (2016) developed sensor nodes with temperature, conductivity, pH, ORP, and flow sensors as part of their research. Their design used microcontrollers to communicate with the sensors through a process of conditioning the signals.

Perumal et al. (2015) demonstrated how to monitor water levels in real time and inform local authorities and the general public using social media in case of potentially hazardous occurrences, such as flash floods or storms. In combination with the ultrasonic sensors, a wireless gateway was employed. A sensor system monitors five water quality parameters: temperature, pH, turbidity, conductivity, and dissolved oxygen (Vijayakumar and Ramya 2015). Microsensors are placed on the site and connected to sensor end device nodes that transmit the data, and mobile communication was used to suggest a low-cost and easy-to-install wireless network system (Cao et al. 2014). Rasin and Abdullah (2009) also developed a wireless sensor network-based system (WSNs). They employed affordable sensors that can measure pH, temperature, and turbidity and conduct signal conditioning to verify their accuracy.

The transceiver transmits information from the node to the central monitoring station when it has been properly conditioned.

6.7.16 OVERVIEW OF WIRELESS SENSORS

Because of the growing need for environmental monitoring and management, wireless sensor networks have emerged as a significant resource for gathering and transferring data to the appropriate location (Engmann et al. 2020; Martinez Paz et al. 2021). There are several reasons sensors are utilized in monitoring water domain. Cross-connections of polluted water entering the channel through defective pipes in a low-pressure area and microbial growth in pipes are all concerns that must be solved (US EPA 2007). Because of the physical characteristics of the pipe networks and the lack of monitoring and surveillance, nationally known water security experts have categorized distribution systems as vulnerable to exploitation (Water Sentinel). Threats or signs of prospective contamination that may be identified are a management problem (US EPA 2007). Therefore, data generated from sensors are used to make management decisions on a range of concerns. The absolute solution is, firstly, to recognize compliance with monitoring water quality criteria and, secondly, detecting noncompliance from major key users in the relevant sites across the system. The third purpose is to verify water quality modeling and, more importantly, to put in place a pollution alert system (Grayman 2013).

The chemical, physical, and microbiological inputs define water quality. Even the smallest change in these features may have a harmful impact on people and companies that rely on water. Monitoring water physicochemical characteristics, such as conductivity, pH, salinity, temperature, dissolved oxygen, residual chlorine, and turbidity, is critical to maintaining its quality. Consequently, water quality monitors have become prevalent in most contemporary distribution systems. There are two major techniques for employing water quality sensors. They are either used to directly measure physicochemical concentrations of water, which serve as indicators of unanticipated contaminants in the water, or they are used to indirectly measure physicochemical concentrations of water, which serve as indicators of unanticipated contaminants in the water.

A water quality monitoring system consists of several of sensors, like pH sensors, turbidity sensors, temperature sensors, conductivity sensors, humidity sensors, and several others. The smart water quality monitoring system is depicted schematically in Figure 6.3. The core controller is the system's heartbeat. All other sensors are linked to a core controller, which controls the process by collecting data from other sensors, comparing it to standard values, and then transmitting the results to the end user through wireless modules. Water quality sensors come in a variety of sizes and designs. The typical examples are enumerated in the following passages.

6.7.16.1 Residual Chlorine Sensor

The measurement of residual chlorine in portable water treatment facilities and distribution networks is a standard procedure that has been required for as long as water has been treated with chlorine. Chlorine is mostly used as disinfectant due to its low cost and efficacy. Free chlorine, monochloramine, and total chlorine are all

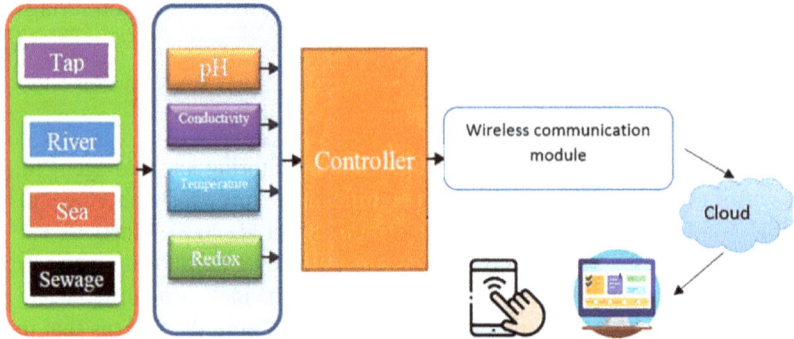

FIGURE 6.3 Schematic diagram of smart water quality monitoring system using sensors.

Source: Adapted from Mamun et al. (2019).

measured through chlorine sensors. Notwithstanding the assertion that total chlorine is regularly determined in treated wastewater, including domestic wastewater, the main use is the treatment and disinfection of portable water.

6.7.16.2   The Total Organic Carbon (TOC) Sensor

Total organic carbon is a significant parameter in water quality evaluation (TOC). There are two types of TOC measuring equipment available. Instrument precision is required for regulatory reporting, the maintenance of a key process control variable, real-time release, and other quality-critical product attributes. If the intended use is for general TOC monitoring rather than making essential quality judgments, other aspects may be more important than accuracy. Sensors are often used to monitor a process, and the data they acquire is solely utilized for information.

6.7.16.3   Turbidity Sensors

Turbidity sensors detect suspended solids or particulate matter in water by measuring the quantity of light that passes through them. The applications of the sensors are in the areas of river and stream assessment, wastewater and evaluation of effluent water, treatment of portable water, treatment process and control, control instruments, sediment transport studies, and laboratory analysis and measurements.

6.7.16.4   Conductivity Sensor

Conductivity assessments are used in industrial operations to obtain information on ionic concentrations, especially dissolved chemicals or compounds in aqueous solutions. Water purification and the monitoring of concentration levels in solutions are all common uses. The measuring system is composed of a suitable inline sensor that is either directly inserted or housed, a cable that connects to a transmitter, and either a Direct Conductivity Sensor, or the received signals are converted into a measurement result.

6.7.16.5   pH Detector

pH is a critical water quality parameter that must be measured, checked, and controlled. It is used to determine the acidity or alkalinity of a solution. A pH sensor's components are usually consolidated into a single device known as a "combination pH electrode." The electrode used in the analysis is often composed of glass, making it rather sensitive. In recent advances, glass has been replaced by more durable solid-state sensors. When conducting pH control, the pH analyzer or transmitter provides a user interface for calibrating the sensor and configuring outputs and alarms.

6.7.16.6 Oxygen-Reduction Potential (ORP) Sensors

Sensors that detect the oxygen-reduction potential (ORP) of a solution are known as ORP sensors. When combined with a pH sensor, the ORP measurement provides information on the degree of redox reactions processes occurring in the solution. The ORP sensor requires a compatible interface and software to collect data. These sensors offer an indicator of water quality for numerous system applications. A well-designed system for water pollution monitoring can provide operators and engineers with valuable data that can assist in calibrating hydraulic models, predicting the formation of regulated substances, providing compliance data, and tracking quality changes over time, all of which can assist system operators in making significant results about marine treatment section methods and effective working conditions. These sensors are used in combination with a contemporary supervisory control and data acquisition system. The supervisory control and data acquisition (SCADA) system provides system operators with real-time data that may be used to sustain and enhance water quality in distribution systems.

6.7.17 ROLE OF ARTIFICIAL INTELLIGENCE IN WATER SCARCITY MITIGATION

Water is a natural resource that should be accessible and available to everyone in an adequate quantity and quality (Gleick 1998). But it is confirmed that around 785 million people on the globe are bereft of portable drinking water, and it is estimated that at least two billion people are drinking water that has been contaminated with human waste, raising the risk of contracting diseases such as diarrhea, cholera, dysentery, typhoid, polio, etc. The World Health Organization (WHO) has proposed that in the 2025, global population will face serious portable water challenge. The collaboration for water, peace, and security is developing novel technologies and services to detect the emergence of water-related security threats. These cutting-edge technologies and services can depict short-term water scarcity changes and consequences, as well as relate them to social, political, economic, and hydrological aspects. Artificial intelligence (AI) was utilized by experts from this organization to identify where worldwide water-related disputes might occur. According to their predictions, the warning systems will be able to check the sources of these water disputes and target assistance to the places where it is needed most. As a consequence, applying artificial intelligence algorithms to monitor quality of water environment–related data and suggest scientific processing procedures is vital for boosting regional water management productivity. Many organizations have begun to employ artificial intelligence to govern water in many areas, such as minimizing water loss or gaining more information about how this resource is used, in order to attempt to address this water constraint. The goal is to alleviate the worldwide water issue, and the good news is that several initiatives have already been proven to be successful.

Predicated on the water and AI application characterizations, there still are unique applications that are presently in use or may be employed in the near future that might have a significant influence on society and the natural environment. While the scientific literature is clearly preoccupied with the most cutting-edge AI approaches, the industry is already using more traditional data science techniques on a wider scale (Doorn 2021).

In India, Dr. Yogesh Simmhan is utilizing IoT in promoting and guaranteeing that people will have access to adequate and safe potable water. The goal of this initiative is to have an in-depth knowledge of the imbalance in the assessment of water in specific neighborhoods, and AI may be used to enhance water scheduling or identify leaks.

The team is able to anticipate peak demand and identify problem locations by combining data from many sources, including reservoir flow, seasonal climate, and residential consumption.

Finding new sources of water and maintaining the current reserves and water systems in a sustainable way are the two most pressing issues confronting the water problem at present. Large datasets can be handled indefinitely by AI.

Researchers are investigating if water plants can be developed using algorithms that can offer current information regarding resource use as well as data for future forecasts. Artificial intelligence (AI) can devise strategies or forecast models for the future to enhance consumption and localization of this fundamental resource. It is impossible, however, to develop policies that govern its proper use.

Artificial intelligence was founded on the principle that human intellectual progression and development can be computerized, and this assumption dates back to primeval Chinese, Indian, and Greek philosophers (Berlinski 2000). While ancient and primitive philosophers were particularly interested in legitimizing reasoning in order to reduce argumentation, computation development of programmable computers only started early in the nineteenth century. Advancement in fields such as mathematics, philosophy, engineering, entrepreneurship, and neuroscience, as well as the evolution of massive code-breaking machines during World War II, as well as analytical work that allowed us to explain what it means to think of a machine as "thinking," led to the establishment of artificial intelligence as an academic field (Crevier 1993; Buchanan 2005). The use of artificial intelligence (AI) patterns is focusing on identifying typical issues in the wastewater collection system.

6.7.18 APPLICATION OF ARTIFICIAL INTELLIGENCE IN PREDICTION AND FORECASTING

This chapter has typically depicted the potential of artificial intelligence (AI) to develop water quality models that do not rely on site-specific qualities or that are applicable to a large number of environmental factors. Although there is a clear overlap with the aforementioned modeling category and quite a few publications are identified which refer to modeling and prediction/forecasting, all prediction and forecasting studies contain a time dimension, while not all modeling publications do (Zakaria et al. 2021). Singh and Gupta (2012), for example, developed numerous prediction models to predict the development of trihalomethanes (THMs) in chlorine-contaminated waters, which pose a significant threat to humans. THM levels in water must be known for adequate regulation, but doing so in the laboratory is too expensive and time-consuming. THM production may be predicted using AI-based prediction models based on more conveniently determined variables, such as water pH, temperature, and the concentration of chemical components such as bromide. The study showed that AI-based models may effectively capture the complex and nonlinear correlation between water conditions and THM development. Additional research investigated the effectiveness and applicability of machine learning models in predicting water quality indicators in coastal regions supported by contributing streams containing potentially polluted water (Alizadeh et al. 2018). In another scenario in Hawaii, USA, an online data obtained from Wailuku River was combined with hourly water quality indicators of salinity, temperature, and turbidity in Hilo Bay. From the present to two hours ahead, the impact on the mentioned parameters in the river was explored using a variety of machine learning techniques. According to the researchers, water performance indicators may be accurately forecasted many hours in advance, which might be valuable information for ecological assessment and tracking in coastal regions. The research by Arnon et al. (2019) provides a good example of AI being used to identify inadvertent water pollution. It focuses on organic water contamination, which is often identified by inferring "recurring" water patterns from measurements of indicator or surrogate physical and chemical characteristics of ostensibly fresh water. The deviation from these patterns is consequently considered an indication of contamination. Turbidity, redox potential, free chlorine, and water conductivity are fundamental physical and chemical criteria used to describe these trends. Activities that do not lead to contamination—for example, alterations in maintenance and operating methods—might cause changes in physical and chemical characteristics, making "genuine" contamination harder to detect. As a result, this technique relies on consistent background circumstances. In this study, a mechanism for predicting water contamination based on its spectroscopic features was established (UV absorbance spectra). The goal is to choose the most important characteristics for the given categorization issue, such as "contaminated" or "portable." An algorithm which combines Pearson correlation and Euclidean distance between the experimental absorption spectrum and the original spectrum of the various water domains online was used to find abnormalities in the water. The application of a basic selection method, which reinforced the significant distinctions between portable and polluted water, was the new aspect.

Gram-based amplification was applied by the researchers, which allows the application of basic selection using computer. The detection system was trained using a data bank of portable water sources. With rapidly variable water properties, the algorithm was effective in identifying pollutants at very low concentrations with a very low percentage of false-positives (Ramona et al. 2012).

Researchers such as Mewes and Schumann (2019) developed a watering planning model with a machine learning–based module that can detect up-to-date hydrological processes and adjust irrigation and cropping strategies in the model. AI techniques had been used to investigate the design of a water quality monitoring network using a different type of optimization: a classic yet complex optimization problem that involves the optimal configuration of sensors to ensure maximum information retrieval from the collected water quality data (Strobl and Robillard 2006). Learning algorithms, artificial neural networks, genetic algorithms, and fuzzy logic systems were all explored as potentially valuable AI techniques in their study. There were a lot of comparable applications in the evaluation that utilized AI to design a system or a water network. The investigation of contaminants and biological markers in fresh wastewater by means of biochemistry and bioinformatics methods is a good example of wastewater-based or sewage-based epidemiology (WBE) (Lorenzo and Picó 2019). It can be used to provide real-time data on the exposure to pesticides or persistent organic pollutants. The previous evidences demonstrate that sustainable AI in water system management entails not just regulating specific categories of usage or applications but also maximizing their potential (Taddeo and Floridi 2018). AI may be utilized to address some of the current challenges in the water domain, as well as lay the foundation for a future distinguished by profiling and one-sided optimization. As a result, sustainable AI may have an impact on current discussions in the domain of water ethics. Paradoxically, the regulatory legislation on human water use is already structured in accessibility to crucial data, which may become even more important as water consumption expands (Braga 2003).

6.8 CONCLUSION

This chapter presented a review of monitoring parameters for quality water, applications of sensors, modeling techniques, and artificial intelligence for detection of pollution in water domains. The subject area of water management has expanded tremendously in recent years due to a global quest for potable, healthier, and improved drinking water. The management of water parameters is an indispensable strategy in maintaining a balance between human activities and the wholesomeness of aquatic species and ecosystems. The pollution of water bodies emanates from natural factors, such as geological, topographical, meteorological, hydrological, biological, etc. The contamination of the water domain by the introduction of nutrients (phosphorus and nitrogen) has grossly distorted and even led to the death of aquatic species through the process of eutrophication. This phenomenon is a dangerous global effect that has attracted attention due to the deleterious diseases it causes.

In light of the prior statement, there is a serious need to observe the quality of these water bodies in order to strike a balance between humans, aquatic species, and eco-habitants. However, the monitoring models, sensors, and artificial intelligence are adopted to analyze and forecast the contamination levels of these water sources, either in short-term or long-term processes. The lists of models, sensors, and artificial intelligence techniques for water quality monitoring are many and varied, but few of them were treated in this chapter. In addition to the application of these monitoring techniques, we suggest that a harmonized blueprint (calibration) be adopted by governments and regulatory agencies to implement the best methods and techniques to provide an international framework for assessing water pollution control in line with the United Nations Sustainable Development Goals (UNSDGs) to restore aquatic environments.

ACKNOWLEDGMENTS

We sincerely appreciate the professional and academic mentorship of Professors Gloria Ukalina Obuzor and Leo Osuji of the Department of Pure and Industrial Chemistry, University of Port Harcourt, Rivers, Nigeria. We are deeply grateful to authors whose books, materials, and research papers were consulted.

REFERENCES

Adelagun, R.O.A., Etim, E.E. and Godwin, O.E., 2021. Application of Water Quality Index for the Assessment of Water from Different Sources in Nigeria. In Wastewater Treatment. IntechOpen.

Adu-Manu, K.S., Katsriku, F.A., Abdulai, J.D. and Engmann, F., 2020. Smart river monitoring using wireless sensor networks. *Wireless Communications and Mobile Computing,* 2020, https://doi.org/10.1155/2020/8897126.

Ahmed, M., Rauf, M., Mukhtar, Z. and Saeed, N.A., 2017. Excessive use of nitrogenous fertilizers: An unawareness causing serious threats to environment and human health. *Environmental Science and Pollution Research,* 24(35), pp. 26983–26987.

Akhtar, N., Syakir Ishak, M.I., Bhawani, S.A. and Umar, K., 2021. Various natural and anthropogenic factors responsible for water quality degradation: A review. *Water,* 13(19), p. 2660.

Akhter, F., Siddiquei, H.R., Alahi, M.E.E. and Mukhopadhyay, S.C., 2021. Recent advancement of the sensors for monitoring the water quality parameters in smart fisheries farming. *Computers*, 10(26), pp. 1–20.

Aldhyani, T.H., Al-Yaari, M., Alkahtani, H. and Maashi, M., 2020. Water quality prediction using artificial intelligence algorithms. *Applied Bionics and Biomechanics*, 2020. doi: 10.1155/2020/6659314.

Alizadeh, M.J., Kavianpour, M.R., Danesh, M., Adolf, J., Shamshirband, S. and Chau, K.W., 2018. Effect of river flow on the quality of estuarine and coastal waters using machine learning models. *Engineering Applications of Computational Fluid Mechanics*, 12(1), pp. 810–823.

Arnon, T.A., Ezra, S. and Fishbain, B., 2019. Water characterization and early contamination detection in highly varying stochastic background water, based on Machine Learning methodology for processing real-time UV-Spectrophotometry. *Water Research*, 155, pp. 333–342.

Arun, P.M., Harish, K., Sachchidanand, S., Chaitanya, B.P., Raj, S. and Shardesh, K.C., 2021. An assessment of in-situ water quality parameters and its variation with Landsat 8 level 1 surface reflectance datasets. *International Journal of Environmental Analytical Chemistry*, doi: 10.1080/03067319.2021.1954175.

Bartram, J. and Balance, R., Eds., 1996. Water quality monitoring-A Practical Guide to the Design and Implementation of Freshwater. Quality Studies and Monitoring programs, United Nations Environment Program and the World Health Organization (UNEP/WHO), CRC, pp. 172–177.

Berlinski, D., 2000. The Advent of the Algorithm: The 300-year Journey from an Idea to the Computer. Houghton Mifflin Harcourt, Boston.

Birje, S. V., Bedkyale, T., Alwe, C. and Adiwarekar, V., 2016. Water pollution detection system using pH and turbidity sensors. *International Journal of Advanced Research in Computer and Communication Engineering,* 5(4), pp. 530–53.

Braga, B.P., 2003. The role of regulatory agencies in multiple water use. *Water Science Technology*, 47(6), pp. 25–32.

Buchanan, B.G., 2005. A (very) brief history of artificial intelligence. *Ai Magazine*, 26(4), pp. 53–53.

Cao, F., Jiang, F., Liu, Z., Chen, B. and Yang, Z., 2014. Application of ISFET Microsensors with Mobile Network to Build IoT for Water Environment Monitoring. In 2014 International Conference on Intelligent Environments. IEEE, China, pp. 207–210.

Chapman, D.V., ed., 1996. Water Quality Assessments: A Guide to the use of Biota, Sediments and Water in Environmental Monitoring. Chapman and Hall, London.

Cloete, N.A., Malekian, R. and Nair, L., 2016. Design of Smart Sensors For Real Time Water Quality Monitoring Organized by Institute of Electrical and Electronics Engineers (IEEE), South Africa, Volume 4, pp. 3975–3990.

Crevier, D., 1993. AI: The tumultuous History of the Search for Artificial Intelligence. Basic Books, Inc.

Doorn, N., 2021. Artificial intelligence in the water domain: Opportunities for responsible use. *Science of the Total Environment*, 755, p. 142561.

Dotaniya, M.L., Meena, V.D., Saha, J.K., Dotaniya, C.K., Mahmoud, A.E.D., Meena, B.L., Meena, M.D., Sanwal, R.C., Meena, R.S., Doutaniya, R.K., Solanki, P., Lata, M. and Rai, P.K., 2022. Reuse of poor-quality water for sustainable crop production in the changing scenario of climate. *Environment, Development and Sustainability*. doi: 10.1007/s 10668-022-02365-9.

Duru, C.E., Njoku, V.O. and Obi, C., 2008. Heavy metal concentration in urban water wells: A case study of Owerri municipal. *Journal of Chemical Society of Nigeria*, 33(2), pp. 89–93.

El Din Mahmoud, A. and Fawzy, M., 2016. Bio-Based Methods for Wastewater Treatment: Green Sorbents. In: Ansari, A.A., Gill, S.S., Gill, R., Lanza, G.R., Newman, L. (Eds.) Phytoremediation: Management of Environmental Contaminants, Volume 3. Springer International Publishing, Switzerland Cham, pp. 209–238.

Encinas, C., Ruiz, E., Cortez, J. and Espinoza, A., 2017. Design and Implementation of a Distributed IoT System for the Monitoring of Water Quality in Aquaculture. Wireless Telecommunications Symposium, IEEE, USA, pp. 1–7.

Engmann, F., Katsriku, F.A., Abdulai, J.D. and Adu-Manu, K.S., 2020. Reducing the energy budget in WSN using time series models. *Wireless Communications and Mobile Computing*, 2020(1), pp. 1–15.

Ferguson, J.E. and Redish, A.D., 2011. Wireless communication with implanted medical devices using the conductive properties of the body. *Expert Review of Medical Devices*, 8(4), pp. 427–433.

Gazzaz, N.M., Yusoff, M.K., Aris, A.Z., Juahir, H. and Ramli, M.F., 2012. Artificial neural network modeling of the water quality index for Kinta River (Malaysia) using water quality variables as predictors. *Marine Pollution Bulletin*, 64(11), pp. 2409–2420.

Geissen, V., Mol, H., Klumpp, E., Umlauf, G., Nadal, M., Van der Ploeg, M., Van de Zee, S.E. and Ritsema, C.J., 2015. Emerging pollutants in the environment: A challenge for water resource management. *International Soil and Water Conservation Research*, 3(1), pp. 57–65.

Gholizadeh, M.H., Melesse, A.M. and Reddi, L., 2016. A comprehensive review on water quality parameters estimation using remote sensing techniques. *Sensors (Basel)*, 16(8), p. 1298.

Gleick, P.H., 1998. The human right to water. *Water Policy*, 1(5), pp. 487–503.

Grayman, W. M., 2013. Contamination of Water Distribution Systems. In International Seminar on Nuclear War and Planetary Emergencies—45th Session: The Role of Science in the Third Millennium, pp. 389–398.

Groen, J., Schuchmann, J.B. and Geirnaert, W., 1988. The occurrence of high nitrate concentration in groundwater in villages in Northwestern Burkina Faso. *Journal of African Earth Sciences (and the Middle East)*, 7(7–8), pp. 999–1009.

Gupta, S.C., 1981. Evaluation of quality of well waters in Udaipur District. Indian *Journal of Environmental Health*, 23, 195–202.

Gwenzi, W., Selvasembian, R., Offiong, N.A.O., Mahmoud, A.E.D., Sanganyado, E. and Mal, J., 2022. COVID-19 drugs in aquatic systems: A review. *Environmental Chemistry Letters*, 20, pp. 1275–1294.

Helmer, R., 1994. Water quality monitoring: National and international approaches. *IAHS Publications-Series of Proceedings and Reports-Intern Assoc Hydrological Sciences*, 219, pp. 3–20.

Ibezim-Ezeani, M.U. and Obi, C., 2017. Fundamentals of Solution Thermodynamics, 1st ed., pp. 95–97. M & J Grand Orbit Communications Ltd, Diobu, Port Harcourt, Nigeria.

Jamie, B. and Richard, B., 1996. Water Quality Monitoring: A Practical Guide to the Design and Implementation of Freshwater Quality Studies and Monitoring Programs, World

Health Organization & United Nations Environment Program. E & FN Spon. https://apps.who.int/iris/handle/10665/41851.

Latiu, G., Cret, O. and Vacariu, L., 2013. Graphical user Interface Testing Optimization for Water Monitoring Applications. In 2013 19th International Conference on Control Systems and Computer Science, IEEE, Romania, pp. 640–645.

Lorenzo, M. and Picó, Y., 2019. Wastewater-based epidemiology: Current status and future prospects. *Current Opinion in Environmental Science & Health*, 9, pp. 77–84.

Madhusmita, G., Deba, P. S. and Narasimham, M. L., 2020. A brief overview of water quality models. *International Journal of Advanced Research in Engineering and Technology*, 11(10), pp 535–544.

Mahmoud, A.E.D., Fawzy, M. and Radwan, A., 2016. Optimization of Cadmium (CD2+) removal from aqueous solutions by novel biosorbent. *International Journal of Phytoremediation*, 18, 619–625

Mahmoud, A.E.D. and Fawzy, M., 2021. Nanosensors and Nanobiosensors for Monitoring the Environmental Pollutants. In Waste Recycling Technologies for Nanomaterials Manufacturing, Springer, Cham, pp. 229–246.

Mahmoud, A.E.D., Umachandran, K., Sawicka, B. and Mtewa, T.K., 2021. Water Resources Security and Management for Sustainable Communities. *In Phytochemistry, the Military and Health*, pp. 509–522.

Mahmoud, A.E.D., 2020. Graphene-based nanomaterials for the removal of organic pollutants: Insights into linear versus nonlinear mathematical models. *Journal of Environmental Management*, 270, p. 110911.

Mahmoud, A.E.D., Al-Qahtani, K.M., Alflaij, S.O., Al-Qahtani, S.F. and Alsamhan, F.A., 2021. Green copper oxide nanoparticles for lead, nickel, and cadmium removal from contaminated water. *Scientific Reports*, 11(1), pp. 1–13.

Mahmoud, A.E.D., Hosny, M., El-Maghrabi, N. and Fawzy, M., 2022. Facile synthesis of reduced graphene oxide by Tecoma stans extracts for efficient removal of Ni (II) from water: batch experiments and response surface methodology. *Sustainable Environment Research*, 32(1), pp. 1–16.

Mahmoud, A.E.D., Franke, M., Stelter, M. and Braeutigam, P., 2020. Mechanochemical versus chemical routes for graphitic precursors and their performance in micropollutants removal in water. *Powder Technology*, 366, pp. 629–640.

Mahmoud, A.E.D., Stolle, A., Stelter, M. and Braeutigam, P., 2018. Adsorption Technique for Organic Pollutants using Different Carbon Materials. Abstracts of Papers of the American Chemical Society, American Chemical Society NW, Washington, DC.

Mahmoud, A.E.D. and Kathi, S., 2022. Chapter 7 — Assessment of Biochar Application in Decontamination of Water and Wastewater. In: Kathi, S., Devipriya, S., Thamaraiselvi, K. (Eds.). Cost Effective Technologies for Solid Waste and Wastewater Treatment. Elsevier, pp. 69–74.

Mamun, K.A., Islam, F.R., Haque, R., Khan, M.G.M., Prasad, A.N., Haqva, H., Mudliar, R.R. and Mani, F.S., 2019. Smart water quality monitoring system design and KPIs analysis: Case sites of Fiji surface water. *Sustainability*, 11(24), p. 7110.

Mansoor, S., Farooq, I., Kachroo, M.M., Mahmoud, A.E.D., Fawzy, M., Popescu, S.M., Alyemeni, M.N., Sonne, C., Rinklebe, J. and Ahmad, P., 2022. Elevation in wildfire frequencies with respect to the climate change. *Journal of Environmental Management*, 301, p. 113769.

Mariola, K., 2020. Regional response to global warming: Water temperature trends in semi-natural mountain river systems. *Water*, 12(1), p. 283.

Martinez Paz, E.F., Tobias, M., Escobar, E., Raskin, L., Roberts, E.F., Wigginton, K.R. and Kerkez, B., 2021. Wireless sensors for measuring drinking water quality in building plumbing: Deployments and insights from continuous and intermittent water supply systems. *ACS ES&T Engineering*, 2(3), pp. 423–433.

Mewes, B. and Schumann, A., 2019. The potential of combined machine learning and agent-based models in water resources management. *Hydrologie und Wasserbewirtschaftung*, 63(6), pp. 332–338.

Mulholland, P.J., Houser, J.N. and Maloney, K.O., 2005. Stream diurnal dissolved oxygen profiles as indicators of in-stream metabolism and disturbance effects: Fort Benning as a case study. *Ecological Indicators*, 5(3), pp. 243–252.

Najafzadeh, M. and Ghaemi, A., 2019. Prediction of the five-day biochemical oxygen demand and chemical oxygen demand in natural streams using machine learning methods. *Environmental Monitoring and Assessment*, 191(6), pp. 1–21.

Obi, C. and Eigbiremonlen, S., 2016. Biosorption characteristics of water hyacinth (Eichhornia crassipes) in the removal of nickel (II) ion under isothermal condition. Pakistan Journal of Scientific & Industrial Research Series A: Physical Sciences, 59(2), pp. 118–120.

Obi, C., Ngobiri, N.C., Agbaka, L.C. and Ibezim-Ezeani, M.U., 2020. The application of monkey Cola Pericarp (*Cola lepidota*) in the removal of toluene from aqueous medium. *Asian Journal of Applied Chemistry Research*, 5(2), pp. 53–67.

Onojake, M.C., Ukerun, S.O. and Iwuoha, G., 2011. A statistical approach for evaluation of the effects of industrial and municipal wastes on Warri Rivers, Niger Delta, Nigeria. *Water Quality, Exposure and Health*, 3(2), pp. 91–99.

Perumal, T., Sulaiman, M.N. and Leong, C.Y., 2015, October. Internet of Things (IoT) enabled Water Monitoring System. In 2015 IEEE 4th Global Conference on Consumer Electronics (GCCE), IEEE, pp. 86–87.

Praveen, S., Babitha, R. H. and Ranjith, S., 2020. A review on water quality model QUAL2K. *International Journal of Creative Research Thoughts*, 8(5), pp 426–431.

Qin, Y., Kwon, H.J., Howlader, M.M. and Deen, M.J., 2015. Microfabricated electrochemical pH and free chlorine sensors for water quality monitoring: recent advances and research challenges. *RSC Advances*, 5(85), pp. 69086–69109.

Ramona, M., Richard, G. and David, B., 2012. Multiclass feature selection with kernel gram-matrix-based criteria. *IEEE Transactions on Neural Networks and Learning Systems*, 23(10), pp. 1611–1623.

Rasin, Z. and Abdullah, M.R., 2009. Water quality monitoring system using zigbee based wireless sensor network. *International Journal of Engineering & Technology*, 9(10), pp. 24–28.

Refsguard, J.C. and Henriksen, H.J., 2004. Modelling guidelines—terminology and guiding principles. *Advances in Water Resources*, 27(1), pp. 71–82.

Sanders, J.M. and Nee, V., 1987. Limits of ethnic solidarity in the enclave economy. *American Sociological Review*, 52(6), pp. 745–773.

Singh, K.P. and Gupta, S., 2012. Artificial intelligence-based modeling for predicting the disinfection by-products in water. *Chemometrics and Intelligent Laboratory Systems*, 114, pp. 122–131.

Snoeyink, V. L., and Jenkins, D., 1980. Water Chemistry, John Wiley & Sons, New York.

Standard methods for the examination of water and wastewater, 1998. American Public Health Association (APHA): Washington, DC. Federation, W.E. and APH Association.

Strobl, R.O. and Robillard, P.D., 2006. Artificial intelligence technologies in surface water quality monitoring. *Water International*, 31(2), pp. 198–209.

Taddeo, M. and Floridi, L., 2018. How AI can be a force for good. *Science*, 361(6404), pp. 751–752.

Tan, K.L. and Hameed, B.H., 2017. Insight into the adsorption kinetics models for the removal of contaminants from aqueous solutions. *Journal of the Taiwan Institute of Chemical Engineers*, 74, pp. 25–48.

U.S. EPA., 2007. Water security initiative: Interim Guidance on Planning for Contamination Warning System Deployment. EPA-817-R-07–002, Office of Water, Office of Ground Water and Drinking Water, U.S. EPA, Washington, D.C.

Umair, A., Rafia, M., Hirra, A., Sadaf, M., and Ali, M. Q., 2020. Water quality monitoring: From convectional to emerging technologies. *Water Supply*, 20(1), 28–45.

United Nations, 2015. Resolution adopted by the General Assembly 22 September 2015, Geneva: Transforming our World: The 2030 Agenda for Sustainable Development.

United States Environmental Protection Agency (USEPA), 2012. The Facts about Nutrient Pollution. Www.epa.gov/waters/ir.

USEPA (U.S. Environmental Protection Agency), 1991. Watershed Monitoring and Reporting for Section 319 National Monitoring Program Projects. U.S. Environmental Protection Agency, Office of Water, Office of Wetlands, Oceans, and Watersheds, Washington, DC.

USEPA (U.S. Environmental Protection Agency), 1994. EPA Requirements for Quality Assurance Project Plans for Environmental Data Operations. Interim Final, EPA QA/R5. U.S. Environmental Protection Agency, Quality Assurance Management Staff, Washington, DC.

USGS, 2019. Geological Survey Office of Water Quality. USGS Water-Quality Information: Water Hardness and Alkalinity. usgs.gov.

Vijai, P. and Sivakumar, P.B., 2016. Design of IoT systems and analytics in the context of smart city initiatives in India. *Procedia Computer Science*, 92, pp. 583–588.

Vijayakumar, N. and Ramya, A.R., 2015. The Real Time Monitoring of Water Quality in IoT Environment. In 2015 International Conference on Innovations in Information, Embedded and Communication Systems (ICIIECS). IEEE, India, pp. 1–5.

Wilson, N., 2020. Soil Water and Ground Water Sampling. CRC Press.

World Bank Group, 1999. Water Quality Models. Pollution Prevention and Abatement Handbook. Washington, DC.

World Health Organization (WHO), 1984. Guideline for Drinking Water Quality, USA, pp. 1–60.

World Health Organization (WHO), 2004. Guideline for Drinking Water Quality, 3rd edition, Volume 1, Summary, Geneva.

Zakaria, M.N.A., Malek, M.A., Zolkepli, M. and Ahmed, A.N., 2021. Application of artificial intelligence algorithms for hourly river level forecast: A case study of Muda River, Malaysia. *Alexandria Engineering Journal*, 60(4), pp. 4015–4028.

Zorigto, N., Anna, M., Anastasia, K., Vasily, I., Anastasia, R. and Evgenii, I., 2020. Algal bloom occurrence and effects in Russia. *Water*, 12(1), p. 285, doi: 10.3390/w12010285.

7 Monitoring of Contaminants in Aquatic Ecosystems Using Big Data

*Miraji Hossein, Asha Ripanda, Mureithi Eunice,
Othman Chande, Faustin Ngassapa, and
Alaa El Din Mahmoud*

CONTENTS

DOI: 10.1201/9781003260455-7

7.1   INTRODUCTION

7.1.1   MEANING OF DATA

Data are statistical collections of facts and information about an object, person, or something under study. The terms *data* and *information* are used synonymously, despite their significant differences. While data contains raw facts without clear meaning, the information contains processed, organized, and structured data that portrays a meaning. Data (singular: *datum*) do not focus on specificity; rather, it is a pile of collected facts that contain coded information (Erickson and Rothberg 2015; Taylor 2022). Raw data are unprocessed data before cleaning to remove unnecessary facts. Facts collected in situ under uncontrolled conditions are field data, while facts collected in a controlled environment are experimental data. Primary data involves the firsthand collection of facts in the field/experiments, while secondary data relies on the facts collected by other sources (Taylor 2022).

7.1.2   TYPES OF DATA AND DATA PROCESSING

There are four principal types of data: nominal (categorical), ordinal, discrete, and continuous data, as indicated in Figure 7.1. Nominal (categorical) data are the facts used to label variables without giving quantitative values. They are name-like facts that enable identification without measurement or evaluation. It contains unique groups without common elements—for example, names of people, hair color, ethnicity, and sex. Ordinal data is nominal data that contain scale or set orders. The scale can range from 1 to 10 or based on researcher preferences (Sharma 2020).

For example, the Likert scale intends to understand satisfactions, and the interval scale shows each response in its interval (child, teenagers, youth, etc.). Open data is a form of freely accessible data that can be reused and redistributed without limits—for example, government, global, and health data (Murray-Rust 2008). On the contrary, attribute data contain facts necessary to establish identification. For example, area codes, clothing sizes, and pass or fail assessment. Quantitative data is a form of facts containing or describing the number of facts obtained through measurements, such as mass and length. Discrete, a subset of quantitative data, contains collected facts that can be counted and have a set of limit values. For example, number of

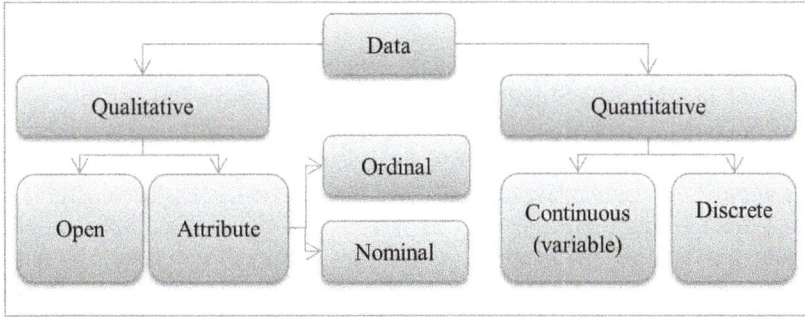

FIGURE 7.1 Schematic types of data.

TABLE 7.1

Different Types of Data Processing

Type of Data Processing	Descriptions
Commercial data processing	• In making the use of *large input data* to generate large output data.
	• Most works are automated to minimize errors.
Batch processing	• Several similar cases/facts are processed simultaneously.
	• Used for *large quantities of data*.
Scientific data processing	• It uses computational operations to manage low-volume input data to get low-volume output.
	• It takes more time, with minimized errors that may affect the interpretation.
Real-time processing	• It involves the processing of data in a short period.
Online processing	• It involves automated continuous data collection and processing on a real-time basis.

students in a class, shoe sizes, and questions you answered correctly (Jozova and Nagy 2021). Continuous data are numerical data types that refer to the unspecified number of possible measurements between two realistic points. Continuous data give no limit of measurements, such as concentration, turbidity, flow rate, temperature, and volume. Traditionally, any form of data validation, sorting, classification, calculation, interpretation, organization, and transformation (Zheng et al. 2018).

Data processing involves the manipulation of collected facts to give meaningful information (Runkler 2012). There are different methods for data processing, which is guided by intended information from the data, as tabulated in Table 7.1.

For insight to the types of data and data processing approaches, there are two classical forms of data: normal or small-volume data processed through scientific data processing. Real-time processing and online processing can process both small and large input data. Commercial and batch data processing mainly deals with a big volume of data, known as "*big data*" (BD) (Kambatla et al. 2014; Bai et al. 2021).

7.1.3 Qualifications and Sources of Big Data

Big data are a huge volume and very highly complex set of collected facts at a very high speed through which traditional data processing techniques cannot process them. It contains structured, semistructured, and unstructured facts (Davenport et al. 2012). Structured data are well-arranged data containing names, identification codes, phone numbers, education levels, emails, and physical addresses. Their entry is well organized, such that the reader may extract needed information with less effort. Unstructured data are qualitative data that cannot be analyzed using conventional tools. These data cannot be grouped into categories because they lack a predefined data model—for example, social media charts, video clips, text, audio files, satellite and surveillance imagery. Semistructured data are intermediate data forms, which means there are no clear organization models, such as zipped files, web pages, and emails. Big data are characterized by the 6Vs (volume, variety, velocity, veracity, variability, and value), originating from social media, machine data, and transactional data, as indicated in Figure 7.2. Examples of big data include customers' data, business processes, industrial analytics, students' information, satellite images, and meteorological station data (Santos et al. 2018; Tang and Liao 2021).

These data are controllable if they originate from internal sources, such as students' academic performance and financial transactions. They may be uncontrollable if they originate from outside an organization (From the Editors 2014). Big data are characterized by receiving large volumes of data, implying they exist in very large amounts of 2.5 quintillion bytes (Sowmya and Suneetha 2014). The speed at which they are received is high, up to zettabytes within a few seconds (Rathore et al. 2016). Generally, the varieties in which they are received may be structured, semistructured, or unstructured data. The degree to which big data can be trusted refers to veracity. Big data can be used and formatted in different ways, known as variability. Finally, well-processed data are important in their corresponding organization, which implies its value.

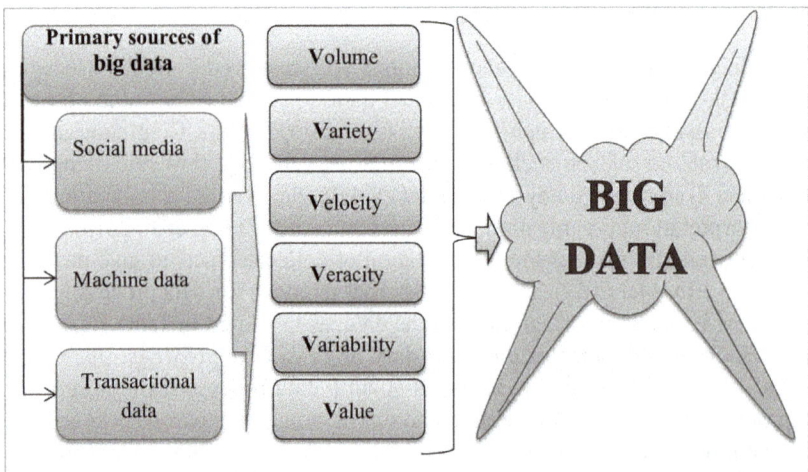

FIGURE 7.2   Characteristics of big data.

Media such as YouTube, Google, Twitter, Instagram, and Facebook are potential sources of a wide range of data, including customers' feedback, individual needs, and most followed update in the media. Data-on-the-cloud is another potential source of big data. Public and private servers and networks make easy access to huge data on the cloud, thus serving storage challenges. Websites and internet services such as Wikipedia and local and international websites give a range of simple to complex data whenever needed. The IoT, or the Internet of Things, involving collective networks of devices and technologies, simplifies communication among devices, clouds, and users (Stone et al. 2014; Jiang and Fu 2018). A database such as for medical needs, tourism, online marketing, and weather gives a potential big data source.

7.1.4 THE FUTURE OF BIG DATA

The majority of global nations, companies, and institutions experience an exponential growth of big data. Due to their complexity, seeking a data analytics expert will be necessary to harness the hidden information for the adoptive environment. These experts will require advanced machinery for more than 175 zettabytes by 2025 of big data, particularly cloud storage. Data storage will shift from customer-based storage to public cloud storage (Lee 2021). This kind of storage will require high security against hackers to protect personal and organizational data. Security issues principally on the sensitive organization privacy will introduce a hybrid environment for indoor and cloud storage (Bazzaz Abkenar et al. 2021; Maheshwari et al. 2021). Again, the cloud storage must have a multisecurity environment that prohibits unauthorized personnel access to some cloud data.

The advancement of big data–processing machinery is expected to work beyond real-time recording of events to detect and predict frauds, traffic, climatic changes, storms, and possible environmental pollution (Bamiah et al. 2018). Transformation of data management may trend, as shown in Figure 7.3.

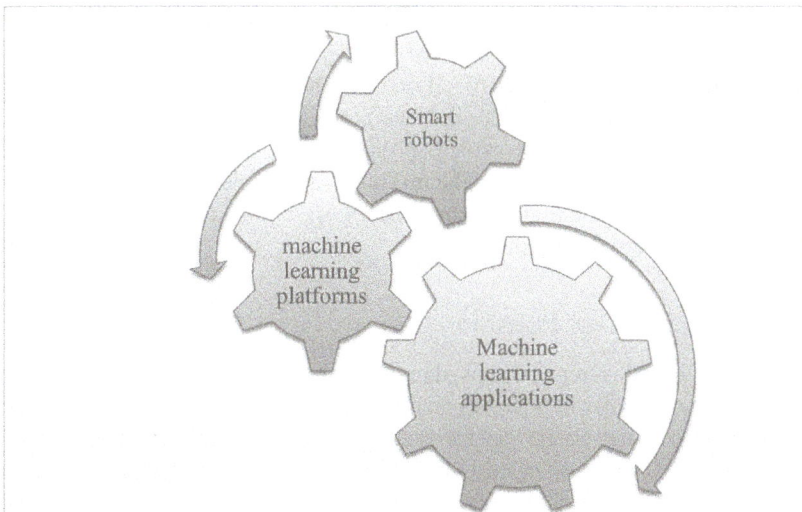

FIGURE 7.3 Trends of big data management in the future.

Unforeseen machinery advancements for big data processing will increase access to open data sources to bring services closer to customers. At the same time, demand for big data analysts with skills in management and designing tools for handling and controlling big data will increase. These analysts must handle privacy and security matters; thus, the need for dynamic security against cyber-attacks and data breaching will be necessary (El Haourani et al. 2020; Tawalbeh and Saldamli 2021). Finally, a transformation from big data to fast data will be necessary so that customers and users can customize big data into personalized mode, store them on the cloud, and access them instantly. Therefore, speed of data access, flexibility of personalized data, customer experience, security, and trustworthiness of data are among factors that will determine the future of big data. Nonetheless, all these factors must predict customer demand, fluctuations, climatic change, and indications of environmental safety, particularly pollution and storm threats.

7.2   INTEGRATED AQUATIC POLLUTION CYCLES

The aquatic ecosystem comprises the inner and surrounding environment of water bodies. The definition excludes isolated water sources, such as water wells, water percolation, and sources whose water is in limited quantities. Traditionally, water bodies can have typically stagnant, saturated waters, known as wetlands; slow-moving water, such as lakes and pools, known as lentic; or fast-moving waters, such as rivers and streams, known as lotic. The aquatic ecosystem is classified into marine and freshwater ecosystems (Weithoff and Beisner 2019). The marine ecosystem covers a high-salt-containing saline oceanic water (Miraji 2018). The latter covers freshwater ecosystems, such as rivers, lakes, streams, ponds, rivers, springs, and wetlands (He et al. 2019). There are exceptions, such as Lake Manyara in Tanzania and Lake Magadi in Kenya, whose water contains high amounts of soda originating from volcanic effects, thus not typically a freshwater ecosystem. Instead, by considering the type of dissolved minerals, it is therefore suggested to be classified as a *soda-ash ecosystem*. There are natural soda-ash ecosystems, such as the lakes Manyara, Natron, and Magadi in East African Rift Valley channels (Scoon 2018; Janssens de Bisthoven et al. 2020), and artificial ones, such as the anthropogenic islands in Poland and, likely, Onondaga Lake in New York City (Effler and Matthews 2003). Interactive aquatic ecosystems explain possible interactions or mixing of waters from the three named ecosystems. The mixing is inevitable since water exists in all spheres of the Earth, including the atmosphere (air), lithosphere (land), hydrosphere (water), and biosphere (living things).

7.2.1   SURFACE AQUATIC POLLUTION CYCLE

Surface waters include rivers, streams, dams, lakes, oceans, and partially, swamps and wetlands. In these bodies of water, contaminants can be dissolved to form a solution freely suspended in the water. Surface water is either stagnant or flowing from high lands to lower lands with suspended and dissolved contaminants (Moiseenko et al. 2019, Mahmoud et al. 2021). The lithosphere, in this case, remains as a platform on which these processes occur. Figure 7.4 indicates the existing surface contaminants cycle.

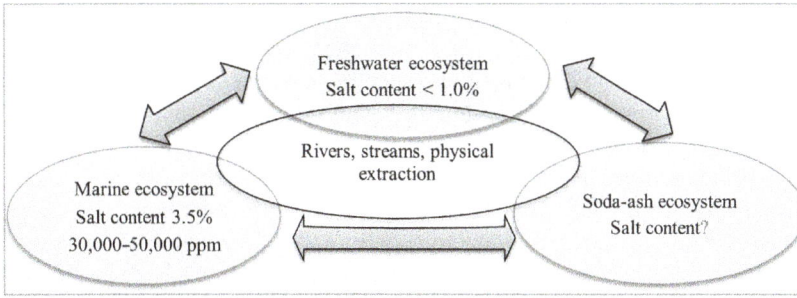

FIGURE 7.4   Interactive mixing of the three aquatic ecosystems.

The cycle starts with introducing pollution-causing agents at a terrestrial or directly to a water body. Terrestrial pollutants are physically dropped or washed in water bodies like rivers and streams. As rivers and streams contain moving water, they transport pollutants to destined points, being dams, lakes, ocean/sea, or swamps (Chaudhary and Walker 2019). In these receivers, pollutants may accumulate from small to huge concentrations, resulting in the soda-ash ecosystem (Wagaw et al. 2019; Lwanyaga et al. 2020), which may terminate the cycle. Some of the worlds closed lakes include Manyara in Tanzania, Lake Chad Basin (Cameroon, Chad, Niger, Nigeria), Lake Rukwa (Tanzania), Lake Vanda (Antarctica), Lake Torrens (South Australia), Torey Lakes (Russia), Sambhar Lake (India), and Zuni Salt Lake (New Mexico). Alternatively, receivers such as Lake Victoria in Tanzania and the Great Lakes in the USA, open lakes, further transport these contaminants via rivers to other receivers, thus making continuous marine or freshwater ecosystem cycles (McSweeney et al. 2017; Polónia and Cleary 2019). Dilution theory is less applicable when marine water and freshwater meet, due to floating less-dense freshwater, slowing the mixing process (Ben-Zur et al. 2021). Nevertheless, applying the concept of partition theory, dissolved or freely suspended pollutants may distribute themselves among the two zones. The distribution may establish a local pollutant transport equilibrium between saline water and freshwater, representing a new local pollutants cycle.

7.2.2   Subsurface Aquatic Pollution Cycle

The subsurface cycle involves the movements of water-containing pollutants, the bulky movements of pollutants in the form of a solution, or the leaching of pollutants in the lower zones of the soil (Ryder and Vink 2007; Chaukura et al. 2022). Natural sources, such as rainwater, or anthropogenic sources, such as industrial wastewater, get in the subsoil through infiltration or penetration to unsaturated soil zones (Zeinali et al. 2020). These processes that allow water from the soil surface to the subsurface carry pollutants that affect groundwater quality (Høisæter et al. 2019). However, they are the potential for recharging underground water. In this case, water aquifers become pools for the accumulation of contaminants. The confined aquifer is the least-affected zone, while the unconfined aquifer is highly affected by this cycle (Burbey 2001). In the presence of earth fissures, this cycle may be partially completed by the horizontal and vertical movement of water with pollutants to lower ground levels or meeting root zones, absorbed by plants followed by transpiration.

Once the unconfined aquifer is saturated with infiltrated water, subsurface water may automatically return on the surface through percolation and spring or manually through boring to form a water well (Goldman and Pabari 2021). In most cases, the return process moves with the least amount of pollutants since soils and sands are natural water filters. Nonetheless, percolation of water, either carrying contaminants or not, becomes a secondary source of surface water flowing back as streams or rivers to its destination. Groundwater again gets to terrestrial and aquatic animals, thus involving the biosphere cycle. The overall subsurface cycle mainly relies on the solution-state movements of contaminants.

7.2.3 Adsorptive-Based Aquatic Pollution Cycle

The superficial concept of the adsorptive-base concept relies on the interaction between pollutants with soil, sediments, and solid organic matters. Physical or chemical interaction allows contaminants to physically or chemically attach to the surface of solid matters (Mahmoud et al. 2018a), moving with the bulk movement of water or remaining stagnant in the water's surface for the organic ones at the bottom of the water body for the case of sediments. This cycle is complicated, as the solid carrier matters may physically move to another point. For example, moving sludge from the wastewater treatment center to the farm or any terrestrial point transfers contaminants (Anne and Paulauskiene 2021, Schnaak et al. 1997). Usually, moving water carries suspended matter; concurrently, sediments move slowly through rolling beneath water surface (Eggleton and Thomas 2004). Rolling movements make sediments finer, increasing the surface area for more attachment of pollutants. In turn, adsorbed contaminants are transported from one ecosystem to another (Liu et al. 2019). Except for sands from rivers and streams, it is not common to withdraw sediment from lakes and oceans, thus preventing direct movements of adsorbed contaminants among the ecosystems. Ruther, the desorption process occurs in the least-contaminated water, whereby sediments releases pollutants in the oceans or lakes, thus making a continuous aquatic pollution cycle, and the food chain makes pollutants move within the ecosystem compartment (Badr et al. 2020).

7.3 AQUATIC AND MARINE POLLUTION

Aquatic pollution refers to the presence of pollutants in surface waters. In contrast, marine pollution addresses pollutants in the oceanic environment. Aquatic and marine pollution is a combination of chemicals and trash, most of which come from land sources and are washed or blown into the surface waters or ocean (Hernando et al. 2006; Shokoohi et al. 2020; Zhi et al. 2020). Eutrophication results from the presence of excessive nutrients, such as nitrogen and phosphorous, in aquatic environments (Karydis 2009). Pollution depletes aquatic habitats and causes an uncontrolled spread of phytoplankton in lakes, leading to the destruction of biodiversity (Kumari et al. 2008).

On the other hand, industrial effluent reused in crop production should be treated prior to application in crop field. Hence, safe reuse of wastewater for cultivation of food material is necessary to fulfill the demands of a growing population across the globe in the changing scenario of climate (Dotaniya et al. 2022; Soni et al. 2020; Patel et al. 2021). Contamination of aquatic and marine areas may lead to a lack of potable water and therefore increased diseases in the ecosystem (Gwenzi et al. 2022; Hernando et al. 2006). Therefore, ecosystem sustainability may be impaired by the presence of these pollutants (Mahmoud et al. 2018b). The use of big data to monitor these areas may lead to understanding parameters' potential for ecosystem sustainability and, therefore, monitoring and managing aquatic and marine ecosystems for sustainability.

7.3.1 Sources of Aquatic Pollution

All aquatic and marine ecosystems are potentially vulnerable to pollution due to the interconnection of aquatic ecosystems via very effective marine and atmospheric transport channels (Islam and Tanaka 2004). Currently, anthropogenic activities include heavy metal mining and processing, fossil fuel burning for transportation, heating, electricity generation, pesticide application, radioactive wastes, food processing, plastic manufacture and disposal, and several other industrial processes that contaminate the aquatic environment (Berninger et al. 2011; Golovko et al. 2014; Shokoohi et al. 2020; Zhi et al. 2020; Sawicka et al. 2021; Singh and Yadav 2021; Mahmoud et al. 2022). Radioactive substances may cause pollution, but mostly in their natural environment, the effect is minimal. Their effect is magnified by anthropogenic activities, such as mining. Figure 7.5 represents sources of aquatic pollution, mainly anthropogenic or amplified by anthropogenic activities.

A recent study by assessed the biological impact of anthropogenic discharge in a river. It utilized gene expression as a biomarker for the combined effects of

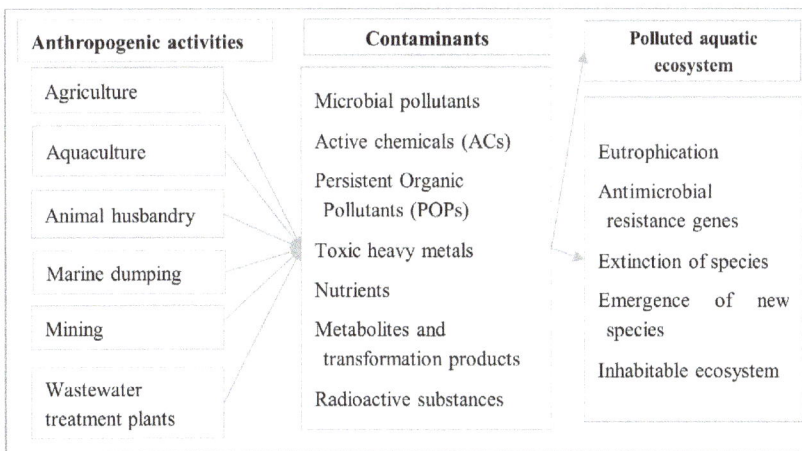

FIGURE 7.5 Sources of aquatic/marine pollution, mainly anthropogenic or amplified by activities.

potentially harmful substances in the aquatic environment (Valenzuela-Nieto et al. 2021). In this case, adult male rainbow trout (*Oncorhynchus mykiss*) were employed as the monitoring organisms. Results indicated that RT-qPCR analyses of biomarker gene expression, vitellogenin (vg), metallothionein (mt), and cytochrome 1A (cyp1A), in liver samples with primers amplifying specific sequences previously verified by cloning and sequencing were used to assess in vivo effects (Valenzuela-Nieto et al. 2021). In two different study areas, such as in Chile and Germany, the modulation of expression of marker genes involved in metal stress, reproduction, and detoxifying systems in the liver of male rainbow trout revealed organismal response to anthropogenic contamination, indicating a potential risk to aquatic ecosystems (Valenzuela-Nieto et al. 2021).

7.3.2 TYPES OF AQUATIC POLLUTION

7.3.2.1 Micro- and Macroplastics

On top of the 150 million tons currently present in our aquatic and marine environments, an estimated 8 million tons of plastic garbage enter them yearly (Okunola et al. 2019). While bigger bits of plastic can harm coral reefs or entangle fish and mammals, they inevitably degrade into much smaller fragments over time (Álvarez-Ruiz et al. 2021; Fendall and Sewell 2009). Microplastics are possibly even more hazardous, as they are more likely to be mistaken for food by species of all sizes (Álvarez-Ruiz et al. 2021). They can injure an animal's internal organs and impair its immune system after consumption, not to mention filling its stomach with plastic debris that has no nutritional value. The presence of microplastics in oysters (wet weight) is in the range of 0.14–7.90 n/g, seawaters in the range of 10.00–27.50 n/L, and sediments in the range of 0.053–0.26 n/g (Wang and Zhang 2021). Microplastics are categorized as significant ingredients based on their fiber and fragment shape, black color, size range of 101–500 m, and polyethene composition (Wang and Zhang 2021). Investigation of microplastic densities in beach sediment has resulted in determining the range between 688.9 and 3308 particles·m^{-2}, whereas in the water column, it varies between 257.9 and 1215 particles·m^{-3} (Nel et al. 2017). With a few exceptions, there is no significant spatial patterns in either the sediment or water column microplastic densities, with little differences in density between bays and the open coast (P N 0.05) (Nel et al. 2017). Another study investigated 218 microplastic samples, with 1,516 from wash water, 447 broken tubes, and 155 from digested tissue. Polyethene, polypropylene, and polyethylene terephthalates were chemically identified as three forms of microplastics, using Raman spectroscopy (Prata et al. 2021).

7.3.2.2 Chemical Pollutants

The use of pharmaceutically active chemicals to mitigate diseases to the whole ecosystem is important. Pesticides (Miraji et al. 2021), fertilizers, and herbicides are used to increase yields and defend against pest attacks. Unfortunately, heavy rain can wash these chemicals into gutters, streams, and rivers, eventually ending in marine ecosystems (Makaye et al. 2022; Makokola et al. 2020) and the food chain.

7.3.2.3 Organic Pollutants

An increase in organic pollutants in living tissue indicates a risk to aquatic and marine ecosystems. This can further affect humans through the food chain, requiring intervention (Mahmoud 2020). Organic compounds such as α-HCH, p,p′-DDE, and polychlorinated biphenyls are the most predominant compounds, detected at ranges of 0.54–4.90 ng/l water, 0.75–2.41 ng/g d. wt. sediment, and 2.19–28.11 ng/g fresh wt. biota. β and γ-HCHs, endosulfan compounds, heptachlor, and heptachlor epoxide have low detection frequencies (Abbassy 2018). The organochlorine pollutants are at high levels and abundances in *Donax* spp. than in *Tilapia* spp., followed by sediment and water (Abbassy 2018).

About 181 of the 484 tested EOPs, and 44 other analytes, were found at all sample locations (Peng et al. 2018), the majority being industrial compounds, with the greatest amounts of 1H-benzotriazole and organophosphate flame retardants (PFRs) (Peng et al. 2018). Maximum amounts of atrazine and isoproturon were found to be higher than Europe's annual average environmental quality criteria for 21 chemicals, predominantly herbicides (Peng et al. 2018). In the YRD area, the most common pharmaceuticals and personal care products (PPCPs) were amantadine and DEET (Peng et al. 2018). Chemical profiles from industrial locations were more complex than those from urban areas (primarily in the Qinhuai River). Industrial operations most likely influence the composition of chemical combinations in surface water from the YRD area. 4-methylbenzylidene camphor and the ISO E Super identified are bioaccumulating (Peng et al. 2018).

7.3.2.4 Inorganic Pollutants

Inorganic chemicals, including of natural or anthropogenic origin, represent a wide range of pollutants that may originate from terrestrial activities such as agriculture, mining, and transportation (Mahmoud 2022; Mahmoud et al. 2021). The atmosphere may be a secondary source through deposition of suspended matters such as dust containing inorganic chemicals. Marine activities such as volcanic eruption, offshore oil drilling, and construction are direct sources of inorganic pollutants in water bodies. Transboundary sources of inorganic pollution include combustion, whose deposit may occur far from the origin, in rivers and oceanic tides. Inorganic contaminants may be washed by rainwater or municipal wastewater to the aquatic ecosystem, thus polluting aquatic and marine ecosystems. Through continuous deposition of inorganic pollutants in marine sediments, the concentrations of different metals build up beyond quality guidelines. The occurrence of Cu at a range between 2.8 and 17 μg/L and Zn at a concentration of 151 μg/L in the water exceeds the chronic and acute effect levels (Groffen et al. 2021). The presence of metals such as Cd, Cu, Pb, Cr, and Zn in mollusks at higher concentrations has been observed, which suggests a threatened ecosystem. Occurrence of these inorganic elements in aquatic organisms implies that, apart from deposition in the sediments, inorganic chemicals can enter the food chain. Heavy metals have shown various trophic transmission patterns with increasing trophic levels magnification. The presence of these metals in the marine ecosystem at higher concentrations affects the aquatic ecosystem's health, and therefore, terrestrial organisms are directly or indirectly affected too.

7.3.2.5 Microbiological Pollutants

Microbiological pollution is a type of water contamination that occurs naturally. Bacteria, protozoa, and viruses can contaminate water supplies, resulting in diseases like bilharzia and cholera. Humans are particularly vulnerable to this type of contamination in areas where proper water treatment facilities are lacking. Oil pollution has increased the relative abundance of Actinobacteria and the putative TM7 phylum (*Saccaribacteria*), decreasing the relative number of Bacteroidetes (Galitskaya et al. 2021). Regardless of the extent of petroleum pollution or soil type, the gene abundance (number of OTUs) of oil-degrading bacteria (*Rhodococcus* sp., candidate class TM7–3 representative) became dominant in all soil samples (Galitskaya et al. 2021). Marked phytoplankton biomass decrease (40–76%) and production rates (22–96%) were observed (Ochieng Okuku et al. 2019). The abundance of heterotrophic bacterial production increased (4–68% and 17–165%) (Ochieng Okuku et al. 2019). Alongside, amplicon sequencing of the 16S rRNA gene revealed that oil-degrading bacteria became abundant 48–96 h post-oil addition, while the cosmopolitan *Synechococcus* and SAR11 lineages were significantly reduced (by 78–98% and 59–98%) (Ochieng Okuku et al. 2019).

The maximum counts (MPN per 100 mL) were observed on the site close to the city center throughout the sampling period, where values ranged from 1,700 to 4,240,000 total coliforms (TC), 200 to 92000 fecal coliform (FC), and 11 to 4,900 enterococci (ENT) (Lyimo 2009). Other sites showed less-predictable results, with a range of values from 0 to 920, 0 to 540, and 0 to 46 for TC, FC, and ENT, respectively. Furthermore, the fecal indicator bacteria concentration varied significantly with sampling time ($P < 0.05$) and between sampling points ($P < 0.05$).

7.3.2.6 Nutrient Pollution

Nutrient pollution results in algal blooms that deplete oxygen, thus leading to oxygen scarcity. This leads to species that rely on oxygen to survive being eliminated and anaerobic microbes surviving, and some can produce hazardous poisons, making the water even more dangerous to the whole ecosystem. Evaluation of fecal indicator bacteria values indicates consistent and significant correlation with each other, reflecting the presumed human fecal pollution (Lyimo 2009). The maximum counts (MPN per 100 mL) were observed on the site close to the city center throughout the sampling period, where values ranged from 1,700 to 4,240,000 total coliforms (TC), 200 to 92,000 fecal coliform (FC), and 11 to 4,900 enterococci (ENT) (Lyimo 2009). Other sites showed less-predictable results, with a range of values from 0 to 920, 0 to 540 and 0 to 46 for TC, FC, and ENT, respectively. Furthermore, the fecal indicator bacteria concentration varied significantly with sampling time ($P < 0.05$) and between sampling points ($P < 0.05$). Similarly, nutrients were significantly higher ($P < 0.05$) at Ocean Road, where concentration (mmol L^{-1}) ranges were 0.2–54 (NO_3), 0.0–20 (NO_2), and 0.3–45 (PO_4). The levels of fecal indicator bacteria and nutrients were higher during the rainy seasons than in the dry seasons, showing the inclusion of rain runoff as a source of contamination. The fecal indicator bacteria correlated positively with nutrients in both one-year and daily datasets ($P<0.01$) (Lyimo 2009). Groundwater is an important source of drinking water in arid and semiarid areas. The limited availability of surface water in arid and semiarid areas led to significant dependence on groundwater as a source of drinking water. The quality of groundwater is threatened by increased pollution, including nutrient pollution.

7.4 APPLICATIONS OF BIG DATA IN AQUATIC POLLUTION

Aquatic pollution integrates several factors since this sphere is not a standalone sphere; rather, whether marine or freshwater ecosystem, they interact with the atmosphere, lithosphere, biosphere, and hydrosphere. This interaction makes the aquatic ecosystem a complex system that cannot be isolated and treated separately (Snaphaan and Hardyns 2021). Direct aquatic data from oceans, lakes, rivers, streams, estuaries, and wetlands may include the type of water body, volume of water, quality of water, sources of water, quality status, residing organisms, uses of water, recharging and exploitation rate, type of soils, economic activities, population around the sources, threat factors, management system, geographical values such as boundary, spatiotemporal variations, and sunken vessels (Brannon 2013). Satellite images are also an important big data source in managing environmental activities. The applications of selected big data about aquatic pollution are presented in Table 7.2.

Big data are only useful after the interpretation of the hidden information. They are interdependent, originating from vast sources. Agriculture, hunting, biodiversity, natality ratio, morbidity, and mortality are prospective sources of big data for prediction, signaling, and evaluation of prospects of aquatic pollution.

7.5 ESSENTIAL OF BIG DATA AND ALGORITHM

7.5.1 ESSENTIAL AQUATIC ECOSYSTEM BIG DATA

Aquatic species are sensitive to environmental changes and serve as good indicators of the health of the ecosystems. It is essential to collect information for building up the database during spring, summer, and fall seasons (Wang and Zhang 2021). Therefore, the aquatic organismal sampling is complemented by aquatic biogeochemistry, hydrology, morphology, and continuous sensor measurements for quality data that can aid in addressing the characteristics of the aquatic and marine ecosystems. In aquatic systems, periphyton, phytoplankton, and aquatic plants (vascular and nonvascular), bryophytes, lichens, and macroalgae, as well as bacteria, provide the foundation of the food web. Flooding, scouring, wave activity, water level, light attenuation, and nutrient availability are all elements that significantly impact the algal and aquatic plant populations (Murphy et al. 2021; Vadeboncoeur et al. 2021). The most important environmental elements affecting these ecosystems are light amount and quality, water depth, wave activity, current velocity, and flow regime. Invasive species, changes in nutrient loads, toxicity, and land cover in watersheds can be tracked by monitoring primary producer populations.

Plant communities serve as a haven for invertebrates and fish and a substrate for algae. Therefore, collecting information on the trends of these phenomena in preparation for a big data database will help give insights about aquatic and marine environments for assessment and monitoring to ensure a sustainable ecosystem.

TABLE 7.2

Examples of Big Data and Their Applications in Aquatic Pollution

Examples of Big Data	Applications in Aquatic Pollution
Insurance	• Insurance data from different water bodies can indicate the feasibility of human-related activities that are predictors of environmental pollution. Also, big data can be the source of understanding the behavior and lifestyle of people for insurers, which can enable them to predict their contribution, competitors, and associated risks to environmental pollution (Keller et al. 2018).
Education	• Education data, such as levels of education, syllabus taught, theoretical and practical education, public awareness, inclusive and exclusive education, participatory education, are major indicators of welfare of the society, awareness on the aquatic pollution issues, management of aquatic pollution, and community understanding of the effects and measures to be taken upon aquatic pollution (Thakuriah et al. 2017).
Government	• These may include population, budget, priorities, policies, and available preserved areas. • This information is crucial in understanding the government's efforts to manage and prevent aquatic pollution (Kitchin 2014).
Transportation	• Transportation data may include the number of vehicles using fuels, import and export of gasoline, marine transportation vessels, and port information. • These data tell the extent of air pollution, coastal pollution, wastewater managing facilities, car washes, and destruction of natural resources to facilitate transport, directly related to aquatic pollution (Ben Ayed et al. 2015).
Energy and utilities	• Hydroelectric power, wastewater treatment, water supply, and sanitation data are directly connected to aquatic pollution. • Data representing the well-being of the aquatic ecosystem can be used as an indicator of pollution (Marinakis et al. 2020).
Health-care providers	• Water quality plays a significant contribution to the well-being of a community. • Medical records such as diarrhea, pneumonia, and gastrointestinal cases are directly linked to polluted water bodies. • Weak community is less effective in managing the environment, and they are pioneers of environmental pollution (Fleming et al. 2022).
Banking, retail, and wholesale trade	• Well-established banking facilities with high security are related to the medium- to the high-income community, indicating a high population. • High population and income imply utilizing large quantities of water and releasing more wastewater to the aquatic ecosystem. • Banking data showing the extent of investment in aquatic-based ventures can predict the related extent of aquatic pollution (Fleming et al. 2022).
Manufacturing and natural resources	• The presence of manufacturing facilities requires raw materials for production. • This may involve harvesting natural resources, clearing land, and producing wastes channeled to the rivers, resulting in aquatic pollution (Song et al. 2017).
Communications and media	• These are channels for communication among each other. • The aquatic ecosystem's awareness, safety, and status are shared through various media, thus adding value to the aquatic ecosystem (Wang and Zhang 2021).

Aquatic macroinvertebrates and zooplankton are two types of organisms that are diverse and abundant. Aquatic macroinvertebrates and zooplankton can be easily tested, are found in all but the most polluted waters, and are heavily influenced by water quality (Galitskaya et al. 2021; Mishra et al. 2022; Murphy et al. 2021; Vadeboncoeur et al. 2021). Building a big data database of these organisms can aid in addressing biodiversity problems since they are found in practically all freshwater bodies, leading to conservation.

7.5.2   THE BIG DATA ART AND ALGORITHM

The drastic increase in classical to advanced sensors and internet access devices, such as smartphones, smartwatches, and cars, has tremendously increased the need for art for data computation. The art of using a single device in the management of data was a traditional way of data processing (Lederer and Mendelow 1986). The daily regenerated 2.5×10^{12} GB data require thousands of normal personal computers for their storage, complicating the management of all those computers at once (Singh et al. 2020). The role of a human being is not to extract valuable information from big data, as it may take centuries, but rather, machine learning algorithms with computational tools capable of diagnostic data and data mining to simplify understanding of information in the data.

The art of big data management begins with data engineering that involves designing and building infrastructures capable of handling, storing, processing, and sending feedback immediately (Sun and Scanlon 2019). Machine learning is a form of artificial intelligence that uses software to predict outcomes by previous computing data to predict new data. In this case, the computer has been programmed to use available data to predict the outcome (Zhou et al. 2017; Sun and Scanlon 2019). For example, image recognition relies on the intensity of the pixel in the data. This can enable the labelling of x-ray as cancerous or not, tagging names to a photographed face, and handwriting recognitions (Zerdoumi et al. 2018). It is useful in speech recognition, voice search, and voice dialing (Chen and Lin 2014). Medical diagnosis, predictive analytics, and data extraction from unstructured data are also achieved through machine learning (Siuly and Zhang 2016). From big data, data science can be applied to acquire information, then information is transformed to knowledge, and the knowledge will generate wisdom on decision-making. Three basic methodologies used in data science include classification, regression, and similarity matching. The whole transformation simplifies data from descriptive analytics to system automation level. Table 7.3 indicates some potential big data infrastructures for handling and processing them.

Various models are required in a business to manage customers' requirements and create customers' analytic platforms.

The selection of these models/algorithms depends on the problem at hand. Examples of selected algorithms are shown in Table 7.4. The algorithm covers the rules or procedures for solving a problem.

7.5.3   TRADITIONAL VERSUS MODERN DATA MANAGEMENT

Data management covers all processes involved, from data sources to decision-making. It covers the collection, storage, security, retrieval, and implementation costs. It is

TABLE 7.3

Selected Technologies to Work with Big Data

Big Data Infrastructure	Descriptions
Streams: Apache Kafka (Hiraman 2018)	• Kafka Streams is a client library build to store, process, analyze, and stream historical and real-time data.
	• It connects distributed components in the system and allows subscription and publication of data in real time.
	• Sources of databases, sensors, cloud services, mobile devices, and software applications.
	• It ensures a continuous flow and interpretation of data to enable access to the right information at the right time.
Indexing: Elasticsearch (Voit et al. 2017)	• Elasticsearch is a search engine–like interphase that allows users to store, search, and analyzes a large volume of data in real time within milliseconds.
	• Examples: Netflix and eBay.
	• The index is a database structure/layout that improves the performance of a database.
Visualization: Kibana, Grafana (Cruz et al. 2021)	• Kibana is an Elasticsearch platform that allows visualization of real-time graphs, charts, and maps.
	• Advanced form of Kibana is Knowi.
Cloudera, Hortonworks (Bhathal and Dhiman 2018)	• Cloudera is a hybrid shared-data cloud platform for any data, anywhere and anytime.
	• Hortonworks is an open-source data software for storage, distribution, and processing of large multiple-source data.

TABLE 7.4

Examples of Algorithms Used in the Management of Big Data

Types of Algorithms	Descriptions
Linear regression algorithms	This is a computational technique that shows the linear relationship between dependent and independent variables.
K-means clustering algorithm	This is an unsupervised learning algorithm that groups unlabeled "k" datasets into different clusters.
Logistic regression algorithms	This is a supervised learning algorithm used to predict the probability of a target variable. For example, spam/not spam email, fraud/not fraud transaction.
C4.5	This a data mining tool using existing sample data to predict outputs.
Naive Bayesian	This is an algorithm with a working assumption of independence among predictors.
Association rule mining algorithm	This is an algorithm with a working assumption of the probability of co-occurrence of items in the collection.
Apriori	This is a kind of algorithm using frequent item-sets to generate association rules.
Support vector machine (SVM)	This is a kind of algorithm used in both regression and classification challenges.
AdaBoost	This is a short form for Adaptive Boosting; it is a boosting technique used as an ensemble method in machine learning.

Source: Vedder and Naudts (2017); Yang et al. (2020); Awan et al. (2021); Lee (2021).

extended to creating, accessing, and updating data, in-house and cloud storage, algorithms involved, existing backup systems, architecture, security, and policy. Data management ensures access to prepared data with acceptable quality, enables visualization of integrated data, data governance, and streaming. Information technology is one of the most important components in interpreting data/facts to a clear message for taking action (Bakshi 2012; Chen et al. 2013; Silva et al. 2016). There is no clear boundary between traditional and big data since the volume alone can't qualify the difference. The only reference for big data is its exponential growth, massive flow, and the existence of heterogeneity in structured and unstructured data. Apart from volume, velocity, and variety (Ishwarappa and Anuradha 2015), the main differences between traditional and big data are presented in Table 7.5 (Wang and Wu 2009; Furht and Villanustre 2016; Ali et al. 2019).

The existing differences outweigh traditional data due to the ability of big data to provide detailed insight into data, fast processing, and high efficiency. However, traditional data are easily stored without involving a third party that may risk security; data can be processed using conventional software with normal configuration and can be quickly interpreted even by a nonexpert person. Institutions are not forced to move to the big data era instantly. However, strategies to be dynamic with the transformation from traditional to big data become a promising approach. Big data and

TABLE 7.5

Comparison between Traditional and Big Data

Differentiation Category	Traditional Data	Modern/Big Data
Storage	Fixed-on-premises systems	Dynamic cloud-based systems
Open-data access	Structured database	Variety of data format
Logical construct	Placed in physical servers	Cloud warehouse
Functionality	Reactive analytics	Predictive analytics
Metadata-driven	Requires IT team to manually process data, update, change codes, and waste time	Integrates and analyzes data in real time and quickly exploits them
Resources utilization	Requires time to move data, storage, updating, etc.	Optimized resource consumption through automation
Highly secure	Difficult to limit access to data levels, thus the security threat	Only required and authorized data can be accessed
Data sources	Within the enterprise level	At the enterprise level and outside
Volume	From GB to 103 GB (TB)	From petabytes to zettabytes
Frequency of data production	Per hour or per day	Continuously, per second
Management	Centrally managed	Distributed management
Data integration	Simple approaches	Complex approaches
System requirements	Normal database tools	Special kind of database tools
Data model	Static	Dynamic
Stability	Stable with relationship among data	Unstable and unknown relationship
Time delay	Requires time to process after the occurrence of the events	Real-time analytics

Source: Disk Storage: 1 ZB = 10^{12} GB = 10^9 TB.

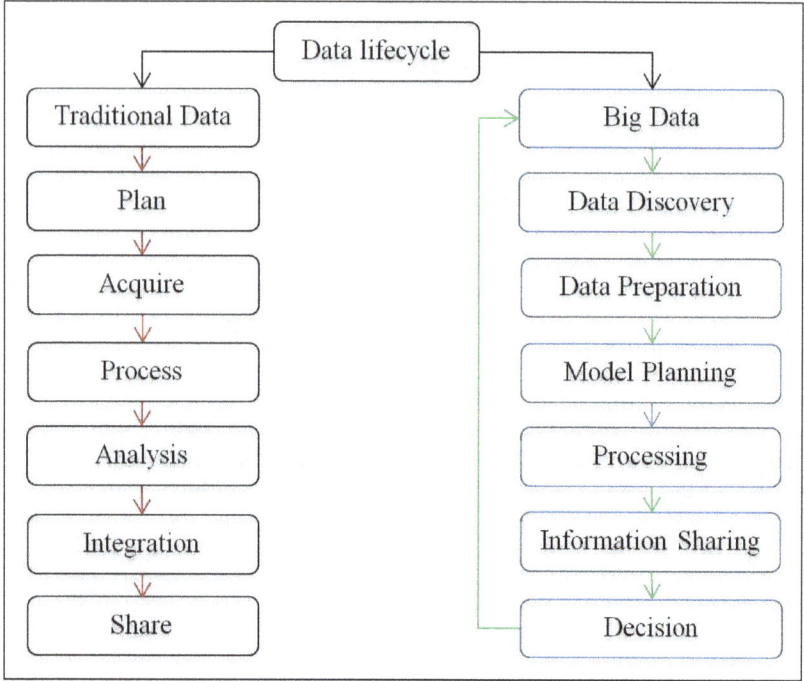

FIGURE 7.6　Comparative life cycle of traditional and big data.

the traditional analytics life cycle are presented in Figure 7.6, where traditional data are terminated in the storage level, while big data are dynamic.

7.5.4　ROLES OF BIG DATA ON MONITORING AQUATIC POLLUTION

Some areas, especially semiarid regions, endure the water crisis for longer. Water scarcity impacts more than 4% of the global population, and more than two billion people lack access to clean drinking water (Turley et al. 2021; Ndiritu 2021). On the other hand, the COVID-19 pandemic increased the need for safer water, magnifying the problem. Increased anthropogenic activities and urbanization lead to increased release of pollutants to the aquatic ecosystem, decreasing safer water that requires intervention. In this case, the management of aquatic pollution using big data is of great importance (Murphy et al. 2021; Vadeboncoeur et al. 2021). The significance of big data can be achieved by taking data from any source and analyzing it properly, and this can give the possibility to manage the aquatic ecosystem, improve operational efficiencies, generate revenue, provide growth, give the possibility to evidence-based decision-making and sustainability opportunities as a source of fresh water.

Further, the utility of big data can be enhanced by combining it with high-performance analytics. This will determine the root cause of aquatic pollution in near real time. On the other hand, it can aid to sport variations of aquatic ecosystem faster and more accurately than by human eye. That can further aid in rapidly converting environmental data into insights and calculating the entire risk in minutes. The creation and sharpening of deep learning models can help accurately classify and react to changing variables and aid in detecting fraudulent activities before it affects the aquatic ecosystem for sustainability.

7.5.5　Big Data in Attaining Sustainable Development Goals

World leaders set global Sustainable Development Goals (SDG) in 2015 to transform the world into a better and fair world by 2030 (UN 2015). They focused on ending poverty and inequality and addressing climatic changes as a global agenda. Among the 17 SDGs, big data plays a significant role in evaluating achievements of the goals and prediction of the future. Big data on education and sanitation address access to education and aquatic ecosystem pollution, thus serving as an indicator for SDG1 (no poverty). Sustainable food production is controlled by climatic changes, irrigation practices, availability of agro-vet services, and affordability of agricultural services, import, export, and national food storage reserve. All these data are an essential indicator for SDG2 (zero hunger). The availability of big data on health services, system, and life expectancy gives clues on the status of a nation's health and well-being, representing SDG3 (good health and well-being). Data on the number of schools, universities, teachers, and student enrollment are key indicators for SDG4 (quality education). Unbiased service data on sex equality and equality in representation are the key indicators for SDG5 (gender equality). Access to big data on clean water and sanitation serves in monitoring of SDG6 as it is not a standalone goal since SDG3 is directly connected.

The availability of smart metering across countries controls the water supply, gas, and electricity. Thus, their big data enables affordable and clean energy availability monitoring, SDG7. The continuous flow of employment information, per capita income, rights of works, and sustainable income is necessary big data for assessing and evaluating SDG8 (decent work and economic growth). The governmental, private and investors empowerment policy, clear regulations, and implemented environmental impact assessment are tools that supply big data to estimate achievements of SDG9 (industry, innovation, and infrastructure). Successful and industrial revolution data are useful in predicting aquatic ecosystem pollution. The economic growth patterns existing among countries and between rural and urban can provide big data upon ensuring the feasibility of SDG10 (reducing inequalities). Implementation and evaluation of SDG require safety and sustainable cities. The absence of these will hinder the implementation of SDG11 (sustainable cities and communities).

The attainment of SDG12 (responsible consumption and production) is the responsibility of everyone with sustainable governmental policy. Big data on the consumption and production of different resources is essential for managing future generations' resources. A supportive environment for organisms to survive requires informed industrial emissions and investment in sustainable energy sources. The SDG13 (climate action) focuses on climatic issues, whereby historical and current big data are useful for the prediction of future patterns as well as the ability to diagnose any possible aquatic pollution threats. SGD14 (life below water) promotes conservation and regulation of marine practices and policies for the sustainability of aquatic lives. All information in SDG14 is a potential source of big data for monitoring, evaluating, and managing pollution in the aquatic ecosystem. Apart from marine life, SDG15 (life on the land) ensures sustainability and enables the environment for terrestrial survival and the breeding of organisms. Data related to deforestation, agriculture, hunting, and urbanization supplement big data necessary to predict aquatic ecosystem pollution. A suitable environment for SDGs to be successful requires big data on SDG16 (peace, justice, and strong institutions), which provides inclusive societies, equal access to justice, and strong institutions. To achieve all SDGs, working together among nations is inevitable. SDG17 (partnerships for the goals) emphasizes working together, through which big data obtained will enable assessment of all SDGs, wherein aquatic pollution will be directly or indirectly addressed.

7.6 CHALLENGES OF AQUATIC BIG DATA

The phrase "big data" provokes a range of emotions in aquatic ecologists, ranging from excitement about the prospects for ecological understanding over a wide range of spatiotemporal scales to dread at the enormous chore of data management. Big data has aided findings from "macrosystems ecology" at regional scales, from climate research to molecular biology (Damerow and Ely 2021). Despite these advantages, hurdles towards big data utilization frequently prevent it from being used to its full potential, requiring intervention. Among the challenges includes how to recognize if you have aquatic ecosystem big data, handling aquatic ecosystem big data, issues with aquatic ecosystem classical analytical techniques, verification of aquatic ecosystem big data, considerations for aquatic ecosystem data sharing, and community development of knowledge infrastructures based on aquatic ecosystem big data, but these challenges are not exclusive to the aquatic ecosystem (Dafforn et al. 2016). Figure 7.7 illustrates the challenges that could face aquatic big data.

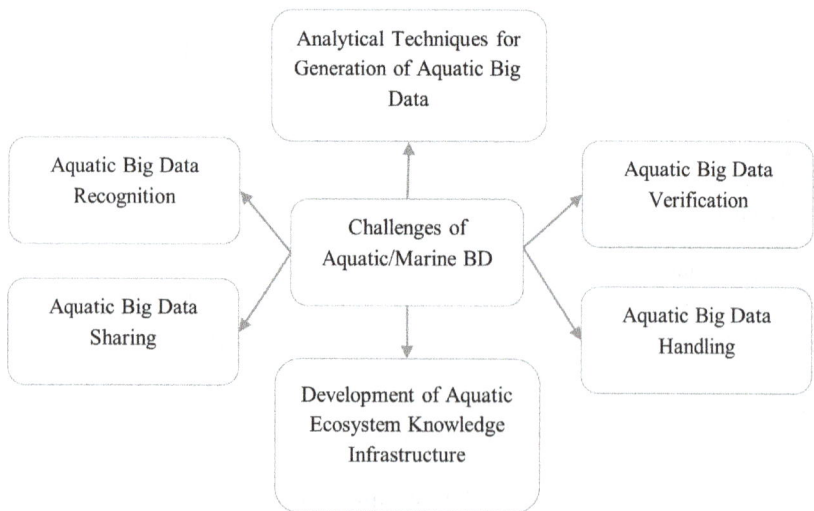

FIGURE 7.7 Illustration of the challenges facing aquatic big data.

7.6.1 Inclusivity and Exclusivity of Aquatic Big Data

Based on the interconnection of earth compartments, there are challenges associated with qualifying typical data that could represent the aquatic ecosystem alone. In most cases, atmospheric, lithospheric, and biospheric pollution implies their presence in the aquatic ecosystem compartments. Likewise, surface- and groundwater contamination on and in the land, respectively, might end up in one of the earth compartments connected to the aquatic ecosystem. In this case, there's no standalone parameter that can only represent one ecosystem. Thus, there is no clear boundary for inclusivity and exclusivity of big aquatic data. However, apart from direct marine and big freshwater data, other big data are a potential predictive indicator for the well-being of the aquatic ecosystem.

7.6.2 Aquatic Ecosystem Big Data Recognition

How do you tell whether you have big aquatic data? It's a subjective concept. For instance, hundreds of thousands of seabed photographs constitute big data. Aquatic ecologists must evaluate their data in the context of other studies in the field (Damerow and Ely 2021). Large collaborations that share instruments and infrastructure across institutions generate more big data than small researchers using local resources (Borgman et al. 2015). Due to ecosystem dynamics, aquatic ecosystems have historically been defined as "little science," in which "artisanal" data is collected using locally made tools and specialized procedures, with limited reproducibility (Bowen and Roth 2007). However, technological advancements favor automated data collecting over artisanal data collection, resulting in increased data volume and velocity.

Recognition systems utilize unique features and patterns to differentiate things. For instance, the behavior of species interaction, breeding, and feeding may tell the quality of the aquatic ecosystem. It is necessary to have appropriate technology capable of recording sounds, images, and any other changes for a long period. Analyzing these recorded data can give information, such as feeding patterns, telling what happened. This information can be used to predict prevailing status and enable decision-making for the stainable ecosystem.

7.6.3 Aquatic Big Data Handling

Scientists must first solve the challenges involved with handling, managing, and manipulating enormous aquatic big datasets before analyzing and understanding big data (Borgman et al. 2015). The need for the aquatic scientist to be computationally literate comes in. In this case, enhanced access to computing platforms, software development tools, and database/archival management is necessary.

7.6.4 Aquatic Big Data Verification

Checking if the information/data recorded is the one required to be recorded is important to ensure the quality of data generated (Borgman et al. 2015). This process may be done at any step and is recurring to guarantee aquatic big data quality.

7.6.5   Analytical Techniques for the Generation of Aquatic Big Data

The big data era requires aquatic scientists to rely on advanced analytics to handle massive sample sizes, false associations among explanatory variables, zero-inflated datasets, nonnormality, and spatial-temporal autocorrelation (Dafforn et al. 2016). A few of the statistical issues that big data brings due to these statistical problems are that traditional basic statistics such as linear regression, analysis of variance, and t-tests are incapable of processing the aquatic ecosystem big data. Therefore, there is a need to focus on a new era of data analysis that can fit big data. These include independence and autocorrelation, significance, normality, and Bayesian approaches (Durden et al. 2017).

7.6.6   Development of Aquatic Ecosystem Knowledge Infrastructure

Knowledge infrastructure, a platform for interresearcher interaction, makes data administration and exchange easier. The most commonly discussed knowledge infrastructure is data repositories (Borgman et al. 2015; Cheng et al. 2021; Damerow and Ely 2021). They serve as information hubs, decreasing the expense and effort of data management for individual researchers and ensuring equal and merit-based data access (Borgman et al. 2015). Rather than being domain-based, they could be institutional or national.

7.6.7   Aquatic Big Data Sharing

The culture of aquatic data sharing is crucial to its success. There is a lack of a culture of public sharing, and a cultural shift towards aquatic data sharing is required before it becomes commonplace, apart from the fact that BD sharing enables cross-domain data fusion and analysis in different applications, including transportation and communication (Borgman et al. 2015). Three present cultural barriers have been identified: lack of incentives and rewards for sharing, a higher risk of sharing for specific groups of academics, such as early career researchers, and no ethical impulse for sharing in the culture.

7.7   CONCLUSION

The aquatic ecosystem represents a sphere among the Earth's spheres, which are vulnerable and always get pollution due to accumulation of all terrestrial and atmospheric washouts. Its complexity arises due to its receiving nature, continuous siltation and depositions, excessive evaporations and damping, and in most cases, its existence in lower lands. These factors make the aquatic ecosystem typically polluted with both intentional and accidental sources. The acquisition of all data related to aquatic pollution starts with understanding the status of atmospheric pollution, followed by data for terrestrial pollution, agricultural, industrialization, and ending up in collective information about water pollution. Other supportive data from metrology, climatic changes, and transboundary effects lead to acquisition of huge amount of data classified as big data. Management of these data requires specialized individuals and software that its interpretation gives necessary multidimensional information for the understanding of the historical, the current, and the ability to predict the future of aquatic pollution. Effective evaluation of big data enables prediction of possible monitoring, management, and a prospective future of aquatic ecosystem. Big data presents a prospective predictive future of the aquatic ecosystem provided that they will be handled properly and purposefully.

REFERENCES

Abbassy MMS 2018 Distribution pattern of persistent organic pollutants in aquatic ecosystem at the Rosetta Nile branch estuary into the Mediterranean Sea, North of Delta, Egypt. *Mar. Pollut. Bull.* **131**:115–121. doi: 10.1016/j.marpolbul.2018.03.049.

Alabi, OA, Ologbonjaye, KI, Awosolu, O, & Alalade, OE 2019. Public and environmental health effects of plastic wastes disposal: A Review. *J Toxicol Risk Assess*, 5(021), 1–13.

Ali W, Shafique MU, Majeed MA, Raza A 2019 Comparison between SQL and NoSQL Databases and Their Relationship with Big Data Analytics. *Asian J. Res. Comput. Sci.* **4**:1–10. doi: 10.9734/ajrcos/2019/v4i230108.

Álvarez-Ruiz R, Picó Y, Campo J 2021 Bioaccumulation of emerging contaminants in mussel (Mytilus galloprovincialis): Influence of microplastics. *Sci. Total Environ.* **796**:10. doi: 10.1016/j.scitotenv.2021.149006.

Anne O, Paulauskiene T 2021 The assessment of the sewage and sludge contamination by phthalate acid esters (Paes) in eastern europe countries. *Sustainability* **13**:1–14. doi: 10.3390/su13020529.

Awan MJ, Rahim MSM, Nobanee H, et al 2021 A big data approach to black Friday sales. *Intell. Autom. Soft Comput.* **27**:785–797. doi: 10.32604/iasc.2021.014216.

Badr NBE, Al-Qahtani KM, Mahmoud AED 2020. Factorial experimental design for optimizing selenium sorption on Cyperus laevigatus biomass and green-synthesized nano-silver. *Alex. Eng. J.* **59**:5219–5229.

Bai Xi, Zhang F, Li J, Guo T, Aziz A, Jin A, Xi F 2021 Educational big data: Predictions, applications and challenges. *Big Data Res.* **26**:100270. doi: 10.1016/j.bdr.2021.100270.

Bakshi K 2012 Considerations for big data: Architecture and approach. *IEEE Aerosp. Conf. Proc.* 1–7. doi: 10.1109/AERO.2012.6187357.

Bamiah MA, Brohi SN, Rad BB 2018 Big data technology in education: Advantages, implementations, and challenges. *J. Eng. Sci. Technol.* **13**:229–241.

Bazzaz Abkenar S, Haghi Kashani M, Mahdipour E, Jameii SM 2021 Big data analytics meets social media: A systematic review of techniques, open issues, and future directions. *Telemat. Inform.* **57**:101517. doi: 10.1016/j.tele.2020.101517.

Ben-Zur E, Gvirtzman H, Shalev E 2021 Haline convection within a fresh-saline water interface in a stratified coastal aquifer induced by tide. *Water* **13**:1–12. doi: 10.3390/w13131780.

Ben Ayed A, Ben Halima M, Alimi AM 2015 Big Data Analytics for Logistics and Transportation. In: 2015 4th IEEE International Conference on Advanced Logistics and Transport. IEEE ICALT 2015, pp. 311–316. doi: 10.1109/ICAdLT.2015.7136630

Berninger JP, Du B, Connors KA, Eytcheson, SA, Kolkmeier, MA, Prosser, KN, Valenti, TW, Chambliss, CK, Brooks, BW 2011 Effects of the antihistamine diphenhydramine on selected aquatic organisms. *Environ. Toxicol. Chem.* **30**: 2065–2072.

Bhathal SG, Dhiman AS 2018 Big Data Solution: Improvised Distributions Framework of Hadoop. In: 2018 Second International Conference on Intelligent Computing and Control Systems (ICICCS), Madurai, India, pp. 35–38. IEEE. doi: 10.1109/ICCONS.2018.8663142.

Borgman CL, Darch PT, Sands AE, Pasquetto IV, Golshan MS, Wallis JC, Traweek S 2015 Knowledge infrastructures in science: Data, diversity, and digital libraries. *Int. J. Digit. Libr.* **16**: 207–227.

Bowen GM, Roth WM 2007 The practice of field ecology: Insights for science education. *Res. Sci. Educ.* **37**:171–187. doi: 10.1007/s11165-006-9021-x.

Brannon MM 2013 Standardized spaces: Satellite imagery in the age of big data. *Configurations* **21**:271–299. doi: 10.1353/con.2013.0021.

Burbey TJ 2001 Stress-Strain analyses for aquifer-system characterization. *Ground Water* **39**:128–136.

Chaudhary M, Walker TR 2019 River Ganga pollution: Causes and failed management plans (correspondence on Dwivedi et al. 2018. Ganga water pollution: A potentiala health threat to inhabitants of Ganga basin. Environ. Int. 117, 327–338). *Environ. Int.* **126**: 202–206. doi: 10.1016/j.envint.2019.02.033.

Chaukura N, Muzawazi ES, Katengeza G, Mahmoud, AED 2022 Remediation Technologies for Contaminated Soil Systems. In: WillisGwenzi (ed). Emerging Contaminants in the Terrestrial-Aquatic-Atmosphere Continuum, Amsterdam, Netherlands, pp. 353–365. Elsevier. doi: 10.1016/B978-0-323-90051-5.00019-5.

Chen J, Chen Y, Du X, et al 2013 Big data challenge: A data management perspective. *Front. Comput. Sci.* **7**:157–164. doi: 10.1007/s11704-013-3903-7.

Chen XW, Lin X 2014 Big data deep learning: Challenges and perspectives. *IEEE Access* **2**:514–525. doi: 10.1109/ACCESS.2014.2325029.

Cheng Y, Zhan H, Yang W, Jiang Q, Wang Y, Guo F 2021 An ecohydrological perspective of reconstructed vegetation in the semi-arid region in drought seasons. Agricultural Water Management, 243, 106488.

Cruz R, Guimarães T, Peixoto H, Santos MF 2021 Architecture for intensive care data processing and visualization in real-time. *Procedia Comput. Sci.* **184**:923–928. doi: 10.1016/j.procs.2021.03.115.

Dafforn KA, Johnston EL, Ferguson A, Humphrey CL, Monk W, Nichols SJ, Simpson SL, Tulbure MG, Baird DJ 2016 Big data opportunities and challenges for assessing multiple stressors across scales in aquatic ecosystems. *Mar. Freshw. Res.* **67**:393–413.

Damerow JE, Varadharajan C, Boye K, Brodie EL, Burrus M, Chadwick KD, . . . & Agarwal D 2021. Sample identifiers and metadata to support data management and reuse in multi-disciplinary ecosystem sciences. *Data Science Journal*, **20(1)**, 11.

Davenport TH, Barth P, Bean R 2012 How "big data" is different. *MIT Sloan Manag. Rev.* **54**: 1–5.

Dotaniya ML, Meena VD, Saha JK, Dotaniya CK, Mahmoud AED, Meena BL, Meena MD, Sanwal RC, Meena RS, Doutaniya RK, Solanki P, Lata M, Rai PK 2022. Reuse of poor-quality water for sustainable crop production in the changing scenario of climate. Environ. *Dev. Sustain*: 1–32. doi: 10.1007/s10668-022-02365-9.

Durden JM, Luo JY, Alexander H, et al 2017 Integrating "Big Data" into aquatic ecology: Challenges and Opportunities. *Limnol. Oceanogr. Bull.* **26**:101–108. doi: 10.1002/lob.10213.

Effler SW, Matthews DA 2003 Impacts of a soda ash facility on onondaga lake and the Seneca River, NY. *Lake Reserv. Manag.* **19**:285–306. doi: 10.1080/07438140309353940.

Eggleton J, Thomas KV 2004 A review of factors affecting the release and bioavailability of contaminants during sediment disturbance events. *Environ. Int.* **30**:973–980. doi: 10.1016/j.envint.2004.03.001.

El Haourani L, El Kalam AA, Ouahman AA 2020 Big Data Security and Privacy Techniques. In: AMC International Conference Proceeding Series, pp. 425–439.

Erickson GS, Rothberg HN 2015 Data, Information, and Knowledge. Cambridge International Examinations.

Fendall LS, Sewell MA 2009 Contributing to marine pollution by washing your face: Microplastics in facial cleansers. *Mar. Pollut. Bull.* **58**:1225–1228. doi: 10.1016/j.marpolbul.2009.04.025.

Fleming L, Medical E, Centre E 2022 Big Data in Environment and Human Health. In: Oxford Encyclopedia of Environment and Human. Oxford Press, pp. 1–23

From the Editors 2014 Big Data and management. *Stud. English Lang. Educ.* 1: i–iii

Furht B, Villanustre F 2016 Big data Technologies and Applications. Springer International Publishing, Switzerland.

Galitskaya P, Biktasheva L, Blagodatsky S, Selivanovskaya S 2021 Response of bacterial and fungal communities to high petroleum pollution in different soils. *Sci. Rep.* **11**:1–18. doi: 10.1038/s41598-020-80631-4

Golovko O, Kumar V, Fedorova G, Randak T, Grabic R 2014 Seasonal changes in antibiotics, antidepressants/psychiatric drugs, antihistamines and lipid regulators in a wastewater treatment plant. *Chemosphere* **111**: 418–426.

Groffen T, Rijnders J, Doorn L Van, et al 2021 Preliminary study on the distribution of metals and persistent organic pollutants (POPs), including perfluoroalkylated acids (PFAS), in the aquatic environment near Morogoro, Tanzania, and the potential health risks for humans. *Environ. Res.* **192**: 110299. doi: 10.1016/j.envres.2020.110299.

Gwenzi W, Selvasembian R, Offiong N-AO, Mahmoud AED, Sanganyado E, Mal J 2022 COVID-19 drugs in aquatic systems: A review. *Environ. Chem. Lett.* 20, 1275–1294.

He F, Zarfl C, Bremerich V, et al 2019 The global decline of freshwater megafauna. *Glob. Chang. Biol.* **25**:3883–3892. doi: 10.1111/gcb.14753.

Hernando MD, Mezcua M, Fernández-Alba AR, Barceló D, 2006 Environmental risk assessment of pharmaceutical residues in wastewater effluents, surface waters and sediments. *Talanta* **69**: 334–342.

Hiraman BR 2018, August A study of apache kafka in big data stream processing. In: 2018 International Conference on Information, Communication, Engineering and Technology (ICICET). IEEE, pp. 1–3. doi: 10.1109/ICICET.2018.8533771

Høisæter Å, Pfaff A, Breedveld GD 2019 Leaching and transport of PFAS from aqueous film-forming foam (AFFF) in the unsaturated soil at a firefighting training facility under cold climatic conditions. *J. Contam. Hydrol.* **222**:112–122. doi: 10.1016/j.jconhyd.2019.02.010.

Ishwarappa, Anuradha J 2015 A brief introduction on big data 5Vs characteristics and hadoop technology. *Procedia Comput. Sci.* **48**:319–324. doi: 10.1016/j.procs.2015.04.188

Islam MS, Tanaka M 2004. Impacts of pollution on coastal and marine ecosystems including coastal and marine fisheries and approach for management: A review and synthesis. *Mar. Pollut. Bull.* **48**: 624–649.

Janssens de Bisthoven L, Vanhove MPM, Rochette AJ, et al 2020 Social-ecological assessment of Lake Manyara basin, Tanzania: A mixed method approach. *J. Environ. Manage.* **267**: doi: 10.1016/j.jenvman.2020.110594.

Jiang M, Fu KW 2018 Chinese social media and big data: Big data, big brother, big profit? *Policy Internet* **10**:372–392. doi: 10.1002/poi3.187.

Jozova S, Nagy I 2021 Use of linear regression to discrete data. *2021 Smart City Symp. Prague, SCSP 2021* 1–6. doi: 10.1109/SCSP52043.2021.9447393.

Kambatla K, Kollias G, Kumar V, Grama A 2014 Trends in big data analytics. *J. Parallel Distrib. Comput.* **74**:2561–2573. doi: 10.1016/j.jpdc.2014.01.003.

Karydis, M., & University of the Aegean, Lesvos I. (Greece). Department of Environmental Studies; University of the Aegean, Lesvos I. (Greece). Department of Environmental Studies. (2009, September). Eutrophication assessment of coastal waters based on indicators: a literature review. In Proceedings of the International Conference on Environmental Science and Technology (Vol. 1). University of the Aegean, Chania (Greece).

Keller B, Eling M, Schmeiser H, et al 2018 Big Data and Insurance: Implications for Innovation, Competition and Privacy. The Geneva Association-International Association for the Study of Insurance Economics.

Kitchin R 2014 The real-time city? Big data and smart urbanism. *Geo J* **79**:1–14. doi: 10.1007/s10708-013-9516-8.

Kumari P, Dhadse S, Chaudhari PR, Wate SR 2008 A Biomonitoring of Plankton to Assess Quality of Water in the Lakes of Nagpur City. In: Sengupta M and Dalwani R, Sengupta M and Dalwani R (Eds.). The 12th World Lake Conference (Taal2007). ILEC, Rajasthan, India, pp. 160–164.

Lederer AL, Mendelow AL 1986 Issues in information systems planning. *Inf. Manag.* **10**: 245–254. doi: 10.1016/0378-7206(86)90027-3.

Lee S 2021 Reducing complexity of server configuration through public cloud storage. *Electronics* **10**:. doi: 10.3390/electronics10111277.

Liu WS, Guo MN, Liu C, et al 2019 Water, sediment and agricultural soil contamination from an ion-adsorption rare earth mining area. *Chemosphere* **216**:75–83. doi: 10.1016/j.chemosphere.2018.10.109.

Lwanyaga JD, Kasedde H, Kirabira JB, Shemi A, & Ndlovu, S. 2020. Beneficiation of salts crystallized from Lake Katwe brine. *Mine Water Solut.* 43–49.

Lyimo, TJ 2009 Microbial and nutrient pollution in the coastal bathing waters of Dar es Salaam. *Aqua. Conserv.* **19**(S1): S27–S37.

Maheshwari S, Gautam P, Jaggi CK 2021. Role of Big Data Analytics in supply chain management: current trends and future perspectives. *Int. J. Prod. Res.* **59**:1875–1900. doi: 10.1080/00207543.2020.1793011.

Mahmoud, AED, 2020. Graphene-based nanomaterials for the removal of organic pollutants: Insights into linear versus nonlinear mathematical models. J Environ Manag 270, 110911.

Mahmoud, AED 2022 Recent Advances of TiO$_2$ Nanocomposites for Photocatalytic Degradation of Water Contaminants and Rechargeable Sodium Ion Batteries. In: Shalan AE, Hamdy Makhlouf AS, Lanceros-Méndez S (Eds.). Advances in Nanocomposite Materials for Environmental and Energy Harvesting Applications, Engineering Materials. Springer, Cham. doi: 10.1007/978-3-030-94319-6_24

Mahmoud AED, Fawzy M, Khairy H, Sorour A 2022 Environmental Bioremediation as an Eco-sustainable Approach for Pesticides: A Case Study of MENA Region. In: Siddiqui, S, Meghvansi, MK, Chaudhary, KK (eds). Pesticides Bioremediation. Springer, pp. 479–494.

Mahmoud, AED, Stolle A, Stelter M 2018b. Sustainable synthesis of high-surface-area Graphite oxide via dry ball milling. *ACS Sustain. Chem. Eng.* 6, 6358–6369.

Mahmoud AED, Stolle A, Stelter M, Braeutigam P 2018a Adsorption Technique for Organic Pollutants Using Different Carbon Materials. In: Abstracts of Papers of the American Chemical Society (Vol. 255). Amer Chemical Soc.

Mahmoud AED, Umachandran K, Sawicka B, Mtewa TK 2021 Water resources security and management for sustainable communities. In: Andrew G. Mtewa and Chukwuebuka Egbuna (eds). Phytochemistry, the Military and Health. Elsevier, pp. 509–522. https://doi.org/10.1016/B978-0-12-821556-2.00011-6

Makaye A, Ripanda AS, Miraji H, 2022. Transport behavior and risk evaluation of pharmaceutical contaminants from Swaswa Wastewater Stabilization Ponds. *J. Biodivers. Environ. Sci.* **20**: 30–41.

Makokola SK, Ripanda AS, Miraji H 2020 Quantitative investigation of potential contaminants of emerging concern in the water: A focus at swaswa wastewater stabilization ponds in Dodoma City. *Egypt. J. Chem.* **63**: 427–436.

Marinakis V, Doukas H, Tsapelas J, et al 2020 From big data to smart energy services: An application for intelligent energy management. *Futur. Gener. Comput. Syst.* **110**:572–586. doi: 10.1016/j.future.2018.04.062.

McSweeney SL, Kennedy DM, Rutherfurd ID, Stout JC 2017 Intermittently Closed/Open Lakes and Lagoons: Their global distribution and boundary conditions. *Geomorphology* **292**:142–152. doi: 10.1016/j.geomorph.2017.04.022.

Miraji H 2018 Brination of coastal aquifers : Prospective impacts and future fit-for-use Remedial Strategies in Tanzania. *World Wide J. Multidiscipl Ina. Res. Dev*. **4**:202–206.

Miraji H, Ripanda A, Moto E 2021 A review on the occurrences of persistent organic pollutants in corals, sediments, fish and waters of the Western Indian Ocean. *Egypt. J. Aquat. Res*. **47**: 373–379.

Mishra B, Tiwari A, Mahmoud, AED 2022 Microalgal potential for sustainable aquaculture applications: bioremediation, biocontrol, aquafeed. *Clean Technol. Environ. Policy*: (2022): 1–13.

Moiseenko TI, Dinu MI, Gashkina NA, Kremleva TA 2019 Aquatic environment and anthropogenic factor effects on distribution of trace elements in surface waters of European Russia and Western Siberia. *Environ. Res. Lett*. **14**:. doi: 10.1088/1748-9326/ab17ea.

Murphy GEP, Dunic JC, Adamczyk EM, et al 2021 From coast to coast to coast: Ecology and management of seagrass ecosystems across Canada. *Facets* **6**:139–179. doi: 10.1139/FACETS-2020-0020.

Murray-Rust P 2008 Open data in science. *Ser. Rev*. **34**:52–64. doi: 10.1080/00987913.2008.10765152.

Ndiritu SW 2021 Drought responses and adaptation strategies to climate change by pastoralists in the semi-arid area, Laikipia County, Kenya. *Mitig. Adapt. Strateg. Glob. Chang*. **26**. doi: 10.1007/s11027-021-09949-2.

Nel HA, Hean JW, Noundou XS, Froneman PW 2017. Do microplastic loads reflect the population demographics along the southern African coastline? *Mar. Pollut. Bull*. **115**: 115–119.

Ochieng Okuku E, Linet Imbayi K, Gilbert Omondi O, et al 2019 Decadal Pollution Assessment and Monitoring along the Kenya Coast. In: Monitoring of Marine Pollution. IntechOpen.

Patel A, Soni S, Mittal J, et al 2021 Sequestration of crystal violet from aqueous solution using ash of black turmeric rhizome. *Desalin. Water Treat*. **220**:342–352. doi: 10.5004/dwt.2021.26911.

Peng Y, Fang W, Krauss M, Brack W, Wang Z, Feilong Li XZ 2018 Screening hundreds of emerging organic pollutants (EOPs) in surface water from the Yangtze River Delta (YRD): Occurrence, distribution, ecological risk. *Environ. Pollut*. **241**: 484–493.

Polónia ARM, Cleary DFR 2019 Archaeal communities in sponge, sediment and water from marine lakes and open water habitats. *Mar. Biol. Res*. **15**:259–274. doi: 10.1080/17451000.2019.1633469.

Prata JC, da Costa JP, Fernandes AJS, et al 2021 Selection of microplastics by Nile Red staining increases environmental sample throughput by micro-Raman spectroscopy. *Sci. Total Environ*. **783**:146979. doi: 10.1016/j.scitotenv.2021.146979.

Rathore MM, Ahmad A, Paul A 2016 Real time intrusion detection system for ultra-high-speed big data environments. *J. Supercomput*. **72**:3489–3510. doi: 10.1007/s11227-015-1615-5.

Runkler TA 2012 Data Analytics: Model and Algorithms for Intelligent Data Analysis, 2012th Edition. Springer Vieweg.

Ryder D, Vink S 2007 Managing Regulated Flows and Contaminant Cycles in Flood Plain Rivers. Salt, Nutrient, Sediment and Interactions: Findings from the National River Contaminants Program, Australian Government, Volume 109.

Santos AFC, Teles ÍP, Siqueira OMP, de Oliveira AA 2018 Big data: A systematic review. *Adv. Intell. Syst. Comput*. **558**:501–506. doi: 10.1007/978-3-319-54978-1_64.

Sawicka B, Umachandran K, Fawzy M, Mahmoud, AED 2021 Impacts of inorganic/organic pollutants on agroecosystems and eco-friendly solutions. In: Andrew G. Mtewa and Chukwuebuka Egbuna (eds). Phytochemistry, the Military and Health. Elsevier, pp. 523–552.

Schnaak W, Kuchler Th, Kujawa M, Henschel KP, Süßenbach, D, and Donau, R 1997 Organic contaminants in sewage slugde and their ecotoxicological significance in the agricultural utilization of sewage sludge. *Chemosphere* **35**:5–11.

Scoon RN 2018 Lakes of the Gregory Rift Valley: Baringo, Bogoria, Nakuru, Elmenteita, Magadi, Manyara and Eyasi. In: Rob Bowell (ed). Geology of National Parks of Central/Southern Kenya and Northern Tanzania. Springer, pp. 167-180. doi: 10.1007/978-3-319-73785-0_15.

Sharma R 2020 4 Types of Data: Nominal, Ordinal, Discrete, Continuous. In: Public Domain. www.upgrad.com/blog/types-of-data/. Accessed on October 13, 2022.

Shokoohi R, Ghobadi N, Godini K, Hadi M, Atashzaban Z 2020. Antibiotic detection in a hospital wastewater and comparison of their removal rate by activated sludge and earthworm-based vermifilteration: Environmental risk assessment. *Process Saf. Environ. Prot.* **134**: 169–177.

Silva YN, Almeida I, Queiroz M 2016 SQL: From Traditional Databases to Big Data. In: Carl Alphonce (ed). Proceedings of the 47th ACM Technical Symposium on Computing Science Education, pp. 413–418. doi: 10.1145/2839509.2844560.

Singh P, Yadav, D 2021 Link between air pollution and global climate change. In: Pardeep Singh S. Rangabhashiyam K. K. Srivastava (eds). Global Climate Change. Elsevier, pp. 79–108.

Singh V, Asari V K, Kumar S, Patel, RB 2020 Computational methods and data engineering. *Proc. ICMDE*: **1**:117–132.

Siuly S, Zhang Y 2016 Medical big data: Neurological diseases diagnosis through medical data analysis. *Data Sci. Eng.* **1**:54–64. doi: 10.1007/s41019-016-0011-3.

Snaphaan T, Hardyns W 2021 Environmental criminology in the big data era. *Eur. J. Criminol.* **18**:713–734. doi: 10.1177/1477370819877753.

Song M, Cen L, Zheng Z, et al 2017 How would big data support societal development and environmental sustainability? Insights and practices. *J. Clean. Prod.* **142**:489–500. doi: 10.1016/j.jclepro.2016.10.091.

Soni S, Bajpai PK, Mittal J, Arora C 2020 Utilisation of cobalt doped Iron based MOF for enhanced removal and recovery of methylene blue dye from waste water. *J. Mol. Liq.* **314**:113642. doi: 10.1016/j.molliq.2020.113642.

Sowmya R, Suneetha KR 2014 Data mining with big data. *Proc. 2017 11th Int. Conf. Intell. Syst. Control. ISCO 2017* **26**:246–250. doi: 10.1109/ISCO.2017.7855990.

Stone ML, Fielden L, Levy DAL, et al 2014 Big data for Media. *Reuters Inst. Study Journal.* **2014**: 1–35.

Sun AY, Scanlon BR 2019 How can Big Data and machine learning benefit environment and water management: A survey of methods, applications, and future directions. *Environ. Res. Lett.* **14**:. doi: 10.1088/1748-9326/ab1b7d.

Tang M, Liao H 2021 From conventional group decision making to large-scale group decision making: What are the challenges and how to meet them in big data era? A state-of-the-art survey. *Omega (United Kingdom)* **100**:102141. doi: 10.1016/j.omega.2019.102141.

Tawalbeh LA, Saldamli G 2021 Reconsidering big data security and privacy in cloud and mobile cloud systems. *J. King Saud Univ.—Comput. Inf. Sci.* **33**:810–819. doi: 10.1016/j.jksuci.2019.05.007.

Taylor D 2022 Difference between Information and Data, GURU99. Accessed on October 13, 2022. https://www.guru99.com/difference-information-data.html

Thakuriah PV, Tilahun NY, Zellner M 2017 Introduction to Seeing Cities through Big Data: Research, Methods and Applications in Urban Informatics. In: Seeing Cities Through Big Data. Springer, pp. 1–9. doi: 10.1007/978-3-319-40902-3_1.

Turley L, Bréthaut C, Pflieger G 2021 Institutions for reoperating reservoirs in semi-arid regions facing climate change and competing societal water demands: Insights from Colorado. *Water Int.* **47**(1): 1–26. doi: 10.1080/02508060.2021.1981636.

UN 2015 Sustainable development goals: Improving human and planetary wellbeing. *Glob. Chang.* **1**:20–23.

Vadeboncoeur Y, Moore MV, Stewart SD, et al 2021 Blue waters, green bottoms: Benthic filamentous algal blooms are an emerging threat to clear lakes worldwide. *Bioscience* **71**:1011–1027.

Valenzuela-Nieto GE, Leal C, Schwaiger J, Ferling H, Vargas-Chacoff L, Kausel G 2021 Aquatic pollution from anthropogenic discharges modulates gene expression in liver of rainbow trout (Oncorhynchus mykiss). *Austral J Vet Sci* **53**: 99–108.

Vedder A, Naudts L 2017 Accountability for the use of algorithms in a big data environment. *Int. Rev. Law, Comput. Technol.* **31**:206–224. doi: 10.1080/13600869.2017.1298547.

Voit A, Stankus A, Magomedov S, Ivanova I 2017 Big data processing for full-text search and visualization with elasticsearch. *Int. J. Adv. Comput. Sci. Appl.* **8**:. doi: 10.14569/ijacsa.2017.081211.

Wagaw S, Mengistou S, Getahun A 2019 Review of anthropogenic threats and biodiversity assessment of an Ethiopian soda lake, Lake Abijata. *African J. Aquat. Sci.* **44**:103–111. doi: 10.2989/16085914.2019.1596065.

Wang K, Zhang X 2021 The effect of media coverage on disciplining firms' pollution behaviors: Evidence from Chinese heavy polluting listed companies. *J. Clean. Prod.* **280**: 123035. doi: 10.1016/j.jclepro.2020.123035.

Wang YH, Wu IC 2009 Achieving high and consistent rendering performance of java AWT/Swing on multiple platforms. *Softw.—Pract. Exp.* **39**:701–736. doi: 10.1002/spe

Weithoff G, Beisner BE 2019 Measures and approaches in trait-based phytoplankton community ecology-From freshwater to marine ecosystems. *Front. Mar. Sci.* **6**:1–11. doi: 10.3389/fmars.2019.00040.

Yang M, Nazir S, Xu Q, et al 2020 Deep learning algorithms and multicriteria decision-making used in big data: A systematic literature review. *Complexity* **2020**:. doi: 10.1155/2020/2836064.

Zeinali M, Azari A, Heidari MM 2020 Simulating Unsaturated Zone of Soil for Estimating the Recharge Rate and Flow Exchange Between a River and an Aquifer. *Water Resour. Manag.* **34**:425–443. doi: 10.1007/s11269-019-02458-7.

Zerdoumi S, Sabri AQM, Kamsin A, et al 2018 Image pattern recognition in big data: Taxonomy and open challenges: survey. *Multimed. Tools Appl.* **77**:10091–10121. doi: 10.1007/s11042-017-5045-7.

Zheng X, Mukkamala RR, Vatrapu R, Ordieres-Mere J 2018 Blockchain-based Personal Health Data Sharing System Using Cloud Storage. In: 2018 IEEE 20th International Conference on E-health Networking, Applications and Services (Healthcom). IEEE, pp. 1–6. doi: 10.1109/HealthCom.2018.8531125.

Zhi S, Shen S, Zhou J, Ding G, Zhang K 2020 Systematic analysis of occurrence, density and ecological risks of 45 veterinary antibiotics: Focused on family livestock farms in Erhai Lake basin, Yunnan, China. *Environ. Pollut.* **267**: 115539.

Zhou L, Pan S, Wang J, Vasilakos AV. 2017 Machine learning on big data: Opportunities and challenges. *Neurocomputing* **237**:350–361. doi: 10.1016/j.neucom.2017.01.026.

8 Mitigation of Water Shortages with Artificial Intelligence

Gloria Ukalina Obuzor, Fredrick Ojija,
Benton Otieno, and Charles Ikenna Osu

CONTENTS

DOI: 10.1201/9781003260455-8

8.1 INTRODUCTION

All types of life on Earth require water to survive. It is necessary for the growth of several industries, including agriculture, education, industry, health care, and employment (UNICEF 2021). Despite this, it is in little supply and distributed unevenly over the world. Climate change, pollution, and drought are only a few of the variables impacting home and industrial water supplies (Kummu et al. 2016; Mitlin et al. 2019; Šulyová et al. 2021). The result of climate alteration is anticipated to result in a shortage of water in the nearest decades.

Humans' use of natural water resources, particularly freshwater resources, has risen throughout time (Hou et al. 2021). Due to continuous population growth and increased water use for agricultural, industrial, and recreational purposes, this picture is unlikely to change in the future years. Furthermore, there are increasing anxieties about the accessibility of moderate supplies of water to satisfy the forthcoming needs of the world. In so many parts of the world, surface-water resource mining is already at the highest level (Mitlin et al. 2019; UNICEF 2021; Agnes et al. 2021).

It is obvious to relate that water insufficiency isn't the only problem; excessive water consumption has lowered water quality. The volume and quality of water available for human consumption are reduced by herbicides, pesticides, and synthetic fertilizers. Contamination of lakes and rivers, for example, as a result of poor sewage disposal and the discharge of untreated industrial wastes reduces water quality and ultimately leads to water shortages. Water shortages have an impact on people's well-being because they make it difficult to obtain clean water for drinking, cooking, bathing, and growing food. As a result of an increasing number of people, specifically in developing countries, there is a deficiency of potable and healthy water. In sub-Saharan Africa, for example, about 42% of people do not have access to healthy water (Agnes et al. 2021). Increased access to healthy and enough potable water is the baseline indicator for reaching the sixth Sustainable Development Goal (SDG), which is based on "water and sanitation" (Agnes et al. 2021).

In the position of the World Health Organization, the water shortage has an impact on water use and sanitation all over the world (WHO 2019). In locations where there is a lack of water, human health is compromised. Due to water shortages, most infectious diseases, such as COVID-19, Ebola, and influenza, which may be managed through sanitary measures, like handwashing, are likely to spread more quickly. First, understand the movers of water shortages and their implications; it may require planning and identifying efficient ways and techniques to reduce water inadequacy. In achieving sustainable water use reducing shortages in the face of population growth and climate change, countries must develop innovative methods, such as desalination, rainwater harvesting, water loss reduction, smart water systems, including irrigation, dew harvesting, runoff capturing, and wastewater reuse (Judeh and Shahrour 2021).

One of the newest smart water resource management methods is artificial intelligence (AI)–based software. Hard-coded (rule-based algorithms/expert systems) algorithms were used in the early software decision support systems to generate outputs or conclusions by gradually reducing the options available. In the face of changing circumstances and opposing interests, these systems, while functional, quickly become unmanageable. This is changing, thanks to AI's new tools. Pattern recognition systems use a combination of weighting factors, inputs, transfer functions, and summations that are dynamically and autonomously changed when new data is provided. Artificial intelligence programming is similar to the human learning process in many aspects. During the "learning phase" of AI programming, input data is connected with known outputs, allowing the algorithms to learn over time. The algorithm enters the "operational phase" when new data is introduced, where it recognizes patterns (Goralski and Tay 2020).

In a variety of fields, AI has produced several useful and practical tools for dealing with complex real-world applications and solving tough challenges. AI technology has been adopted by scholars for its speediness, user-friendliness, and precision without requiring knowledge of physical difficulties (Rajaee et al. 2019). Artificial intelligence is utilized in medicine to prevent, diagnose, and treat ailments. It is used in the financial business to forecast money movement. It's also been used for "supply chain risk management" in determining menaces and instituting immediate strategy to safeguard against major supply chain disruptions and avoid momentous losses. Furthermore, AI is used to handle real-world difficulties, such as wastewater treatment, water quality improvement, river water quality modeling, and water resource recycling across a variety of engineering disciplines. Consequently, in the water sector and in an ever-changing environment, AI is seen as a valuable tool for water resource management (St-Onge et al. 2019; Ezhilarasu et al. 2019).

8.2 WATER SCARCITY AND SHORTAGE

The term "water scarcity" means nonexistence of safe drinking water or simply a lack of water. It occurs when water demand exceeds supply and present water resources have reached or are approaching sustainable limitations, according to UNICEF (2021). On the other side, *water scarcity* refers to a lack of water per person (Kummu et al. 2016). Due to the combination of climate alteration, basin-level water sources, and management systems' adaptive capacities, water scarcity is dynamic and intricate. It occurs when a population, area, country, or region does not have enough water to meet its demands (Kummu et al. 2016). A shortage of absorbed rain in soil (green water) as well as a lack of "liquid water" (blue water) cause limited water (Falkenmark 2013). Water scarcity has negative consequences for agricultural, industrial, commercial, tourism, biodiversity conservation, and residential water users, jeopardizing the global economy, food security, sustainable development, and human well-being (Salehi 2022). Water scarcity and shortages affect the amount of water accessed and the quality of water that users may access. Despite its scarcity, a large portion of it is mismanaged, polluted, and wasted (Adetoro et al. 2020; Agnes et al. 2021).

Water scarcity is a major problem in the world, with roughly four billion people facing acute freshwater shortages (Adetoro et al. 2020). Even though water scarcity is unevenly distributed over the world, many areas are experiencing severe water shortages. Due to effects of climate alteration, population increase, expanding demand, and bad management, scarcity has increased rapidly and is getting more severe (El-Fakharany and Salem 2021; Salehi 2022). Increased water scarcity is causing significant concern around the world. Water scarcity has increased by 58% since the 1900s, from 0.24 billion to 3.8 billion people in the new century (Kummu et al. 2016; Adetoro et al. 2020).

By 2050, at least a quarter of the world's population is predicted to live in areas where freshwater scarcity occurs on a regular or recurring basis (Hou et al. 2021). Currently, 1.42 billion people live in places with significant water scarcity, with 450 million of them being children (UNICEF 2021). Every year, at least 4 billion people, or approximately world population of 75%, face extreme water inadequacy (Mekonnen and Hoekstra 2016). In addition, a population of 3.2 billion have settled in agricultural municipalities with desperate water constraints (FAO 2020). Data on water risk received from World Resources Institute (WRI), United Nations Environment Programme (UNEP), and WHO/UNICEF Joint Monitoring Programme (JMP) were combined with population concentration maps from Gridded World Population version 4 (GPWv4) and WorldPop to reveal that 1.42 billion people (450 million children) live in areas with limited access to clean water (UNICEF 2021).

Access to healthy and potable drinking water is becoming more limited as the world's population increases and climate alteration wreaks havoc on ecology. Due to water shortages worldwide, about 2.2 billion people have no link to healthy drinking water, according to UNICEF and WHO (2019). In sub-Saharan Africa, around 400 million people do not have access to safe drinking water (Kummu et al. 2016; UNICEF 2021). Furthermore, nearly half of the world's population spends a minimum of one month yearly in potentially water-inadequate locations. This figure is predicted to increase to 5.7 billion by 2050, up from 4.8 billion now (Burek et al. 2016). Approximately 73% of individuals affected live in Asia (Burek et al. 2016). In general, water shortage puts the human population in danger in terms of crop output, food security, and a whole lot of economic endeavors. According to the SDGs, water shortages have a detrimental impact on food security, water quality, sanitation, and educational prospects for disadvantaged people (Hou et al. 2021). One of the critical human rights recognized by the UN and goal 6 of the SDGs is availability of affordable, safe, and clean drinking water. As a result, one of the UN 2030 agenda's fundamental goals (SDG6) for sustainable global development is to enhance universal provision to water (Salehi 2022; UN 2015).

Increased water consumption has resulted in water shortages in several countries. Water consumption has risen considerably in some African countries and in the Eastern and Southern Asia continent (Burek et al. 2016; Judeh and Shahrour 2021). According to polls, the United States, Russia, and Pakistan are countries that use most water after China and India (Burek et al. 2016). As a result of the region's swift economic development, domestic water demand is likewise relatively high in sub-Saharan Africa (Burek et al. 2016). Furthermore, countries like China, India, and Pakistan, where water consumption has doubled due to exponential population growth, are likely to see more water shortages (Kummu et al. 2016). Water shortages afflict more than three-quarters of China's 660 cities, making it 1 of 13 countries experiencing a water crisis (Hou et al. 2021). Water scarcity is also getting worse in Africa and the Middle East (El-Fakharany and Salem 2021; Kummu et al. 2016). Egypt is one of driest countries in North Africa (El-Fakharany and Salem 2021).

As a result of global warming, water scarcity and its severe effects, such as drought, famine, and diarrhea, are expected to worsen. As a result, countries experiencing water shortages should develop effective water management strategies that incorporate AI to make a stable provision of clean water (Jenny et al. 2020). Artificial intelligence algorithms can be used to change the standard of water utility service delivery and performance, resulting in reduced water loss and cheaper costs (Doorn 2021). In addition to supporting SDG6, this digital transformation of water delivery helps to achieve SDG13: climate action by increasing climate-related investments to mitigate the effects of climate change (Jenny et al. 2020). Overall, employment of AI numerical tools in the water industry helps solve some of the world's most pressing water issues (Jenny et al. 2020; Doorn 2021). Furthermore, because water shortage is a stumbling block to a society's or country's long-term growth, it is vital to plan and implement plans to protect and conserve water resources.

8.3 CAUSES OF WATER SCARCITY AND SHORTAGES

Water scarcity is caused by both natural and man-made reasons (Agnes et al. 2021). Poor water management, contamination, and overuse, as well as climate change and global warming, energy production, population growth, urbanization, agriculture, and industrial expansion, all contribute to water scarcity (Kummu et al. 2016; Agnes et al. 2021; Šulyová et al. 2021). On the other hand, human influences decline and cause drought of water. Water shortages are increasing because of climate alteration and global warming, or a rise in temperature (Kummu et al. 2016). Climate change, which is accompanied by extreme occurrences such as droughts and floods, exacerbates droughts and floods. As the severity and frequency of these severe storms increase, water scarcity is projected to intensify (WHO and UNICEF 2019; UNICEF 2021). Some of the variables that lead to water scarcity and shortages are listed in Table 8.1.

The expectation of droughts to triple in magnitude as a result of climate alteration will result in severe water scarcity in the next decades. Floods, for example, contaminate and ruin both water and infrastructure. As a result of global warming, evaporation increases and average air temperature also increases; the water in rivers and lakes evaporates at the same vein. Rivers, streams, ponds, boreholes, and wells can all dry up as a result of long-term evaporation. However, when temperatures rise, sea levels rise as well, contaminating fresh water and generating a water crisis. As a result, turbulent sea levels and temperatures are predicted to put the lives of uncountable millions of people who are sustained through freshwater resources in jeopardy. Due to changes in rainfall patterns, river flows, and growing water demand, water shortages are emerging in lockstep with the frequency and severity of droughts. In agriculture, an unpredictable, limited rainfall pattern reduces agricultural products, thereby forcing farmers to depend heavily on irrigation system. Global water demand is anticipated to rise 20% to 30% annually by 2050 as a result of population expansion. The United Nations Children's Fund (UNICEF) living standards will soften the problem by necessitating more food, agriculture, and energy, all of which demand vast amounts of water. Agriculture (i.e., farming and irrigation) is undeniably a big user of water resources all over the world (Burek et al. 2016; Adetoro et al. 2020).

TABLE 8.1

Some Causes of Water Scarcity and Shortage

S/N	Cause of Water Shortage	Reference
1	Climate change	(Kummu et al. 2016; El-Fakharany and Salem 2021; Judeh and Shahrour 2021)
2	Population growth	(Mitlin et al. 2019; UNICEF 2021; Agnes et al. 2021)
3	Global warming and drought	(Kummu et al. 2016; WHO and UNICEF 2019; UNICEF 2021)
4	Agricultural irrigation, misguided agricultural development, and policies	(Kummu et al. 2016; Adetoro et al. 2020; UNICEF 2021)
5	Pollution and contamination	(UNICEF 2021)
6	Poor water management and misuse	(Agnes et al. 2021; El-Fakharany and Salem 2021)
7	Energy production and industrial development	(Burek et al. 2016; Kummu et al. 2016; Agnes et al. 2021)
8	Change in human consumption behaviour	(Hou et al. 2021; Cheng and Li 2021)
9	Rising sea levels	(UNICEF 2021)
10	Increase in water demand	(UNICEF 2021; Cheng and Li 2021; El-Fakharany and Salem 2021)
11	Urbanization	(Mitlin et al. 2019; Agnes et al. 2021)
12	Conflict and migration	(UNICEF 2021)

In the Republic of South Africa, for example, agricultural output requires more than 60% of the country's fresh water, while irrigation consumes 50% of the country's total freshwater consumption (Adetoro et al. 2020). Water is used extensively in the energy sector for power generation, which primarily cools thermal power plants, power hydropower turbines, and extract, process, and transport fuel. Energy is also used in agroforestry to produce fertilizer, for irrigation, to cultivate and harvest crops, as well as to dry and process products (Burek et al. 2016). Water is necessary for animal, plant, and human life, so developing an innovative strategy and employing AI to manage water resources is critical to each country's long-term development (Doorn 2021).

Cities now accommodate almost above one-half of the populace in the world (WHO and UNICEF 2019; UNICEF 2021; Agnes et al. 2021). As large populations congregate in cities, water supplies are poisoned and depleted. As a result, some cities in the Global South are experiencing water shortages (including Addis Ababa, Dar es Salaam, Kampala, Nairobi, Mogadishu, and Yaoundé) (Mourad 2020; Agnes et al. 2021). By 2050, cities are predicted to accommodate around 2.5 billion people, or roughly 68% of the global population (Mitlin et al. 2019). Around 45 large cities with populations of more than three million people are projected to confront acute water shortages (Mitlin et al. 2019). According to Agnes et al. (2021), water shortages in Cameroon and other developing countries are mostly caused by the development of large cities. Water will be needed for home, industrial, agricultural, and recreational purposes as people and economies grow, putting a strain on limited freshwater resources.

Pollution and contamination of water are also major drivers of water scarcity (UNICEF 2021). Pollutants such as oil, certain chemicals, and feces can contaminate water, making it unfit for human consumption. In agriculture, synthetic fertilizers and other harmful substances can pollute and taint the water, causing water shortages. Furthermore, violence and migration increase water scarcity (UNICEF 2021). Conflicting parties restrict, contaminate, or destroy water and sanitation infrastructure in countries suffering from political crises and conflicts (UNICEF 2021). Water scarcity can also be caused by misuse and poor water management as a result of unregulated agriculture and a lack of laws. Water waste is growing increasingly widespread by the day, with people using considerably more than is necessary (Agnes et al. 2021). Water is occasionally misused, even unnecessarily, for recreational purposes, with little regard for the consequences of water scarcity. This is since most underdeveloped countries lack effective law enforcement to prevent water waste.

Governments and stakeholders must implement rules and regulations to reinforce strategies to prevent water shortages and improve water utilities as water scarcity grows in many locations throughout the world. To reduce water loss and management expenses, these must be combined with cutting-edge technologies, such as AI algorithms (Doorn 2021).

Some of the causes of water scarcity and shortage are further explained next.

Water overuse. Water overuse is a key issue that restricts water supply. This contributes to agricultural product scarcity and hunger.

Water pollution. Water contamination is a big problem, especially in areas where the sewage system is unreliable. Some noticeable examples of pollution are embedded in oil, carcasses, chemicals, faces, etc., then it creates a plethora of issues for persons who want to use it (Hutton et al. 2013).

Conflict. If a piece of land is the subject of a legal dispute, accessing water on that piece of land may be problematic. It may result in the loss of key local infrastructure, causing widespread water supply issues, and if dwellers within try to get water in the place, it may worsen the situation, which may result to violence from the pollution point of view.

Distance. Because they are not close to a supply of water, water scarcity affects several rural areas around the world. In deserts or isolated places, some water sources may be unavailable.

Drought. This is defined as an area that is unusually hot and dry, with inadequate rainfall to sustain life. Some locations are chronically afflicted with drought, while others may experience droughts from time to time. Droughts are common over the world, which diminishes groundwater accessibility, thereby making it unfit for human consumption, and nothing cannot be done to avoid such (Hutton et al. 2013)

Governmental access. People in power in some nations, particularly in dictatorships, may impose stringent laws on water usage which may affect people living in those parts of world. It is used by these regimes to impose control over those in positions of power, which might be a major issue.

Global warming. Heating is a principal factor in water shortage, resulting to air temperature increases from rivers and lakes, which evaporate more speedily, adding to dryness of water bodies. The global warming activities on water make the water contaminated and unfit for consumption, which could have a negative influence on local water supplies.

Illegal dumping. This is a pattern of waste elimination that reduces the volume of water in supply. The dumping of trash into water bodies (sea) as an inexpensive method of evacuating waste contaminates the water substance, thereby resulting in extremely limited water, which impacts people that rely on it as a water source for drinking.

Groundwater pollution. It is the introduction of harmful substances that move from the soil into groundwater in the form of pollution, which can bring soil adulteration. People who think good groundwater is limited would have a serious problem with water inadequacy if an alternative is not provided.

Natural disasters. Due to the loss of needed public goods, floods might result in severe water shortages for the people. A severe natural disaster has the potential to entirely shut down the area's water supply. During floods, large amounts of soil are carried, potentially causing substantial pollution of local water bodies. Local rivers may become unsuitable for drinking for some time following floods.

8.3.1 Impacts of Water Scarcity and Shortage

Many countries and big cities around the world, both rich and poor, have experienced increasing water shortage in the twenty-first century as a result of population expansion, overuse, increasing pollution, and changes in global warming–related weather patterns. Physical and financial scarcity are the two main types of scarcity. Water scarcity, whether natural or absolute, occurs when an area's demand exceeds its limited water resources. About 1.2 billion people live in areas are subjected to natural disasters, according to the Food and Agriculture Organization of the United Nations (FAO). Many of these people live in arid or semiarid regions. Natural water shortage can be periodical. It is estimated that two-thirds of the people in the globe are living where shortage of water dwells at least once a year; as population increases and the weather becomes more unstable and harder, the number of people that are affected by natural water scarcity is likely to increase.

Generally, economic water inadequacy is culminated by poor presence of water infrastructure or ill-management of water resources where there is infrastructure. FAO asserted that more than 1.6 billion people are suffering from economic water deficiency. In places with economic water inadequacy, there is always a large body of water to enhance human and environmental needs, but access is limited. Available water could also be contaminated or unfit for human usage due to poor management or lack of development. Unlimited use of water for agriculture or industry, often to the detriment of the general public, can also cause economic water shortage. Finally, significant inefficiencies in water consumption can lead to water inadequacy, mainly due to the economic devaluation of water as a finite natural resource. Distributing fair water is becoming increasingly difficult as water resources becomes limited. Some groups win at the expense of others when governments are forced to settle between agricultural, industrial, municipal, or environmental interests. Geopolitically, vulnerable locations affected by adverse water shortage can cause forced movement of people or regional wars. Water shortage chronic areas are particularly prone to water difficulties, which the source of water reduces to imaginable low levels. The Day Zero threat faced by the dwellers of Cape Town in South Africa in 2018, on a day the municipal taps ran empty, is a serious primarily city-wide water problem. The instantaneous danger passed without occurrence of unexpected conservation efforts and hence the accidental onset of rain. However, because humans can only choose a few days of unwater, a water shortage can quickly become a complex human problem.

Water issues were tagged "third most serious global risk" from the angle of mankind impact in "2017 World Economic Forum Global Risks Report," after mass destruction weapons and chronic weather occurrence.

8.3.2   Effects of Water Scarcity

Lack of access to drinking water. When water is scarce, it usually has an impact on people's life, like in getting fresh, good water for consumption. The normal body would not function without water, and a shortage of beverages can cause a spread of additional problems, which we explore further in the following text.

Hunger. Poor supply of adequate water will affect the growth of crops; as such, hunger will be on the land, animals will die, and shortage of meat will occur. Summarily, shortage of water tenses both people and animals with mass famine in the area.

Lack of education. Shortage of water affects child growth pattern as a result of insufficient liquid in his growth process; this could cause problem in the brain and sensory organs. The child can also help his parent look for where to get water for family usage, at the expense of education.

Diseases. Water, as an important liquid used for bathing, cooking, and washing of clothes, hands, and other things, if scarce, will collapse the body because of microorganisms' infections. These microorganisms are preventable from the use of water.

Sanitation issues. Unsanitary situations can result from unavailability of water used for daily living (drinking, cooking, washing, or bathing) and can cause health problems. Improper sanitation can cause diseases, which contribute to psychological state problems, like depression and anxiety.

Poverty. Overall, individuals suffering from water shortage are frequently poor, and these people cannot get the resources required to thrive and are living in these hard moments.

Migration. Scarcity of water could result to people living from one place to another in search of greener pastures (farming). In the course of migration for livelihood, lives are lost because of the hotness of the soil.

Destruction of habitats. Water being the commando of life, if water shortage persists, it may lead to extinction of entire life (humans, animals, and plants) on Earth.

Loss of biodiversity. Experience of significant shortage of water in an area could cause hunger to plants and animals, which could eventually call for extinction of biodiversity (Hutton et al. 2013).

8.4 WATER RESOURCE MANAGEMENT

Artificial intelligence–based water technology solutions can help in facilitating the uptake of solutions that are off-grid and more localized for wastewater and water resource management. The AI-based water technologies can also provide strategies for building hybrid decentralized-centralized water systems. Existing centralized water systems currently struggle to cope with the increased urbanization, rapid population growth, limited financing, climate variations, and lack of political will. Shifting to decentralized water network systems will fulfill future water needs and improve urban/city planning and preparedness. Several innovations, such as adopting power sector microgrid strategies for the water system, are currently in place. Also, the monitoring of water quantity, quality, and system performance in real time through AI can significantly increase the adoption of off-grid water systems (World Economic Forum 2018). Digitalizing water systems improves the relationship between users and utility providers through real-time AI-based technologies to monitor water quantity.

8.4.1 WATER LOSS REDUCTION THROUGH AI

About 126 billion m^3 of water (amounting to 39 billion US dollars) is lost through nonrevenue water (NRW) globally. Water losses are mainly caused by too high pressure, resulting in network failures and water pipe material usage in old networks (Rojek and Studzinski 2019). It is crucial to detect and locate water leaks to reduce water loss in water supply systems and promote continuous water usage. The amount of water lost from leakages generally depends on time taken between occurrence and reporting of the detected leak by the water service provider (Hu et al. 2021). The losses can be managed by optimization of pump operation and pumping stations and pipe network revitalization. The optimization process requires the application of complex algorithms and thus is computationally demanding, while revitalizing the network can be expensive (Rojek and Studzinski 2019). In tackling water losses, many attempts have taken the advantage of data (particularly big data) obtained from installed sensors, advanced numerical techniques, and water supply network modeling. Data-driven and physical-based methods are the two major approaches used for network analysis. The physical methods combine statistical tools with hydraulic modeling, enabling the "digital mirroring" of physical water supply networks for testing and checking real-time scenarios. The physically based methods provide water utilities with the starting point for their digital transformation. The data-driven methods apply machine learning algorithms or AI techniques to large sets of data for extracting information and detecting patterns without using network equations (Donghwi and Joong 2018; Vrachimis et al. 2018; Wu and Liu 2017; Li et al. 2015). Through access to big data, AI can provide water loss reduction functionalities, as summarized in Table 8.2.

Digitalization and connectivity are two new technologies that could have positive impact on promoting continuous development in almost all parts of globe. The rapid connection of the world's population through mobile phones and the internet presents an opportunity for exploiting digitally enabled technologies to sustainably manage

TABLE 8.2
Functionalities of AI Algorithms for Water Loss Reduction and Management

Functionality	Description
Optimized design for network monitoring and control	• Provision of objective-based information criteria by AI algorithms to locate several selected sensors in a given network for extracting maximum information on the entire system with the least capital investment.
	• The number of control points is minimized while prioritizing pressure gauge installation instead of the more expensive flowmeters.
Numerical detection of water losses (apparent and physical)	• AI algorithms give spatial information on water losses (type and amount) and then provide the status of the water network.
	• AI algorithms perform probabilistic and continuous calibration of the network, thus allowing for the analysis of the error structure at each control point.
	• The numerical location of losses saves money and time for deploying staff to detect leaks.

Source: Beh et al. (2017); Jung et al. (2019).

TABLE 8.3
Major Areas for Internet of Things Devices for Mapping in Water Management

Weather forecasting and water resources mapping
- GIS—geographical information systems
- Systems for in situ terrestrial sensing
- Remote sensing from satellites
- Sensor networks and the internet

Water distribution network asset management
- Electronic tagging and buried asset identification
- Smart piping systems
- Real-time risk assessment/just-in-time repairs

Meeting future water demand in cities and setting up early warning systems
- Stormwater/rain harvesting
- Flood management
- Aquifer recharge through managed systems
- Smart water metering
- Knowledge of process systems

Landscaping and just-in-time irrigation in agriculture
- GIS
- Sensors networks and the internet

Source: Xiang et al. (2021); Tiyasha et al. (2020); AL-Washali et al. (2020); Dawood et al. (2020).

water resources. The production and efficient distribution of water can be facilitated by advances in AI (Xiang et al. 2021; Tiyasha et al. 2020). Internet of Things (IoT) devices, including sensors, mobile phones, and meters, can be used to improve water management (Table 8 3). Applying geospatial technology for managing water infrastructure can significantly improve water supply and reduce water losses (United Nations 2015). Pipes, pumps, valves, nodes, and tanks for storage are all essential parts of the water supply. The integration of geographical information systems (GIS) with the water infrastructure network can help establish an efficient and continuously maintained water supply (Xiang et al. 2021). The GIS technology, when integrated with the water supply network, allows for spatial network data analysis in the form of water pipe parameters, including diameter, length, coefficients, and roughness. Improved water infrastructure network enhances assets factor productivity and tasks automation, such as pressure management, technical maintenance, leak detection, and improved water quality control (Dawood et al. 2020).

In South Africa, municipalities are turning to advanced metering infrastructure to give accurate data in real time, thus reducing nonrevenue water losses and achieving accurate billing. The advanced metering infrastructure (AMI) is composed of integrated communication networks, smart meters, and data management, enabling two-way communication involving customers and utilities. The communication enables the smart meters to be controlled remotely and managed on a fixed network over a private network or through cloud computing. The prepaid meters allow the consumers to have financial control and accurate billing by municipalities. Moreover, some systems allow for free daily basic water provision with an option for a top-up for consumers wishing to purchase more water above the allowable amount allocated daily (INTELLIGENT CIO 2019; AL-Washali et al. 2020).

8.4.2   AI AND SMART WATER SYSTEMS FOR WATER RESOURCE MANAGEMENT

The conventional methods for water quality analyses are labor-intensive, time-consuming, and costly, thus hindering a synoptic view of the water sources. Data collected weekly, monthly, or sometimes seasonally prevent accurate decision-making informed by data for effective management of water resources or rapid response to operational accidents such as toxic pollution. Involving the citizenship's dispersed collective intelligence can be aided through AI to ensure the rational and optimal use, conservation, and protection of the water resource management regime (Mapani et al. 2019; Ingram and Memon 2020; EU 2020). Water resource management and protection require new paradigms that can combine new systems and intelligent technologies to be formulated. The combination should be capable of increasing the performance and efficiency of the networks, resources, and treatment plants within the territory. Also, the combinations should enable the development of easy-to-access modern water resource monitoring systems for widespread quality control (World Economic Forum 2018). In all cases, AI has prominent work on the massive amounts of data involved. The advancement of interoperable advancement capable of promoting huge amounts of information dissemination and exchange between citizens, decision makers, and managers can result in the creation of widespread knowledge that can feed AI systems and promote environmental protection, with an impact on the behavioral and educational sides (Dawood et al. 2020, 2021; Shamsudin et al. 2016; Schultz et al. 2018; Park et al. 2020).

Through advanced data collection systems, artificial intelligence can significantly improve real-time monitoring and management of water resources, flood assessments, and pollution control (Park et al. 2020). Automated sampling can be used for pollutant load estimation in selected water streams (Park et al. 2020; Alizadeh et al. 2018). Advanced water superiority monitoring can be achieved through in situ sensing and AI-based systems. Real-time smart water quality monitoring systems are used to manage potable water (drinking water) supply to maintain the required quality standard. Smart monitoring can also prevent several unexpected accidents, such as water treatment system malfunction and raw water contamination (Xiang et al. 2021). A system of monitoring based on AI technologies allows for effective control actions targeted on diffused loads formed because of a network overflow as well as production loads, reducing the presence of metal pollutants and organics while maximizing the recovery of nutrients. AI, when combined with modern equipment for monitoring and efficient urban sewer control, allows for quick intervention, thereby lowering the possibility of groundwater contamination (Elhoseny et al. 2014). In response to what citizens, administrators, and managers want, AI uses early warning indicators for identifying and suggesting mitigation measures on a local scale for extreme events caused by natural or anthropogenic factors.

Water in natural systems such as rivers, lakes, and the sea can be managed effectively by systems based on in situ real-time monitoring, whereby multispectral sensors are applied for data collection from a wide range. Recent advances in sensing technologies and AI have enabled low-cost, reliable measurements, transmission, and management of massive data on the environment. The expansion of mobile network connectivity in the developing world in rural and urban areas has created platforms for the emergence of innovative digital solutions that have reduced the costs associated with the provision and accessibility of basic services, including water, energy, sanitation, transport, and waste management (García et al. 2020).

Manufacturing plants and industries can manage their water usage more efficiently by employing AI. Water is used in almost every manufacturing industry during operations for several purposes, such as cooling and cleaning. Cooling systems in industries could be properly managed to reduce water and energy-related operational costs and minimize wastewater and chemical discharge. Process automation and the use of AI in control systems can improve the performance of the plant and optimize water use (Goralski and Tay 2020). Areas of AI application for improved water security are presented in Table 8.4.

8.4.3   AI FOR GROUNDWATER MANAGEMENT

Groundwater level (GWL) measurement and analysis in aquifers is a critical and essential activity in groundwater resource management. The knowledge of GWL changes can be utilized to assess the availability of groundwater. The GWL changes in aquifers/wells allow for a direct assessment of the groundwater development impact, and the constantly recorded GWL time series often contains vital information about aquifer dynamics (Butler et al. 2013). As a result, water managers and engineers must model and predict GWL to quantify and define groundwater resources and achieve the required balance between the supply of water and the demand for water (Rajaee et al. 2019).

The permit/license system is an important component of groundwater sustainability. Effective groundwater management is made possible by groundwater limitations

TABLE 8.4

AI Application for Water Security

Water Security Aspect	AI Application
Water supply	• Management of water supply
	• Monitoring of water supply
	• Water quality simulation
	• Self-adaptive water filtration
	• Maintenance of assets
Water efficiency	• Water use (residential/domestic) management and monitoring
	• Industrial water use optimization
	• Informed maintenance approaches of water plants
	• Systems for early warning
	• Detection of underground leaks
	• Smart metering in homes
Catchment control	• Detection and monitoring of harmful algal blooms
	• Forecasting streamflow
	• Automated flood-centered infrastructure

and rights. Permits are issued to protect the quality of groundwater resources and keep track of how long groundwater abstractions last. Permits also assure that distribution rates, sizes, and magnitudes are within politically, socially, and environmentally acceptable, as well as technically achievable, restrictions. Economic mechanisms, demarcating groundwater rights, and groundwater licensing are the most important aspects of permits in terms of groundwater sustainability.

Physical or conceptual-based models have, over the time, been the primary tool for GWL modeling; nevertheless, they have significant practical constraints, such as the need for a substantial amount of input parameters and data. On the one hand, data is scarce in many situations, while on the other hand, exact forecasts are more crucial than comprehending underlying systems; hence, AI models can be a good option. Groundwater abstraction monitoring can be aided by AI with IoT. Wireless sensor networks (WSNs) that have become critical components of IoT provide detecting abilities to gather data about the physical condition, as well as remote communication that enables spontaneous system management without a preexisting physical framework (Rajaee et al. 2019). Groundwater levels and many hydrological variables can be calculated without the use of a conceptual model. These models include numerous essential phases, including consideration of input data, model characteristics regulation, input data division, training, and testing. Careful construction of all the stages will result in good model performance. Some of the models for GWL modeling include support vector machine (SVM), genetic programming (GP), adaptive neuro-fuzzy inference system (ANFIS), and artificial neural networks (ANN) (Beale et al. 2010; Suryanarayana et al. 2014; Tang et al. 2018; Rajaee et al. 2019).

8.4.4 WATER QUALITY EVALUATION MODEL

Some current water treatment projects and wastewater treatment plants have inadequate results, which are detrimental to the environment and the long-term viability of the economy, as well as major hidden risks to the health and safety of residents (Lee and Oh 2016). As a result, enhancing the effectiveness of aquatic governance requires the application of artificial intelligence algorithms to monitor ecological aquatic environmental data and provide scientific processing methods. Using SO_2 (sulfite) to accelerate wastewater oxidation, Muerdter et al. (2018) developed a sustainable technique to enhance the rate of pollutant removal in the aquatic environment. Muerdter et al. (2018) examined how total suspended solids, nitrogen, phosphates, toxic metals, hydrocarbons, pathogens, and emerging pollutants affect performance and pollutant removal processes in urban rainfall. Although their study examined the effect of plants on pollution, they made no recommendations for immediate treatment. In order to minimize contamination in rural domestic water, Hashimoto et al. examined the use of a sludge replacement titanium oxide (TiO2)–type photocatalyst in the decomposition of pesticides in water (Hashimoto et al. 2016). The integrated pollution index method, gross water quality index technique, and brown water quality index method are a number of water quality assessment methodologies. The following are some of the advances that have resulted from the widespread use of AI algorithms in water quality assessment models:

(i) Developed a set of sensors with ideal functions for real-time data acquisition on dissolved oxygen content, water temperature, turbidity, temperature and humidity, and smoke concentration in the aquatic environment.
(ii) The model of water quality forecasting is made using dynamic variable time forecasting approach, which can firmly forecast water quality and provide a rapid treatment strategy.
(iii) Support vector machine (SVM) and the decision tree supported by the SVM classification method are used to teach and categorize data samples. As shown in equation 1, the water quality assessment model is made using the particle swarm optimization approach, which can accurately classify water quality.

$$g(x) = \mathrm{Sgn}\left[\sum_{i-1}^{i} r_i m_i \{x_i, x_j\} + n \right] \tag{1}$$

Where m_i is Lagrange multiplier, x_i is sample vector, r_i is classification identifier, and n is classification threshold, Sgn is objective function, $g(x)$ is vectors, while x_j and x_i are number of jth and ith particles.

For more than 50 years ago, water quality index (WQI) models have been in existence. In the 1960s, the first WQI model was evolved by Horton based on ten water quality parameters very crucial in some water environments. Support from the National Sanitation Foundation with Brown evolved a stricter version of Horton's WQI model, the NSF-WQI, for which a team of 142 water quality experts informed the selection and weighting of parameters (Abbasi and Abbasi 2012).

Other several WQI models have since relied on NSF-WQI. In 1973, the Scottish Research Development Department (SRDD) came up with SRDD-WQI, which was also anchored on Brown's model, and used it to assess river water quality. The Bascaron index (1979), House index (1986), and Dalmatian index are all later derivatives of SRDD-WQI. The model of environmental quality index for assessment of water quality in ecosystems of the Great Lakes was later evolved by Steinhart et al. (1982). Another important development was the British Columbia WQI (BCWQI), developed by British Columbia Department for Environment, Land, and Parks in the mid-1990s and used to assess the standard of many bodies of water in the province of British Columbia, Canada (Saffran et al. 2001). Said et al. (2004) addressed that BCWQI was discovered to have the maximum level of sensory technique in sampling design and highest dependence on specific application of water quality objectives.

The Canadian Environment Council Working Group on Water Quality Guidelines developed the CCME WQI in 2001 (Saffran et al. 2001), following a review and revision of the BCWQI model (Lumb et al. 2011). To date, more than 35 WQI models have been introduced by various countries and/or bodies to assess surface water quality worldwide (Abbasi and Abbasi 2012; Dadolahi-Sohrab et al. 2012; Kannel et al. 2007). Table 8.5 shows that, although WQI models have been applied to all

TABLE 8.5

Summary of WQI Model Applications (in Total and by Study Area) Found in Literature Published from 1960 to 2019

WQI Model	Number of Applications	Type of Study Area		
		River	Lake	Marine/ Coastal/Sea
CCME	36	28	5	3
NSF	18	17	1	—
FIS	12	10	1	1
MWQI	8	6	1	1
Horton	7	6	—	1
SRDD	6	6	—	—
Bascaron	4	3	—	1
EQI	2	1	1	—
Oregon	2	2	—	—
Smith	2	2	—	—
Almedia	1	1	—	—
BCWQI	1	1	—	—
Dalmatian	1	—	—	1
Dojildo	1	1	—	—
Dinius	1	1	—	—
Hanh index	1	1	—	—
House index	1	1	—	—
Liou index	1	1	—	—
Said	1	—	—	1
WJWQI	1	—	—	1

(Md. Uddin et al. 2021)

major water body types, 82% of applications were for river water quality assessment. In addition, the table shows that the CCME and NSF models have been used in 50% of the studies examined.

The model of WQI parameters was carefully chosen because of statistical information availability to expert opinion or the environmental importance of a water quality parameter. Debels et al. (2005) commented that a lot of WQI models used only basic water quality parameters due to lack of availability of other parameter measurement data (Cude 2001; Banerjee and Srivastava 2009). Scholars have improved the parameter lists of model based on accessibility and data availability, and sometimes it is not possible to infuse the critical water quality parameter to a model for this reason (Ma et al. 2020, Naubi et al. 2016).

Abbasi and Abbasi (2012), Abrahão et al. (2007), Lumb et al. (2011), and Sutadian et al. (2018) outlined the general structure of WQI models, indicating the four most main steps of WQIs:

(i) **Selection of water quality parameters**. For inclusion in the assessment, one or more water quality parameters are chosen.
(ii) **Generation of the parameter subindices**. Parameter concentrations are changed to units of fewer subindices.
(iii) **Assignment of the parameter weight values**. Parameters are assigned weightings depending on their relevance to finding.
(iv) **Computation of the water quality index using an aggregation function**. The individual parameter subindices are amalgamated using the weightings to give one an overall index. A rating index is usually adopted to categorize/classify water quality based on overall index value.

8.5 ALTERNATIVE WATER SOURCES TO MITIGATE WATER SHORTAGES

8.5.1 RAINWATER HARVESTING

The technique involving the collection and storage of rainwater instead of allowing it to flow away is identified as "rainwater harvesting." Rainwater is gotten from a roof-zinc surface that is directed towards a tank, cistern, deep pit, aquifer, or reservoir through percolation, where it seeps down and is restored to earth. Rainwater collection is now a widely used technique for supplying both drinkable and nondrinkable water supplies, specifically in growing nations where drinkable water is limited to satisfy society's growing needs as a result of speedy commercialization and development, so also population explosion (Ikhioya et al. 2015; Olaoye and Olaniyan 2012). Nowadays, rainwater has become the only mainstay of water collection for families and also eliminates the suffering of water gathering from other sources (Vikaskumar et al. 2007).

The simplest form of rainwater collection is rainwater runoff obtained from residential roofs. Rooftop runoff has less rainfall runoff contamination than in a ground catchment system, in addition to being less expensive and easier to maintain because of the durability of the corrugated sheet. Roof catchment gives water for drinking (Gould and Nissen-Peterson 1999). The collected water is critical in places where there is heavy rain but no good water delivery facility. The amount of effective roof area and the materials used in construction have an impact on collection efficiency, water quality, and water quantity (Ikhioya et al. 2015).

AI-related technologies can be used to support water management using geographic information system (GIS) technology. According to research undertaken by the International Water Management Institute (IWMI), up to 6,000 traditional water tanks could be restored using GIS and Landsat data to absorb 15–20% of local rainfall. When compared to the construction of a normal irrigation dam, the restoration could capture 1.7 km3 of water, providing for a 50% increase in irrigation capacity and a 75% cost-per-hectare reduction (UNCTAD 2011).

8.5.2 DESALINATION

Water scarcity is becoming a serious concern around the world due to a poor supply of stream resources and highly expensive transferring of fresh water from local sources to water-demand areas. As a result of this problem, there has been a renewed interest in creating drinkable water from seawater and brackish water. In addition, both the energy consumption of water infrastructure and the carbon impact of water consumption have emerged as serious issues. Total dissolved solids (TDS) concentration is the best predictor of water quality in relation to salinity, which is the sum of all minerals, metals, cations, and anions that can be dissolved in water. (The Environmental Protection Agency (EPA) 2009).

A high TDS level in water is detrimental to human beings and can lead to pipe scaling and pipeline and fixture corrosion, among other problems. The USEPA has recognized a secondary maximum contaminant level (SMCL) or aesthetic standard of less than 500 mg/L for TDS in potable water; a TDS concentration of <200 mg/L is better in drinking water. Desalination is the process of treating high-salinity water so that it can be utilized for drinking and/or other uses. Modern desalination techniques include thermodynamic (distillation) and membrane filtering, as shown in Figure 8.1.

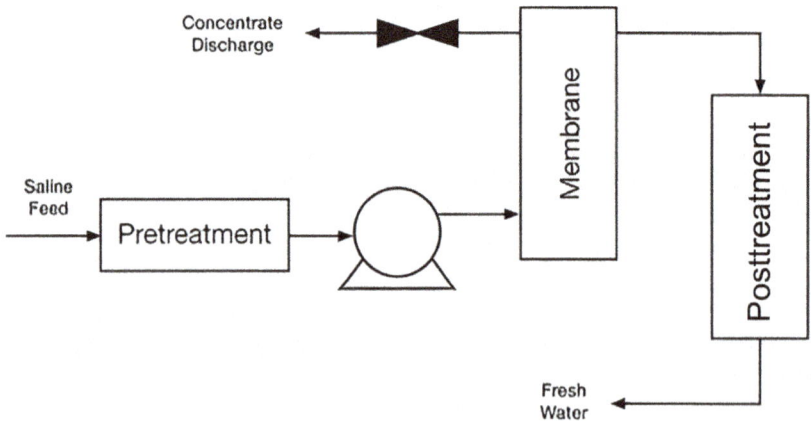

FIGURE 8.1 Modern desalination technique.

8.5.2.1 Energy Use Efficiency

A lot of several parameters decide desalination plant's energy usage, including desalination process, TDS concentration, temperature of source water (feedwater), plant capacity, and plant location concerning water source intake and concentrate discharge. From the angle of water–energy position, three approaches—collocated plants, cogeneration generators, and hybrid generators—provide opportunities for use of energy efficiency in desalination facilities (Charisiadis 2018).

Plants that have been grouped include the following: desalination and power generators in an industrial area are coupled in this approach, with the power plant drawing seawater for cooling. The desalination engine applies hot seawater from cooling water loop of power plant as feedwater, which requires small power for desalination. The desalination plant discards the concentrated by-product of high TDS in cooling water effluent of power plant, allowing it to dilute before being discharged back into the ocean. By "piggyback" on cooling water loop of power plants, the desalination unit can decrease energy intake. TECO's Big Bend power plant, for example, takes up to 1.4 billion gallons of seawater a day from Tampa Bay and uses it as cooling water for the power plant. The Tampa Bay seawater desalination plant "catches" up to 44 million gallons of hot seawater daily for desalination. The diluted concentration is delivered to Tampa Bay (see Figure 8.2).

Cogeneration Plants. A regular power plant produces high-pressure, high-temperature steam, which is typically ejected as waste in this technique. To reduce fossil fuel use, a cogeneration facility inhabits steam as an additional power medium during the desalination process. By selling waste steam to desalination plants, the power plant makes more money, while the desalination plant avoids having to spend on the development and operation of its energy medium (EPA 2009; Charisiadis 2018).

Hybrid Plants. Plants of a hybrid nature apply a combination of treatment approaches to help the plant recover energy and perform better. For example, the Cape Hatteras desalination plant in North Carolina uses brackish water to power a hybrid RO/ion exchange engine. Water is delivered to the power plant from two distinct wells: high-salinity water from well 1 is treated with RO, while high-organic-constituent water from well 2 is treated using a natural process. The final product water is made from RO and natural process water that has been processed. This plant's RO treatment process also incorporates an energy recovery turbine (EPA 2009; Charisiadis 2018).

8.5.2.2 Renewable Energy Use

Since the mid-1990s, there has been talk about using renewable energy for desalination. A few conventional water treatment plants in the United States have integrated solar PV for water treatment, such as many drinking water treatment plants in Massachusetts (EPA 2009; Charisiadis 2018).

8.5.2.3 Environmental Impacts

The developments of the infrastructure for water intake and, therefore, the pipe network conveying the feedwater to the desalination plant, which can harm environmentally sensitive areas, are two risks linked with plant construction. However, concentrate (reject salt) remains the foremost serious environmental issue, which

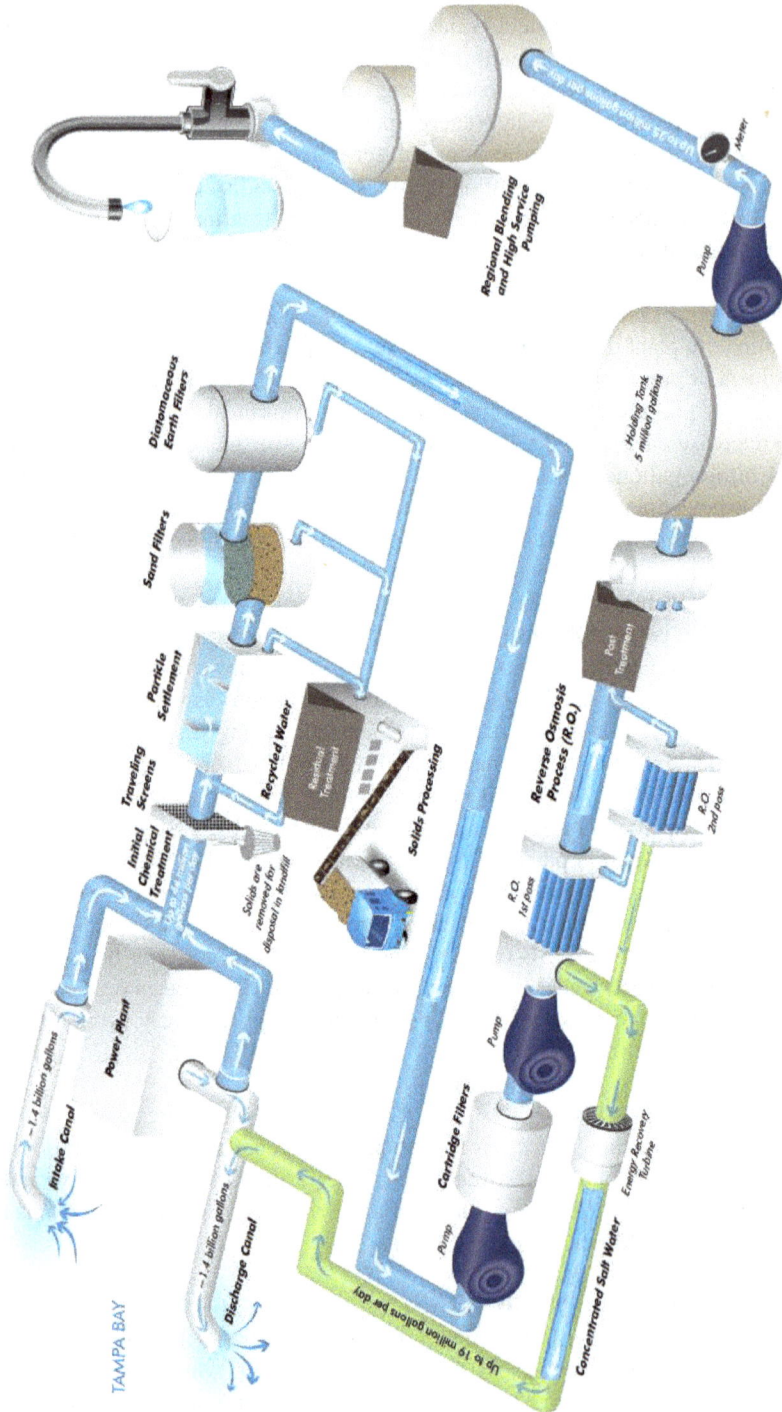

FIGURE 8.2 Tampa Bay seawater RO desalination facility.

has an impression on desalination's cost-effectiveness. High TDS concentrations (>65,000 mg/L) are produced by RO facilities, which can also contain some harmful chemicals utilized during the pretreatment and posttreatment (cleaning) procedures. With the quality and quantity of concentrate and the situation of the plant, therefore, this is applicable for environmental requirements that influence the selection of a concentrate management method.

Surface water disposal. The foremost typical way of concentrate disposal is in coastal waters, like oceans, tidal rivers and streams, estuaries, and bays close to the plant. Environmental issues include the long-term influence on coastal aquifer quality and therefore the negative impact on the ecosystems of receiving waters.

Submerged disposal. The concentrate is transferred from generator to an estuary and ocean area via underwater pipes. The possible impact of sinking considers benthic marine creatures dwelling on seafloor as a source of environmental concern.

Deep well injection. Deep groundwater aquifers are immediately injected with the concentrate. Thanks to geologic circumstances and therefore the possibility of contamination of beverage sources, this method is site-specific.

Evaporation ponds. Water evaporation from the concentrate is allowed by constructing ponds with liners, while the residual salts accumulate within the pond bottom. In places with weather and high evaporation rates, evaporation ponds are an economical choice. They, however, require a sizeable land area and produce critical evaporation of the elemental water resource.

Land application. Infiltration trenches, spray irrigation, and percolation ponds are all utilized in this method. The land application method allows for the efficient utilization of the concentrate. The concentrate can, for instance, be used to irrigate crops and grasses that are salt-tolerant. The viability of land application is decided by the tolerance of local vegetation to salinity, prevailing temperature, groundwater table placement, and land availability.

Integration with a wastewater treatment plant. Concentrate discharge at the top or front of a wastewater treatment plant can be considered. Discharge to the front is not recommended since most wastewater treatment machines are not manufactured to extinct TDS, and a high TDS level can interfere with the biological treatment process. Because concentrated trash mixes with treated wastewater after a wastewater treatment plant, the result is a diluted concentrate. It's worth noting that, despite the dilution, this approach may have a harmful effect. Additionally, establishing a separate pipeline (and possibly utilizing a pump) to move the concentrate to the wastewater treatment plant will incur additional costs. Waterworld may be a website dedicated to all things or anything associated with water.

Brine concentrators. Deaerators, heat exchangers, and vapor compression are utilized in the brine concentrator process to turn the liquid concentrate into the slurry. Utilizing a brine concentrator could allow up to 95% high-purity water (distillate) recovery with TDS of about 10 mg/L. In the crystallizer, the remaining concentrated slurry is often converted into easy-to-handle solid cakes for land disposal. Some chemicals could also be present within the dried solid, so take care.

Zero liquid discharge (ZLD). The ZLD method, which was first designed for solid waste management, is now being promoted as a replacement technology for concentrate management. The ZLD method converts the liquid concentrate (brine) into a dry, hard substance which will subsequently be used for beneficial goals through an evaporation process. Some chemicals could also be present within the dried solid, so take care. Furthermore, the ZLD process is an energy-intensive method. Using thermal technology for concentration control could end in a less-expensive ZLD application.

Desalination technology and research are rapidly improving, and abundant seawater and brackish water sources will provide a major opportunity to deal with present and future water scarcity challenges. This page gives a fast overview of key desalination concerns, like the water–energy nexus, concentrate management, and regulatory/permit needs. The event of ecologically acceptable and economically viable desalination facilities in urban regions, also at smaller coastal towns, is additionally projected to be aided by evolving renewable technologies.

8.5.3 WASTEWATER REUSE AND RECYCLING

With new trends like the zero liquid discharge and, therefore, the reuse of intersector water, like the utilization of urban wastewater for industrial uses, the reuse and recycling of household and industrial wastewater, have become routine practice for several homes and enterprises. Most wastewater minimization solutions are built around the waste hierarchy. The goal of the waste hierarchy is to urge the foremost practical benefits out of things while producing the smallest amount of ultimate wastewater. (www.epa.gov)

The avenue of converting wastewater into sewage that can be reused immediately or returned to the water cycle with minimal environmental impact is known as wastewater treatment. Water recovery is the process of healing wastewater so that it can be used again for a variety of uses. The healing process takes place in a wastewater treatment plant (WWTP), also known as a water resource plant (WRRF). Pollutants in urban wastewater (generated by households and small businesses) are discharged or degraded. Chitosan and an alumina-chitosan nanocomposite made of *Rhynchophorus phoenicis* (palm arm) have been adopted for wastewater redress. Alumina chitosan composite and chitosan have been effectively synthesized and used to remove gentian violet dye from industrial effluents. This could assist in addressing society's water shortage.

8.5.3.1 Basics of Water Reuse

The method of recovering water from various medium, treating and reusing it for beneficial purposes such as agriculture and irrigation, drinking water supply, groundwater replenishment, company activities, and environmental remediation, is known as water reuse. By giving alternatives to existing water supplies, water reuse can help increase water safety, sustainability, and resilience (www.epa.gov). Planned and unintentional water reuse are two types of water reuse. When a water source contains a large amount of previously used water, involuntary water reuse occurs. Involuntary water reuse is common in areas that draw their water from rivers that receive wastewater treated from upstream communities.

Scheduled water reuse refers to water systems designed to recycle recycled water beneficially. Communities will often try to reduce overall water use by reusing water as much as possible within the community before returning it to the environment. Planned reuse includes agricultural and landscape irrigation, treatment water, drinking water supplies, and groundwater supply management.

8.5.3.2 Types of Water Reuse

Water can be used again from urban wastewater, company processes and cooling water, rainwater, agricultural runoff and return streams, and water produced from natural resource extraction activities. These water sources have been properly treated to meet the needs of a particular next use. To avoid damage to plants and soils, to maintain food safety, and to protect the health of farmers, for example, the recovered water for irrigation of crops should be of sufficient quality. In applications where there is a greater risk of human contact, the water may require further treatment. Cities, farms, and businesses use recycled water for a variety of potable and nonpotable purposes (contaminated drinking water) (www.epa.gov). Nonpotable applications include the following:

- Landscape and crop irrigation
- Stream and wetlands enhancement
- Industrial processes
- Recreational lakes, fountains, and decorative ponds
- Toilet flushing and gray water applications
- As a barrier to protect groundwater supplies from seawater intrusion
- Wetland habitat creation, restoration, and maintenance
- Groundwater recharge

For such purposes, used water reduces dependence on increasingly scarce and expensive surface water. It has the ability to reduce groundwater overflows as well as discharge of treated wastewater into rivers and oceans. Different water applications require different amounts of treatment. The amount of treatment—secondary, tertiary, or advanced—is determined by the initial quality of the water, the end use, and the laws of the state.

8.5.3.3 Uses of Recycled Water

Examples of the use of recycled water sources are stated next:

- Irrigation for agriculture
- Irrigation for landscaping, such as in parks, access rights, and golf courses
- Municipal water supply
- Water treatment for power stations, refineries, mills, and factories
- Internal uses, such as flushing the toilet

- Dust control or cleaning of road surfaces, construction sites, and other areas with traffic
- Concrete mixing and other construction processes
- Supply of artificial lakes and inland or coastal aquifers
- Environmental restoration

8.6 WATER DEMAND MANAGEMENT

Local water resources will be scarce in a growing number of regions and countries. In many wet and semiarid areas with low water supplies, irrigation, industry, communities, and the environment must compete for available freshwater resources. As a result, freshwater intakes will increase, leaving the ecosystem with less water (WRI 1999). The amount of water exported for human use has reached a stage where it has significant detrimental effects on ecosystems in some areas (Falkenmark 1997). As a result of the potential for water scarcity, water demand management and effective solutions to meet future requirements have evolved without compromising the long-term viability of current water resources systems. In the 1990s, a new management strategy was developed to address this issue. This new example, known as integrated water resources management (IWRM), has been widely accepted at major international water conferences (ICWE 1992).

It becomes more expensive to build and use when water supplies are scarce. The supply–demand balance is discarded when an individual, farmer, or industrial plant uses water that is not available to other customers in an area with limited water resources. The productive value of water in a farm or industrial facility where it is rare is the alternative cost of water. Water efficiency usually suggests that water contributes more to human happiness. In countries where people are poor and food security is a problem, its use in agricultural production is considered a benefit to human well-being. Water must be distributed to meet the basic needs of all populations, rural and urban. Managing water demand may be the only viable option to meet "now" and "tomorrow" water needs, given the current political shortfalls to water exports and the relatively high cost of desalination.

8.6.1 Efficient Use and Allocation of Water

WDM (water pressure management) is an established policy and practice aimed at encouraging people and businesses to manage the quantity, manner, and price at which they supply, use, and dispose of water, thus reducing the demand for freshwater supplies and maintaining quality. Water demand can be adjusted with various nonfinancial (awareness, technology) and economic (incentives, prices) strategies and practices that are either compulsory (legislation) or selective (pricing) (e.g., purchasing systems).

Practically, water needs management, which is comprised of two interrelated activities: enhancing water efficiency and efficiently allocating available water among competing sectors. In the urban, industrial, and agricultural sectors, water providers and users often strive for water efficiency by meeting individual users' and uses' existing demands with less water. The technical efficiency criterion comes in helpful when comparing different items and procedures. For example, if one showerhead can get the same outcome (i.e., showering) with less water or other inputs, it is more efficient than another (e.g., lower water pressure). Some communities employ drip irrigation instead of channel irrigation, or low flushing toilet size instead of local toilets (pit toilets), to conserve a lot of water without abandoning the fundamental function for which water is used (Rosegrant and Ringler 1998)

Any water saved by conservation can be stockpiled, used for the same purpose, or redistributed to other industries. The redistribution of water savings and the current water resources of each important sector is the second stage of water demand management. The process can become controversial when consumers and water providers reject water redistribution for fear of losing access to water resources. Once the potential for water saving through efficient use is fully exploited and all available water is appropriated, new uses can only be settled, to the detriment of existing users. Irrigated agriculture is responsible for the majority of freshwater intake (approximately 85% worldwide, according to the WRI, 1999), so irrigation water is an obvious target for redistribution in industrial, domestic, and environmental applications (Rosegrant and Ringler 1998).

8.6.2 Demand Management: Measures and Programs

Any action, practice, technical device, regulation, or policy that has potential to reduce water use is referred to as demand (or conservation) management. In literature, there are hundreds of different measurements. A variety of methods can be used to measure individual clustering measurements. A popular strategy is to categorize measures according to the purpose of water use. The following are some examples of water-saving technologies and methods that can be used to reduce water use in primary areas: low-flow showerheads, shower flow restrictors, toilet tank inserts, faucets, dual-flush toilets, flush water heaters, flush water pipes shaft, low-pressure supply connections, pressure-relief valves, water-saving landscape designs, and other appliances are used in home. Water recycling, ozone treatment for cooling towers, countercurrent washing and rinsing systems, process water reuse, cooling water recycling, ozone treatment for cooling towers, and treatment and reuse of blasting water are all examples of industrial applications. Water metering, water supply system restoration, leak detection and repair, and pressure relief in distribution systems are all alternatives to agricultural irrigation.

8.6.3 Water Management Programs

Water systems, especially metropolitan piping systems, are manufactured to provide water "on demand" because system operators do not have direct control over the amount of water consumed by users. Many excellent practices and activities that fall under the direct responsibility of water utilities are included in a management-based strategy. Public and commercial entities that distribute water to cities, farmers, and industries can usually improve water use efficiency by following some water management best practices. The most critical of these is accurate monitoring and recording of water flow. Water demand management strategies and implementation programs (Table 8.6) can help achieve this goal (World Economic Forum 2018).

TABLE 8.6

Demand Management Implementation Strategies and Programs

Strategy	Implementation Program
Public Education	• Primary and secondary school programs
	• Promotional campaigns and events
	• Mass media advertising campaigns
	• Dissemination of information through personal contacts
	• Outreach programs to educate water users and help them install conservation hardware
	• Xeriscape garden demonstration
Water Management Programs	• Meter testing and replacement program
	• Leak detection and repair program
	• Distribution system audit program
	• Tax incentives, subsidies, and rebates for adoption of conservation measures
	• Social conservation incentives and disincentives program
Government Regulation	• National water management laws and policies
	• Water rights and priorities of use statutes
	• Government-enforced performance criteria and standards
	• Local conservation codes and ordinances
	• Water use restrictions and bans during emergencies
Economic Incentives	• Water pricing and rate-making policies
	• Tradable water rights
	• Regional water markets and water banks
	• Subsidies and rebates to water users
	• Cross-subsidization of agricultural conservation
	• Tax credits and incentives
	• Penalties for excessive use (quotas)s
	• Privatization of the water supply sector

Source: Dziegielewski and Baumann (1992), Dziegielewski et al. (1993).

8.7 THE TANZANIA PILOT CASE STUDY OF INTEGRATED WATER RESOURCE MANAGEMENT

The Global Water Partnership (GWP) collaborates with several international partners to address the global water resource management challenges through several projects and activities. The projects are aimed at improving water security concerning the United Nations (UN) Sustainable Development Goals (SDGs), focusing majorly on SDG6 and the related targets and other SDGs dealing with issues related to water, including health, climate change, energy, and hunger. Technological advancements and innovations present great opportunities for improved design and planning of effective and sustainable projects for monitoring water. The Global Water Partnership leverages pervasive technologies to enhance sustainable water resource management. For instance, water data can be collected through smartphones, and local citizens are empowered to play significant roles in monitoring water resources at the locality. With a click, the smartphone application can allow users to take measurements of water levels, control water discharge into receiving streams or systems, and make them fit for the purpose data of other people in different parts of the world. By developing existing water collection, access, and storage mechanisms, the data collected can add value by integrating nontraditional data sources (WMO 2017).

A consortium of international partners, including the Swiss and the GWP, has developed innovative monitoring and modeling (iMoMo) systems to promote and develop new data sources. The new data sources complement the already-existing sources used for observing water resource networks operated and maintained by national/regional hydrometeorological services. Through the approach, the long-standing gaps in data sourcing and management are filled with innovative solutions. Where it is possible, the new data sources are codesigned by relevant experts working in conjunction with the users in the communities (WMO 2017).

The Themi River catchment in northeastern Tanzania had increased water demand for irrigation, drinking, and livestock, thus putting pressure on catchment. The pressure led to water shortages, thus affecting dependent ecosystems, electricity (hydropower) generation, and the local community. The iMoMo initiative, in collaboration with relevant local stakeholders, enabled smart water usage through transparency on water allocation and availability. The project, through hardware and software components integration and improved analytical decision support, has benefited the stakeholders and local communities. Some of the outcomes include (WMO 2017):

- The iMoMo Service Centre supports institution strengthening and building, outreach, and project advocacy.
- Site instrumentation by equipping the catchment with river gauges enables local communities to monitor daily irrigation measurements using smartphone applications.
- Tailored information delivered through SMS for end users, such as weather forecast updates and market prices.
- Data collection to inform the administration, management, and planning for water resources.
- Successful outscaling to other areas, such as Rufiji Basin, also in Tanzania, for compliance monitoring of water abstractions for irrigation by farmers.

8.8 CONCLUSION

Most populations in the developing parts of the world lack appropriate water resources to meet household, environmental, and economic development needs. This is attributed due to natural and artificial factors, such as poor management, contamination and misuse of water resources, climate change and global warming, energy production, population growth, urbanization, agriculture, and industrial development. In such areas, human health and productivity continue to be harmed by a lack of drinkable, clean, and safe water for meeting human demands for drinking and sanitation purposes, as well as environmental and natural ecosystem conservation. In recent years, there has been a push to harness emerging technologies to provide sustainable water treatment, transportation, harvesting, recycling, and reuse options. Artificial intelligence–based technologies and solutions have a lot to offer the water resource management industry. AI-based solutions can efficiently handle real-world water supply challenges while also giving tangible benefits. By combining power of AI algorithms with the scale of big data, water resource managers can use data analytics to optimize the information and data they have to make better decisions on water resource management while improving service delivery and lowering costs.

REFERENCES

Abbasi, T. and Abbasi, S.A., 2012. Water Quality Indices. Elsevier, pp. 353–356

Abrah˜ao, R., Carvalho, M., Da Silva, W.R., Machado, T.T.V., Gadelha, C.L.M. and Hernandez, M.I.M., 2007. Use of index Analysis to Evaluate the Water Quality of a Stream Receiving Industrial Effluents. Water SA.

Adetoro, A.A., Abraham, S. and Paraskevopoulos, A.L., 2020. Alleviating water shortages by decreasing water footprint in sugarcane production: The impacts of different soil mulching and irrigation systems in South Africa. *Groundwater for Sustainable Development* 11, pp. 1004–10064.

Agnes, E., Matchawe, C. and Nsawir, B.J., 2021. The use of alternative water sources as a means of adaptation to water shortages in Nsimeyong, Yaounde city: A quality assessment. *Scientific African* 13.

Alizadeh M.J., Kavianpour M.R., Danesh M., Adolf J., Shamshirband S. and Chau K.W., 2018. Effect of river flow on the quality of estuarine and coastal waters using machine learning models. *Engineering Applications of Computational Fluid Mechanics* 12 (1), pp. 810–823.

AL-Washali, T.M., Elkhider, M.E., Sharma, S.K. and Kennedy, M.D., 2020. A review of nonrevenue water assessment software tools. *Wiley Interdisciplinary Reviews Water* 7 (2), p. e1413.

Banerjee, T. and Srivastava, R.K., 2009. Application of water quality index for assessment of surface water quality surrounding integrated industrial estate-Pantnagar. *Water Science and Technology* 60 (8), pp. 2041–2053.

Beale, M.H., Hagan, M.T. and Demuth, H.B., 2010. Neural Network Toolbox 7 User's Guide. The Mathworks Inc., Natick, MA.

Beh, E.H.Y., Zheng, F., Dandy, G.C., Maier, H.R. and Kapelan, Z. 2017. Robust optimization of water infrastructure planning under deep uncertainty using metamodels. *Environmental Modelling and Software* 93, pp. 92–105.

Burek, P., Satoh, Y. and Fischer, G., 2016. Water Futures and Solution: Fast Track Initiative—Final Report (ADA Project No. 2725–00/2014). International Institute for Applied *Systems* Analysis, Laxenburg, Austria.

Butler, J.J., Stotler, R.L., Whittemore, D.O. and Reboulet, E.C., 2013. Interpretation of water level changes in the high plains aquifer in western Kansas. *Groundwater* 51(2), pp. 180–190.

Charisiadis, C., 2018. Brine Zero Liquid Discharge (ZLD) Fundamentals and Design, Lenntech *Research Report*.

Cheng, B. and Li, H., 2021. Improving water saving measures is the necessary way to protect the ecological base flow of rivers in water shortage areas of Northwest China. *Ecological Indicators* 123, pp. 107347.

Cude, C.G., 2001. Oregon water quality index: A tool for evaluating water quality management effectiveness. *Journal of the American Water Resources Association* https://doi.org/10.1111/ j.1752–1688.2001.tb05480.x.

Dadolahi-Sohrab, A., Arjomand, F. and Fadaei-Nasab, M., 2012. Water quality index is a simple indicator of watersheds pollution in the southwestern part of Iran. *Water and Environment Journal* 26, pp. 445–454. https://doi.org/10.1111/j.1747-6593.2011.00303.x.

Dawood, T., Elwakil, E., Novoa, H.M. and Delgado, J.F.G., 2021. Toward urban sustainability and clean potable water: Prediction of water quality via artificial neural networks. *Journal of Cleaner Production* 291, pp. 1252–1266.

Dawood, T., Elwakil, E., Novoa, H.M. and Delgado, J.F.G., 2020. Artificial intelligence for the modelling of water pipes deterioration mechanisms. *Automation in Construction* 120, pp. 1033–1098.

Debels, P., Figueroa, R., Urrutia, R., Barra, R. and Niell, X., 2005. Evaluation of water quality in the Chill´an River (Central Chile) using physicochemical parameters and a modified Water quality index. E*nvironmental Monitoring and Assessment* https://doi.org/10.1007/s10661-005-8064-1.

Donghwi, J. and Joong, H.K., 2018. State estimation network design for water distribution systems. *Journal of Water Resources Planning and Management* 144 (1).

Doorn, N., 2021. Artificial intelligence in the water domaIn: Opportunities for responsible use. *Science of the Total Environment* 755, pp. 1–10.

Dziegielewski, B. and Baumann, D., 1992. Benefits of managing urban water demands. *Environment* 34 (9), pp. 6–11.

Dziegielewski B., Opitz E.M., Kiefer, I.C. and Baumann, D.D., 1993. Evaluating Urban Water Conservation Programs: A Procedures Manual. American Water Works Association. Denver, CO.

El-Fakharany, Z.M. and Salem, M.G., 2021. Mitigating climate change impacts on irrigation water shortage using brackish groundwater and solar energy. *Energy Reports* 7, pp. 608–621.

Elhoseny, M., Yuan, X., El-Minir, H.K. and Riad, A.M., 2014. Extending Self-Organizing Network Availability using Genetic Algorithm. In: Proceedings of the 5th International Conference on Computing Communication and Networking Technologies, ICCCNT, Hefei, China.

Ezhilarasu, C., Mattuvarkuzhali, Z.S. and Ian, K.J., 2019. The application of reasoning to aerospace integrated vehicle health management (IVHM): Challenges and Opportunities. *Progress in Aerospace Sciences* 105, pp. 60–73.

EU., 2020. Self-Sustaining Cleaning Technology for Safe Water Supply and Management in Rural African Areas—SafeWaterAfrica. https://doi.org/10.14332/svc19.proc.0005

Falkenmark M., 1997. Analytical Summary. In: Water for the Next 30 Years: Averting the Looming Water Crisis. Mar del Plata Anniversary Seminar. Stockholm International Water Institute. Stockhom, Sweden.

Falkenmark, M., 2013. Growing water scarcity in agriculture: Future challenge to global water security. *Philosophical Transactions of the Royal Society A* 371, p. 20120410.

FAO., 2020. The State of Food and Agriculture 2020. Overcoming Water Challenges in Agriculture. Rome www.wri.org/facts/data-tables.html.

García, L., Lorena, P., Jose, M., Jimenez, J.L. and Pascal, L., 2020. IoT-based smart irrigation systems: an overview on the recent trends on sensors and Iot systems for irrigation in precision agriculture. *Sensors* 20 (4), p. 1042.

Goralski, M.A. and Tay, K.T., 2020. Artificial Intelligence and Sustainable Development. *International Journal of Management Education* 18 (1), p. 100330.

Gould J., and Nissen-Peterson E., 1999. Rainwater Catchment Systems for Domestic RaIn: Design Construction and Implementation Intermediate Technology. Publication London. p. 335.

Hashimoto, Y., Hara, K. and Sarangaraja, B., 2016. Decomposition of pesticides in water through the use of a slurry-type TiO2 water treatment apparatus. *Journal of Water and Environment Technology* 14(1), pp. 1–5.

Hou, C., Wen, Y., Liu, X. and Dong, M., 2021. Impacts of regional water shortage information disclosure on public acceptance of recycled water—evidences from China's urban residents. *Journal of Cleaner Production* 278, pp. 1239–1265.

Hu, Z., Tan, D., Chen, B., Chen, W. and Shen, D. 2021. Review of model-based and data-driven approaches for leak detection and location in water distribution systems. *Water Supply* 21(7), pp. 3282–3306.

Hutton, G., Kerstens, S., Van Nes, A. and Firmansyah, I., 2013. Downstream Impacts of Water Pollution in the Upper Citarum River, West Java, Indonesia: Economic Assessment of Interventions to Improve Water Quality. In: Water and Sanitation Program Technical Paper, World Bank, Washington, DC.

ICWE (International Conference on Water and the Environment). 1992. The Dublin Statement on Water and Sustainable Development. <www.wmo.ch/web/homs/ icwedece.html>.

Ingram, W. and Memon, F.A., 2020. Rural water collection patterns: Combining smart meter data with user experiences in Tanzania. *Water*, 12 (4), p. 1164.

Ikhioya, P.E., Osu, C.I. and Obuzor, G.U., 2015. Physiochemical characteristic of Harvested Rain water from Bodo Community in Rivers State, Nigeria. *Journal of Applied Science Environmental Management* 19 (4), pp. 671–677.

Intelligent CIO. 2019. How Technology Can Conserve and Improve Access to Water—Intelligent CIO Africa.

Jenny, H., Wang Y., Garcia, E. and Minguez, R., 2020. Using artificial intelligence for smart water management systems. *Asian Development Bank* pp. 1–10. DOI: 10.22617/ BRF200191–2

Judeh, T. and Shahrour, I., 2021. Rainwater harvesting to address current and forecasted domestic water scarcity: Application to arid and semi-arid areas. *Water* 13, pp. 1–16.

Jung, D., Lee, S. and Kim, J.H., 2019. Robustness and water distribution system: State-of-the-art review. *Water (Switzerland)* 11(5), p. 974.

Kannel, P.R., Lee, S., Lee, Y.-S., Kanel, S.R., Khan, S.P., 2007. Application of Water Quality Indices and Dissolved Oxygen as Indicators for River Water Classification and Urban Impact Assessment. E*nvironmental Monitoring and Assessment* 132, pp. 93–110. https:// doi. org/10.1007/s10661-006-9505-1.

Kummu, M., Guillaume, J.H.A. and de Moel, H., 2016. The world's road to water scarcity: Shortage and stress in the 20th century and pathways towards sustainability. *Scientific Reports* 6, pp. 384–3895.

Lee, L.H. and Oh, J.E., 2016. A comprehensive survey on the occurrence and fate of nitrosamines in sewage treatment plants and water environment. *Science of the Total Environment* 556, pp. 330–337.

Li, R., Huang, H., Xin, K. and Tao, T., 2015. A review of methods for burst/leakage detection and location in water distribution systems. *Water Science and Technology: Water Supply* 15(3), pp 429–441.

Lumb, A., Sharma, T.C. and Bibeault, J.-F., 2011. A review of genesis and evolution of water quality index (WQI) and some future directions. *Water Quality, Exposure and Health.* 3(1), pp. 11–24. https://doi.org/10.1007/s12403-011-0040-0.

Ma, Z., Li, H., Ye, Z., Wen, J., Hu, Y. and Liu, Y., 2020. Application of modified water quality index (WQI) in the assessment of coastal water quality in main aquaculture areas of Dalian. China. M*arine Pollution Bulletin* 157. https://doi.org/10.1016/j.

Galal Uddin, M., Nash, S., and Olbert, A.I., 2021. A review of water quality index models and their use for assessing surface water quality. *Ecological Indicators* 122, (1–21), p. 107218

Mapani, B., Makurira, H., Magole, L., Meck, M., Mkandawire, T., Mul, M. and Ngongondo, C., 2019. Integrated water resources development and management: Innovative in Eastern and Southern Africa. *Physics and Chemistry of the Earth* 112, pp. 1–2.

Mekonnen, M.M. and Hoekstra, A.Y., 2016. Four billion people facing severe water scarcity. *Science Advances* 2, pp. 1500323

Mitlin, D., Beard V.A., Satterthwaite, D. and Du, J., 2019. Unaffordable and Undrinkable: Rethinking Urban Water Access in the Global South. In: Working Paper. World Resources Institute, Washington, DC, pp. 1–24.

Mourad, K.A., 2020. A water compact for sustainable water management. *Sustainability* 12, pp. 7339.

Muerdter, C.P., Wong, C.K. and Lefevre, G.H., 2018. Emerging investigator series: The role of vegetation in bioretention for stormwater treatment in the built environment: pollutant removal, hydrologic function, and ancillary benefits. *Environmental Science: Water Research & Technology* 4(5), pp. 592–612.

Naubi, I., Zardari, N.H., Shirazi, S.M., Ibrahim, N.F.B. and Baloo, L., 2016. Effectiveness of water quality index for monitoring Malaysian river water quality. *Polish Journal of Environmental Studies* 25, pp. 231–239. https://doi.org/10.15244/pjoes/60109.

Olaoye, R.A. and Olaniyan, O.S., 2012. Quality of rain water from different roof materials. *International journal of Engineering and Technology* 2(8), pp. 1413–1414.

Park J., Kim, K.T. and Lee, W.H., 2020. Recent advances in information and communications technology (ICT) and sensor technology for monitoring water quality. *Water* 12, p. 510.

Rajaee, T., Ebrahimi, H. and Nourani, V., 2019. A review of the artificial intelligence methods in groundwater level modeling. *Journal of Hydrology* 572, pp. 336–351.

Rojek, I. and Studzinski, J. 2019. Detection and localization of water leaks in water nets supported by an ICT system with artificial intelligence methods as away forward for smart cities. *Sustainability (Switzerland)* 11 (2).

Rosegrant, M.W. and Ringler, C., 1998. Impacts on food security and rural development of transferring water out of agriculture. *Water Policy* 1 (6), pp. 567–586.

Saffran, K., 2001. Environment, A., Cash, K., Canada, E., 2001. Canadian Water Quality Guidelines for the Protection of Aquatic Life Ccme Water Quality Index 1. 0 User' s Manual. Quality 1–5.

Said, A, D.K. Stevens and G. Sehlke., 2004. An innovative index for evaluating water quality in streams. *Environmental Management.*

Salehi, M., 2022. Global water shortage and potable water safety; Today's concern and tomorrow's crisis. *Environmental International* 158, pp. 106936.

Schultz, W., Shahram, J. and Alla, S., 2018. Smart water meters and data analytics decrease wasted water due to leaks. *Journal—American Water Works Association* 110(11), pp. 24–30.

Shamsudin, S.N., Ahmad, A.A., Nurlaila, I., Mohd, H.F.R., Azizah, H.A. and Mohd, N.T., 2016. Review on Significant Parameters in Water Quality and the Related Artificial Intelligent Applications. In: Proceedings—2015 6th IEEE Control and System Graduate Research Colloquium, ICSGRC 2015, pp. 163–68.

Steinhart, C.E., Schierow, L.-J. and Sonzogni, W.C., 1982. An environmental quality index for the great lakes. *JAWRA The Journal of the American Water Resources Association* 18, pp. 1025–1031. https://doi.org/10.1111/ j.1752–1688.1982.tb00110.x

St-Onge, X.F., Cameron, J., Saleh, S. and Scheme, E.J., 2019. A symmetrical component feature extraction method for fault detection in induction machines. *IEEE Transactions on Industrial Electronics* 66(9), pp. 7281–7289.

Šulyová, D., Vodák, J. and Kubina, M., 2021. Effective management of scarce water resources: From antiquity to today and into the future. *Water* 13, pp. 27–34.

Suryanarayana, C., Sudheer, C., Mahammood, V. and Panigrahi, B.K., 2014. An integrated wavelet-support vector machine for groundwater level prediction in Visakhapatnam, India. *Neurocomputing* 145, pp. 324–335.

Sutadian, A.D., Muttil, N., Yilmaz, A.G. and Perera, B.J.C., 2018. Development of a water quality index for rivers in West Java Province. Indonesia. *Ecological Indicators* 85, 966–982. https://doi.org/10.1016/j.ecolind.2017.11.049.

Tang, Y., Zang, C., Wei, Y. and Jiang, M., 2018. Data-driven modeling of groundwater level with least-square support vector machine and spatial—temporal analysis. *Geotechnical and Geological Engineering* 37, pp. 1661–1670.

The World Resources Institute., 1999. World Resources 1998–99. Chapter 12. *Fresh Water.* New York.

Tiyasha, K.E., Tung, T.M. and Yaseen, Z.M., 2020. A survey on river water quality modelling using artificial intelligence models: 2000–2020. *Journal of Hydrology* 585, pp. 1246–1270.

UN, 2015. Transforming our world: The 2030 Agenda for Sustainable Development. https://sdgs.un.org/2030agenda. Accessed on November 30, 2021.

UNCTAD, 2011. Water for Food-Innovative Water Management Technologies for Food Security and Poverty Alleviation.

UNICEF, 2021. Reimagining WASH: Water security for allwww.unicef.org/media/95241/file/water-security-for-all.pdf. Accessed on December 02, 2021

United Nations., 2015. Technology and Innovation Report; Harnessing Frontier Technologies for Sustainable Development,

U.S. Environmental Protection Agency., 2009. Massachusetts Energy Management Pilot Program for Drinking Water and Wastewater Case Study, EPA-832-F-09–014, U.S. Environmental Protection Agency, Washington, D.C., 2009.

Vikaskumar, G.S., Hugh, D., Phillip, M.G., Peter, C., Timothy, K.R. and Tony, R., 2007. Comparison of water quality parameters from diverse catchments during dry periods and following rain events. *Water Research* 41(16), pp. 3655–3666.

Vrachimis, S.G., Eliades, D.G. and Polycarpou, M.M. 2018. Real-time hydraulic interval state estimation for water transport networks: A case study. *Drinking Water Engineering and Science* 11 (1), pp 19–24.

WHO, UNICEF., 2019. Progress on drinking water, sanitation and hygiene: 2000–2017: special focus on inequalities. https://data.unicef.org/resources/progress-drinking-water-sanitation-hygiene-2019/. Google Scholar. Accessed on November 30, 2021.

WMO., 2017. Innovative and Community-based Sustainable Water Management. World Meteorological Organization (WMO).

World Economic Forum., 2018. Harnessing the Fourth Industrial Revolution for Water. Retrieved from www.weforum.org/reports/harnessing-the-fourth-industrial-revolution-for-water

Wu, Y. and Liu, S. 2017. A review of data-driven approaches for burst detection in water distribution systems. *Urban Water Journal* 14 (9), pp 972–983.

Xiang, X., Li, Q., Khan, S. and Khalaf, O.I., 2021. Urban water resource management for sustainable environment planning using artificial intelligence techniques. *Environmental Impact Assessment Review*, 86, pp. 1065–1115.

9 Implementation of Industry 4.0 in Water Treatment

Alan Guajardo, Abrahan Mora,
Marco A. Mata-Gómez,
and Pabel Cervantes-Avilés

CONTENTS

9.1   INTRODUCTION

The wise use of water is a key factor for sustainable development and keeping the basic quality of life. Water is necessary for many different anthropogenic activities and is vital for humans, animals, plants, and the environment. Because of this, the management and treatment of water is of top importance for any civilization.

Poor water quality in water bodies of developing countries has caused concern about the efficiency of water treatment plants. An inefficient treatment can induce a harmful effect on the health of individuals. Moreover, this has also triggered the inhabitants to increase the consumption of industrial beverages, such as soda and bottled water, which are more expensive, less convenient, as they require several production processes, which end up wasting a large amount of water during filtration process and creating by-products like liquid treatment sludge (Samuel et al. 2019).

DOI: 10.1201/9781003260455-9

Water treatment can be dated back to antiquity, when water was boiled for safer consumption. However industrial water treatment wasn't used until many centuries later, in the first industrial revolution, when, for example, by accident some worker left potatoes inside a boiler to later find out that the tank had less scale than normal, which could be cleaned easier (Wang et al. 2015). With the creation of the steam engine, the study of water treatment became more important, as in the beginning of the twentieth century, higher-temperature engines were more prevalent; as an effect, the scale produced had become an even bigger problem, so by using phosphate and excess carbonate, this was solved (Domoney et al. 2015).

In general, water treatment has been involved in all industrial revolutions—for example, with the discovery of electricity, it had been possible to convert water load or flow to electricity. In the Third Industrial Revolution, it was possible to use electronics to monitor the water treatment process, which leaves us in the beginning of the Fourth Industrial Revolution. This new revolution brings to the water treatment industry a wide scope of tools for monitoring and enhancement of the process.

The latest Industrial Revolution, called Industry 4.0, or I4, started in Germany and centers on the change from machine-dominant production to digitally dominated. When implemented correctly, Industry 4.0 can greatly improve the efficiency of production systems. Applying this philosophy to water treatment plants, we may more easily identify and prevent issues, enhancing water quality and saving money by eliminating any part of the process which does not bring value to the consumer, and, overall, tend to guarantee a better-functioning plant. Based on this, Industry 4.0 technology can be applied to the water treatment and distribution industry to improve it and make it more competitive.

Industry 4.0 and its inherent technologies can provide a positive change that could help the water industry develop better processes for water treatment and supply, either in the industry or in the society. Industry 4.0 can contribute to improve water quality and efficiency during treatment, which is beneficial for consumers and the environment.

9.2 INDUSTRY 4.0 AND THE WATER SECTOR

To understand Industry 4.0 and its importance for water treatment, it is needed to retrieve some details from the previous Industrial Revolutions. The development of mechanical production facilities, which began in the second part of the eighteenth century and accelerated throughout the whole of nineteenth, was the First Industrial Revolution. The Second Industrial Revolution began in the 1870s as a result of electrification and the division of labor (also known as Taylorism). The Third Industrial Revolution, sometimes known as "the digital revolution," began in the 1970s, when sophisticated electronics and information technology furthered the automation of manufacturing processes (Oztemel and Gursev 2018). In the current Industrial Revolution, there are concepts that must be considered, and further described, such as (Lasi et al. 2014):

- Smart factory
- Cyber-physical systems

- Self-organization
- New systems in distribution and procurement
- New systems in the development of products and services
- Adaptation to human needs (New manufacturing systems should be designed to follow human needs instead of the reverse.)
- Corporate social responsibility

These concepts are applied to a product life cycle that contains six phases of *design, fabrication, delivery, use, refurbishment, and disposal* (Labuschagne and Brent 2005), which can be replicated for the water treatment industry. Mass customization models have gained popularity to represent the life cycle due to rising consumer demand for a wide range of items (Cerdas et al. 2017). The market demand, which is currently becoming more customer-focused (Choudhari and Patil 2016), now calls for smaller batch sizes with shorter life cycles. Diversity is typically obtained at different points in the product life cycle. Adjustments made while the product is being used, as well as those made throughout the design, manufacture, sales, and distribution phases, can all be used to achieve this. For instance, rapid prototyping can be utilized to achieve manufacturing process diversity (Mourtzis 2016). Because additive manufacturing systems generate fewer components at a given time than traditional manufacturing techniques, it is essential to continuously monitor the entire process and make the necessary adjustments to improve the number of acceptable components.

Currently, there are two main approaches to enhance the manufacturing systems via Industry 4.0. One is based on sensors and intelligent data collection that are extending the life cycle of any asset, from design to manufacture, distribution, maintenance, and recycling (Errandonea et al. 2020). The other one is the real-time three-dimensional measurement technique, which is becoming important together with artificial intelligence and robotics (Wang 2020). However, applied research is needed to understand all phases of a life cycle during additive manufacturing (Liu et al. 2021).

9.2.1 Smart Factory

Smart factories are a highly digitalized and interconnected places that can automate and self-optimize their processes to improve them. It is up to the manufacturer to carry out the actual processing necessary to transform primary materials and semi-finished items into finished goods. In this process, it includes various physical or informational subsystems, which are part of the production and management inside the limits of a factory. The actuator and sensor, control, production management, manufacturing, and execution, as well as corporate planning levels, are only a few of the hierarchical levels where these subsystems can be found (Wang et al. 2016). Some of these systems are already part of the utilities for water treatment; however, this does not imply that it can address self-optimization processes.

With the internet and network systems' rapid development, smart decision-making is easier; this capability has also prepared the way for the creation of intelligent machines that, in the future, will be able to think, learn, and remember as well as share their knowledge at any given time or respond to specific situations (Hozdić

2015). In spite of this, it is anticipated that in the near future, smart machines will shape occupations, industrial processes, and production systems, including water utilities either for water or wastewater treatment.

9.2.2 CYBER-PHYSICAL SYSTEMS

Recent technological advancements have resulted in the emergence and evolution of digital technologies, such as cyber-physical systems (CPS), Internet of Things (IoT), Internet of Services (IoS), Big Data, cloud computing, semantic web, and virtualization (Mabkhot et al. 2018). CPS is defined as transformative technologies for managing interconnected systems between their physical assets and computational capabilities. This technology is one of many that make up digital twin (DT). The idea of using "twins" can be attributed to NASA's Apollo program, when more than two identical space vehicles were built. This allowed the engineers to have the identical conditions of the space vehicle during the task, with the vehicle remaining on Earth being experimented as a reference or control for the mission's conditions (Boschert and Rosen 2016). Then, the words "digital twin" were first introduced to the public in NASA's Integrated Technology Roadmap for the Technology Area 11: Modeling, Simulation, Information Technology and Processing (Shafto et al. 2012).

Due to the fact that these technologies are constantly evolving, it is believed that DT will also do so (Warke et al. 2021). DT is increasingly being investigated as a way to enhance the performance of physical entities, such as water treatment plants (Jones et al. 2020). This can be potentialized with the use of computational methods made by the virtual counterpart. Moreover, the use of DT represents an ideal way to facilitate remote human interaction with physical machines and get over the problem of distance (Hu et al. 2018).

Virtual or digital representations of systems or objects are also known as digital twins. DT can be as intricate as an entire city, a factory, or even a planetary system. These virtual representations include real-world items, like people, buildings, cars, highways, houses, and transit networks (Canedo 2016). With this in mind, it would be able to develop a digital duplicate of the city's pipeline as well as a water treatment facility.

Using digital representations that can be shared across the product life cycle makes DTs for products and processes more practical. To generate data for digital twin modeling and visualize the outcomes, it can use computer-aided design (CAD), computer-aided manufacturing (CAM), computer-assisted instruction (CAI), and computer numerical control (CNC) technologies (International Organization for Standardization [ISO], 2021). Although these tools may enhance the virtual experience of users exploring and understanding water utilities, the water sector demands the physical operation of many components, such as the interruption of the supply due to unexpected leaks, determination of water quality caused by spills, among others.

Systems that combine computing with physical operations are called as cyber-physical systems (CPS). These systems embed computing and digital networks to monitor and regulate the physical processes with feedback loops, where physical inputs influence calculations (Lee 2008). The economic and societal potential of

CPS has been recognized, and significant investments are being made globally to advance the technology in many sectors, including water. However, the deployment of Industry 4.0 can have difficulties for implementation, especially in developing countries, mainly due to capital expenditure and a highly specialized workforce. Nevertheless, by using CPS in water treatment facilities, the treatment process can be optimized in cost–benefit ratio. According to some authors, to create a water CPS, the following are needed (Wang et al. 2015):

- Providing an accurate and reliable model of the water infrastructure and enabling time-aware and time-critical activities inside a water CPS through the use of sensing, communications, and networking technologies.
- Using computing technologies, such as computational modeling, data management, and machine learning, to overcome the difficulties of water sustainability.
- Creating adaptive and predictive hierarchical hybrid control systems to accomplish coordinated and synchronized actions and interactions in a water CPS.

Implementation of the digital twin in a water treatment process should include the following characteristics, which include those typical for DT (Boschert and Rosen 2016): behavioral descriptions through various simulation models, operational data, and process data; updated information while developing the actual system throughout its life cycle; and coming up with a solution that is relevant to the actual system, such as water flow, pumping energy, chemical reagents dosification, that allow explaination of how things work within water utilities.

In addition to these characteristics, a digital twin must support many different functions (Kostenko et al. 2018) which are in line with water treatment aims, such as prediction of status and asset analysis, a life span reflection of the item in digital form, resource administration, and decision support based on technical and statistical analysis. According to Kostenko et al (2018), these functions allow the provision of real-world items, the ability to grow the hierarchy of the processes and the user interfaces, making a model more complex by adding both hierarchical levels, portraying several process types using a single display composition, showing various visualization scales with various hierarchical levels and levels of detail, a framework that links diverse asset modeling techniques with comprehensive digital models, and finally, a semantic model that organizes all data pertaining to a plant and lucidly explains the logic control of various processes within water treatment plants.

In general, the benefits of using digital twins in the water sector can include parameter optimization offline or in real time, simplifying automatic data flow management, and monitoring of work parameters, as well as anticipating and stopping machine or subsystem due to defects (Corradini and Silvestri 2021).

9.2.3 SELF-ORGANIZATION AND SELF-OPTIMIZATION

Self-organizing in a distributed manufacturing is essential to achieve Industry 4.0 principles, as new technologies and escalating industry competition have driven production to be flexible, reconfigurable, adaptable, and efficient. Inspiring is the idea of self-organizing drawn from nature, which allows systems to operate independently and impulsively without outside influence. It occurs naturally in biology, chemistry, economics, and other complex systems. Examples include the invisible hand in economics and ant or bee foraging strategies. In this regard, biology, nature, chemistry, and evolutionary theories all serve as excellent sources of inspiration for strengthening systems with self-organization (SO), in order to maximize their potential for solving complex issues. A distributed approach was suggested for the so-called self-organizing manufacturing system (SOMS) (Zhang et al. 2017).

With new mechanisms for procurement and distribution, these processes will become more customized. Various different channels will be used to manage connected processes (Lasi et al. 2014).

Based on system state changes and the feedback incentives that accompany actions, reinforcement learning develops an optimal decision that may be applied to an unfamiliar environment. The target system can be dynamically adjusted by the reinforcement of a learning-based system, which lowers the need for manual intervention or threshold-based rules or tactics (Deng et al. 2021). The DT can aid in real-time optimization and evaluation of production planning and process behavior (Semeraro et al. 2021).

In order to generate a framework for digitalization of CPS in water treatment facilities and other potential CPS, there are five tiers of its architecture that can be considered (Josifovska et al. 2019), which include configuration level, collaborative diagnosis and decision-making, twin component and machine models at the cyber level, degradation and performance prediction, and smart connection level (Figure 9.1). This process goes from self-adjusting and self-optimizing up to performance prediction and management of sensor network.

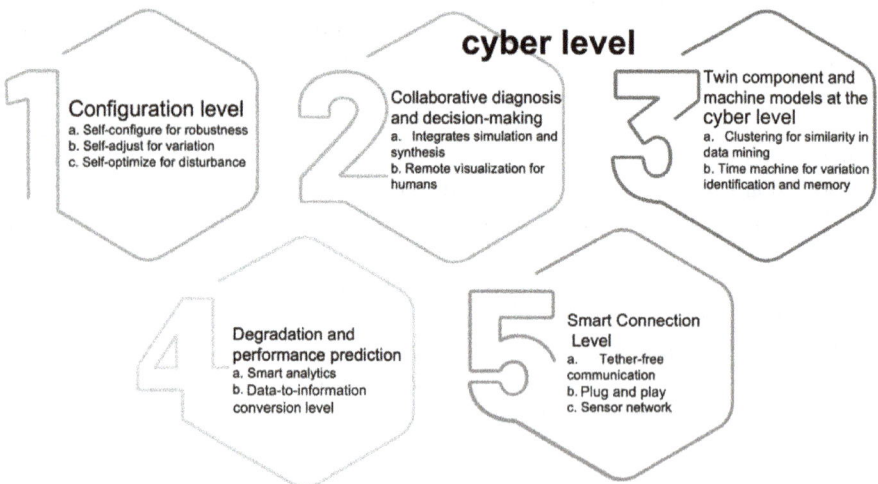

cyber level

Configuration level
a. Self-configure for robustness
b. Self-adjust for variation
c. Self-optimize for disturbance

Collaborative diagnosis and decision-making
a. Integrates simulation and synthesis
b. Remote visualization for humans

Twin component and machine models at the cyber level
a. Clustering for similarity in data mining
b. Time machine for variation identification and memory

Degradation and performance prediction
a. Smart analytics
b. Data-to-information conversion level

Smart Connection Level
a. Tether-free communication
b. Plug and play
c. Sensor network

FIGURE 9.1 General framework of architecture for cyber-physical system according to Josifovska et al. (2019).

9.2.4 SMART CITY

A systematic strategy to integrating data and digital technology to ensure sustainability, citizen welfare, and urban environment's economic growth is known as smart city (Ivanov et al. 2020). As a result of incorporating new technology into their management, it is trendy, and we essentially learn about new smart cities every day. The idea of a "smart city" refers to an area where the essential elements of urban infrastructure, such as power, emergency management, traffic management, and the environment, are integrated in a way that new technologies and other functions can be linked between them through data networks and systems (Nam & Pardo 2011). The quality of life of residents must improve for a city to earn the title of "smart city"; therefore, it is imperative that it utilizes all the technological advancements' potential for cost savings, increased efficiency, new service provision, reduced environmental impact, and innovation (Aguaded-Ramírez 2017).

Today, approximately 55% of the world's population resides in cities, and by 2030, that percentage is projected to rise to 66% (UNESCO 2019). New issues, such as traffic congestion, waste management, pollution, parking distribution, etc., occur as the city expands and resources are in short supply. Therefore, it is important for all society sectors to concentrate on adapting the city to meet the demands of the present and the future (Camero and Alba 2019).

Principles of Industry 4.0 have been applied in a smart city by applying a water monitoring method that makes use of inexpensive water turbines in a smart city (Domoney et al. 2015). In this work, a system was described by Domoney et al. which can measure flow rates and produce electrical power by identifying the key components of a smart city water management solution. With self-powered flow rate monitoring accessible in the household setting, the hydro-generation scheme's full capability was proven.

The various platforms and technologies needed for a smart environment were examined by Shahanas et al. (Shahanas and Sivakumar 2016). It was found that the use of cutting-edge technologies, like the IoT and predictive analytics, in the context of smart cities is affordable to set a plan for a smart water management system. In fact, both approaches can use two alert mechanisms for water stakeholders as warning, such as mail and SMS, described next.

- Mail. Information about excessive water consumption is taken from the Exus table and sent via mail to the relevant individual or group. Python's smtp package is used to deliver mail alerts. Excel files can be created from the tables using the xlswriter program and emailed as attachments.
- SMS. An SMS alert system is set up when a certain amount of water in a tank is reached. Txtweb is used for this service. For this project, 20 APIs were established in order to deliver SMS messages to interested parties.

9.2.5 EMERGING SYSTEMS IN THE DEVELOPMENT OF GOODS

Innovation in goods and services needs arduous and protracted research and development, which Industry 4.0's cutting-edge technologies, including simulation through virtual reality, are enabling. The manufacturing process, with its associated expenses, is the next step, though, and it can operate as a competitiveness hurdle. The lead time for products or services of markets also gets longer towards the end (Alcácer and Cruz-Machado 2019).

With Industry 4.0, the goal is to adapt the machine to the worker rather than the other way around, making labor safer, more effective, more comfortable, and producing higher-quality products. Normally, workers are taught and adapted to use technology in an industry. Although businesses have traditionally worked hard to boost production, it has recently become clear that doing so is insufficient to maintain competitiveness (Yu et al. 2008). As a result, businesses have started to provide solutions aimed at improving both market share and consumer happiness (Beuren et al. 2013)

Historically, most national wealth has increased by 40% to 90% as a result of newly created technology-based items (Campbell 1983). According to Easingwood and Storey (1991), the most important success factors are overall quality, a differentiated product, product fit and internal marketing, and use of technology. Moreover, for a business to be competitive, it is essential to develop new success factors (Brentani 1991). Therefore, many businesses are attempting to create cutting-edge technological products because economic sectors recognize the strategic value of technology commercialization (Cho and Lee 2013).

In addition to dynamic issues of water sectors, which can be complex, changes in the hardware and software sectors, as well as changes in software users' demands and requirements across a range of industries, have an impact on the life cycles of software products and services in water (Aramand 2008). To create novel goods and solutions that satisfy clients' and stakeholders' requests, all working engineering managers must be able to comprehend the instruments at their availability, including the life cycle assessment and tools that can forecast potential social, economic, and environmental threats. This tendency, in turn, has caused to provide customizable products and services which can be applied in the water sector.

The following inquiries should be included as part of the business model to determine whether a product needs to be customized (Pine 1993).

- Who requires my goods or services, and why? How can I give it to anyone who requests it?
- How do customers use my product or service differently? How do I fulfill all their needs and wants?
- In what locations do buyers need my goods or services? What must I do in order to give it to them wherever they request it?
- When should customers modify my product or service to suit their needs? How can I make it available to them whenever they need it? Can I give it right away? Do you mean 24 hours a day?
- Why do clients require my goods or services? How can I add additional value so that they can achieve their goals?
- In what ways do buyers require my goods or services? What can I do to give it to them however they prefer?

9.2.6 CORPORATE SOCIAL RESPONSIBILITY

The history of the idea of corporate social responsibility (CSR) is extensive and diverse. Evidence of the business community's interest for society can be found dating back decades (1999, Carroll). The pyramid proposed by Carroll is the most common example for CSR, which has in the base economic and legal responsibilities, followed by ethics responsibility, and philanthropic responsibility on top of the pyramid.

Currently, CSR can be addressed by ISO 26000 following seven fundamental topics interconnected, such as organization administration, human rights, work practices, environment, fair operating practices, consumer matters, and active participation and community development (ISO 2011). Since these topics follow the standard principles of accountability, transparency, ethical behavior, respecting the interests of parties and legality, the CSR is commonly reported through the global reporting initiative (GRI) through the accomplishment of the series 400.

As demonstrated by the Volkswagen emissions scandal in September 2015, social responsibility is crucial for a business because it is an integral part of its branding. People frequently purchase products based solely on their brand, so a positive company reputation can significantly increase sales and support business expansion. Hence, Industry 4.0 implementation in water sector must fulfill the CSR standard requirements.

9.3 LEAKS: CAUSES AND PREVENTION

The amount of water leaked in water distribution systems varies widely between different countries, regions, and systems, from as low as 3–7% of distribution input in the well-maintained systems in the Netherlands (Beuken et al. 2006) up to 50% in some developing countries and less-well-maintained systems (Mamlook and Al-Jayyousi 2003).

Natural ground movement is one of the many causes of pipeline failures. Wave propagation and permanent ground displacement (PGD) are two seismic dangers that are significant for pipelines. Natural ground movement–related pipeline breakdowns are exceedingly expensive.

Earthquakes like the Northridge earthquake in 1994 have demonstrated that the actual cost of repairing pipes only accounts for a small portion of the overall expenses involved with the failure of lifelines. Direct losses are often dwarfed by indirect losses, which include economic losses from downtime and fire damage that could have been avoided but for the loss of the water supply (Pour-Ghaz et al. 2011).

As we know, prevention maintenance can greatly reduce the number of failures in any system. Pipes are not the exception; if pipelines are left alone with incorrect maintenance, it will only be a matter of time before a leakage occurs. It is of great importance to have periodic maintenance, but the problem is that the pipes in the water treatment industry are mostly underground. As such, periodic maintenance becomes very costly and time-consuming. Because of this, monitorization and forecasting of failures within the pipe systems would help detect the need of maintenance instead of having to apply periodic maintenance to the whole pipeline, which can be kilometers long. With several approaches based on Industry 4.0, users just need to give maintenance to the affected sector at a time, besides to know when and the recovered capacity. Another approach to involve Industry 4.0 is to monitor the potential failure zones in a water treatment and distribution system. This would greatly reduce cost and will be the best option at long term, preventing water shortage due to water supply outages.

There are a wide variety of technological ways for prevention and detection of leaks. Some of them are based in inspection and sensing of surrounding environment, such as soil, air, temperature, and even sound (Table 9.1). For example, leaks from water supply pipes generate noise, which can be used for leak detection and location by amplification of waver systems and reconstructions of the soil compactions and its influence (Gao et al. 2005). However, the accuracy of listening devices like ground microphones for locating leaks depends largely on the user's experience, and the process is time-consuming. When listening devices are couple to noise correlators, which consider the natural noise background, it can be used for leak prevention or as early warning tool. Compared to listening devices, leak noise correlators are more effective, produce more accurate results, and rely less on human experience (Hunaidi et al. 2004).

Another potential technology highly feasible to be implemented to detect leaks in water treatment or water supply system is the electromagnetic inspection. In this approach, a magnet and a switch make up the two components of magnetic sensors that create a magnetic field. Depending on the magnetic field's intensity, which is affected by the switch's and the magnet's closeness to one another, the switch is activated. This can be used to track the development of a crack in the pipes (Pour-Ghaz et al. 2011). Because it costs less to simply seal a fissure than it does to have to replace a pipe that is already leaking, this can significantly lower the cost of maintenance.

Although many of the technologies applied for leaks detection are highly efficient, their application to respond after natural events such as earthquakes and land

---

TABLE 9.1

Application of Technology for Leaks Detection per Pipe Material

Type of Technology	Works With		
Pit depth measurement	Metallic pipes		
Visual inspection	Metallic pipes	Concrete pipes	
Electromagnetic inspection	Metallic pipes	Concrete pipes	
Acoustic inspection	Metallic pipes	Concrete pipes	Poly pipes
Ultrasonic testing	Metallic pipes		
Radiographic testing	Metallic pipes		
Thermographic testing	Metallic pipes		
Pipe condition assessment from soil properties	Metallic pipes		
Other sensor technologies	Metallic pipes	Concrete pipes	

Source: Adapted from Liu and Kleiner 2013.

movements and subsidence, especially in developing countries, is still challenging. Industry 4.0, especially DT and CPS, can deal with this current challenge that will allow recovery of water treatment and supply in shorter times due to real-time monitoring and response.

9.4 WATER TREATMENT: TECHNOLOGIES OUTSIDE AND INSIDE BUILDINGS

In water sector, it is crucial to have as much quality control as possible, because water quality is crucial to a nation's population's health. The majority of nations now have very strict water purification regulations, but its handling and storage leave much to be desired in developing countries. UNICEF notes in its Water Handbook that there are numerous instances in which water that is bacteria-free at the source becomes polluted during transit, storage, and consumption (UNICEF 2016). Any endeavor to supply water, no matter if this is a household or a utility, that ignores this will be ineffective. Because of this, it would be extremely useful to gauge the quality of the water as function of disinfection level as well as the inorganic and organic pollutants content, especially in tap water that is distributed to the population, and in the common water in the pipeline of sectors and districts of the cities. Definitely, Industry 4.0 has a gap to fulfill by using in-pipe real-time sensors which could gauge the concentration of bacteria and other contaminants in the water. The initiatives ensuring the sanitation of water coupled to CPS can be conceivable and would also generate confidence in water utilities and stakeholders.

De-oiling (removal of dispersed oil and grease), desalination, removal of suspended particles and sand, removal of soluble organics, removal of dissolved gases, removal of naturally occurring radioactive materials (NORM), disinfection, and softening (to remove excess water hardness) are the general goals for operators treating water before its supply (Igunnu and Chen 2012).

Conventional water treatment technologies outside buildings are those used in treatment facilities, such as coagulation-flocculation processes, sand filtration, high-speed and pressurized filtration, as well as membrane filtration. However, biocatalyzer (enzymes), biochemical treatments, and even advance oxidation processes based on photochemical or electrochemical reactions are more commonly applied to remove recalcitrant contaminants in water sources.

Through a number of different methods, nanotechnology is being applied to improve traditional ceramic and polymeric membrane materials for water treatment. The most promising of the many ideas put forth so far are membranes coated with zeolitic and catalytic nanoparticles, hybrid inorganic-organic nanocomposites, and membranes inspired by living things, such as membranes made of aligned nanotubes, isoporous block copolymers, and hybrid protein-polymer biomimetic membranes (Pendergast and Hoek 2011; Asif and Zhang 2021).

In developing nations, it is becoming more common for buildings and housing to treat water supplied due to its poor quality. Therefore, filter and disinfectant tap water are highly common for residents that have no confidence in water supplier and sources besides to provide an added benefit. Treatment technologies tend to be those smaller mainly based in membrane filtration and UV radiation, which increases the costs of treatment compared to outside technologies. An example is the case of Atria Vertical Living, where residents can choose to pay extra for this service and the cost depends on consumption. In contrast, in another apartment building in the metropolitan area of San Pedro, Mexico, all residents receive filtered water as it is already included in the maintenance fee. The changes for water treatment outside and inside the buildings open the opportunity for Industry 4.0 to create customizable water treatment based on CPS, which can, in turn, optimize water treatment and reuse.

9.5 ADDITIVE MANUFACTURING

The act of transforming any substance into a product with diverse forms and sizes, whether or not the material attributes of the end product are changed, is known as a manufacturing process (Kaushish 2010). The selection of manufacturing processes and their design are important variables in the quality and economic efficiency of industrial output, including water treatment (Klocke and Kuchle 2009). Traditional manufacturing techniques can be divided into (Rajput 2007):

- Subtractive
  - Machining
  - Grinding and finishing
  - Unconventional machining
- Constant mass processes
  - Casting
  - Metal forming
  - Powder metallurgy process
  - Heat treatment
- Additive
  - Wielding
  - Joining
  - 3D printing

According to ASTM, "a technique of combining materials to produce items from 3D model data, usually layer by layer," is how *additive manufacturing* (AM) is defined (Yossef and Chen 2015). A group of cutting-edge manufacturing techniques known as 3D printing is used to create physical parts in a discrete, line-by-line, layer-by-layer, or additive way from 3D CAD models that have been digitally cut into 2D cross sections (Chen et al. 2019). Additive fabrication (AF), additive layered manufacturing (ALM), rapid casting, rapid manufacturing (RM), rapid prototyping (RP), fast tooling, and solid free-form fabrication are just a few of the many names given to the rapidly evolving subject of AM. In ISO/ASTM 52900, an effort is made to standardize the nomenclature (Buchanan and Gardner 2019). No matter the type or stage of AM, applications are moving fast to fluids control in industries, while water is in the target in the near future to produce components for water treatment, sensors, pipes, valves, lamps for UV disinfections, and even chemical reagents through different methods for AM.

One of the most often-used methods is fused filament fabrication (FFF), or fused deposition manufacturing (FDM), an additive manufacturing process based on the layered deposition of melted thermoplastic (Baş et al. 2019). A heated nozzle is used to extrude a molten filament to create items in the FDM process (Sun et al. 2008). The mechanical characteristics of the fabricated pieces are one of this technique's drawbacks in engineering applications. Programs that enable users to regulate the printing process are required because it directly affects the part's quality (Fernandez-Vicente et al. 2016).

Binder jetting, material extrusion, directed energy deposition, material jetting, powder bed fusion, sheet lamination, and vat photopolymerization are the seven categories into which AM is divided in accordance with the fundamentals of manufacturing (Singh et al. 2020), which are explained in Table 9.2 and must be considered to create components of the water treatment industry.

Construction, aerospace, food, soft robotics, automotive, biomedical equipment, prosthetic implants, health care, printed electronics, biomimetic designing, water treatment, energy harvesting, and desalination are just a few of the industries that have frequently used 3D printing in recent years (Ahmed 2021). There are many different applications for this technology; for instance, 3DP technology is used in the medical sector to create bones and joint grafts as well as anatomical pieces for research and analysis. For their clients, architects use 3DP to generate intricate 3D models. Airfoils are also printed using 3DP in the aerospace sector (Kidwell 2017). In water sector, 3DP has been applied in membrane separation, desalination, and water treatment (Tijing et al. 2020), as well as for sensors of several pollutants (Contreras-Naranjo et al. 2021).

The product development stage of product design and manufacturing has greatly benefited from additive manufacturing technology (Lemu 2012). However, for the AM method to be considered as a true manufacturing process, it is missing crucial information, such as process repeatability and uniformity of the made items (Qattawi and Ablat 2017). To carry out this and apply it to contaminants sensors and components of water treatment plants, a general workflow to create a 3D part can be followed (Figure 9.2).

9.6 SIMULATION VALIDATION

It's crucial to confirm that a new component's mechanical requirements meet the particular requirements of the applications for which they are created. Because experimental testing takes a lot of time and money, finite element analysis (FEA) was developed to handle challenging elasticity and structural analysis problems in civil and aeronautical engineering (Erdemir 2012). FEA can lower the cost of materials and experimental testing (Paul 2021). FEA is a technique that combines elements from the approximative solutions of partial differential equations with a linear combination of polynomial trial functions (Paul 2021).

FEA reduces complicated problems to a manageable number of variables. FEA is used to simulate a variety of boundary conditions, materials, constraints, and loads for constrained parts (Domingo-Espin et al. 2015, Salonitis and Al Zarban 2015).

TABLE 9.2
Description, Advantages, and Limitations of the Additive Manufacturing (AM) Families

Family	Alternative Names	Description	Advantages & Limitations	Typical Materials
Photopolymerization	Scan, Spin, and Selectively Photocure Stereolithography Digital Light Processing Continuous Liquid Interface Production	A laser or projector is used to selectively expose a vat of liquid photopolymer resin to light, causing polymerization to start and turning the exposed areas into solid parts.	Advantages: Good degree of precision and intricacy; finished surface is smooth; accommodates extensive build zones. Limitations: Needs support and postcuring.	UV curable photopolymer resins
Powdered Fusion	Selective Heat Sintering Selective Laser Melting Selective Laser Sintering Direct Metal Laser Multijet Fusion Sintering Electron Beam Melting	Powdered materials are selectively consolidated by melting it together using a heat source, such as a laser or electron beam. The unfused powder surrounding the consolidated part acts as a support material for overhanging features.	Advantages: Substantial complexity as a support material; power a variety of materials. Limitations: Needs postprocessing; the machine is challenging to clean; employ caution while using x-rays.	Powders, plastics, sand, metal, and ceramic
Binder Jetting	3D Printing, Voxel-jet Ex-One	To construct pieces layer by layer, liquid bonding agents are selectively put onto thin layers of powdered material. Both organic and inorganic materials are used as binders. After being printed, pieces made of metal or powdered ceramic are usually burned in a furnace.	Advantages: Enables for printing in colors; extremely productive; uses a variety of materials.	Sand, powdered plastic, glass, metal, and ceramic
Material Jetting	Smooth Curvatures Printing Polyjet Multijet Modeling	To create pieces, material droplets are placed one layer at a time. Both jetting thermally molten materials that afterwards solidify in ambient temperatures as well as jetting photocurable resin and curing it with UV light are common kinds.	Advantages: Excellent precision; enables elements to be fully colored; enables the use of multiple materials in one part. Limitations: Wax-type materials and a lengthy production time are weaknesses.	Polymers, photopolymers, waxes

Sheet Lamination	Ultrasonic Additive Manufacturing Laminated Object Manufacture Selective Deposition Lamination	To create an object, sheets of material are piled and bonded together. Adhesives, chemicals (for paper or plastic), ultrasonic welding, or brazing are all possible lamination techniques (metals). After the object is formed, unused areas are ripped out layer by layer and eliminated.	Advantages: Rapid volumetric growth; relatively affordable (nonmetals); allows for various metal foil combinations, including those with embedding elements.	Metal tapes, paper, and plastic sheets
Material Extrusion	Fused Deposition Modeling Fused Filament Fabrication	A nozzle or aperture is used to extrude the material in tracks or beads, which are then assembled into multilayer sculptures. Heated thermoplastic is a common.	Advantages: Affordable and cost-effective; allows for a variety of hues; suitable for usage in an office setting; good structural qualities. Limitations: A slower process than SLA and a subpar finish.	Pellets, thermoplastic filaments, and liquids
Directed Energy Deposition	Direct Metal Deposition Laser Metal Deposition Laser Engineered Net Shaping	Using an energy source like a laser or electron beam, powder or wire is fed into a melt pool that has been created on the surface of the part, where it binds to the underside part or layers. Essentially, this is a type of automated build-up welding.	Advantages: Not constrained by axis or direction efficient for improvements and feature addition; several materials combined into one component; highest rates of single-point deposition. Limitations: Requires polish, but only for little portions.	Metal powder, metal wire with ceramics
Hybrid	AMBIT	Combining CNC machining with laser metal deposition (a type of DED), which combines additive manufacturing and "subtractive" machining, enabling products to take advantage of both techniques.	Advantages: High productivity and a smooth surface finish; DED's material and geometric freedoms; automated support finishing, inspection, and removal.	Metal powder, metal wire with addition of ceramics

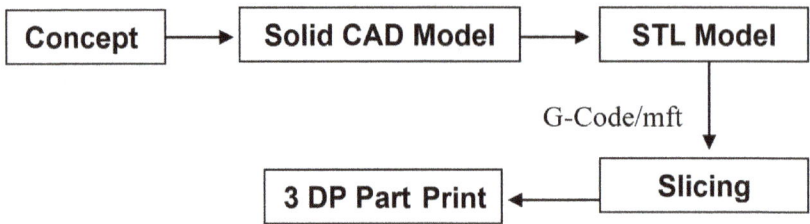

FIGURE 9.2 General workflow to create 3D parts and apply the fused deposition manufacturing (FDM).

The advantages of FEA also include the ability to modify the design once simulation results have been scrutinized and resimulated to ensure compliance. FEA is often used for structural analysis, thermomechanical simulations, thermal analysis, and hydraulics (Paul 2021). FEA can be used to simulate the existing model and predict stress distribution and deformations (Salonitis 2015). Existing FEA simulations do, however, have considerable limitations, particularly when it comes to FDM parts. When designs are developed using CAD software, the sections of the parts are taken into account in FEA as being geometrically and physically homogenous. The tensile behavior of an FDM sample built with variable layer thickness and/or raster orientation, according to Garg and Bhattacharya (2017), is different from the behavior and properties of solid material. An appropriate CAD tool for AM should enable customization of the material distribution and composition inside the part (Qattawi and Ablat 2017).

When using FEA, it is crucial to verify that the conclusions of analysis match up with a physical testing part to get accurate results. Consequently, it is essential to validate experiments (Hambali et al. 2010). Materials that have undergone typical production processes and are isotropic can be found in FEA material databases. They cannot, however, faithfully imitate layered generated anisotropic materials (Domingo-Espin et al. 2015). When employing FEA methodologies, parts made from molded or machined stock material are meant to be practical (Hopkinson and Dicknes 2003). Instead of being insufficient, material characteristics are currently constrained because they are not sufficiently understood (Baikerikar and Turner 2017; Hopkinson and Dicknes 2003).

When making sensors, which often exhibit lesser stiffness and strength than metal due to substantial deformation, it might be difficult to evaluate the qualities of thermoplastics. Since the strength of the bonding and fusing between the extruded infill and the air spaces separating them impacts the part's stiffness and strength, simulating FDM materials in FEA is a challenging task (Qattawi 2017).

For insights into determining whether the objects can fulfill the mechanical criteria unique to the applications for which they are created, the understanding of the interdependence of process parameters, such as the mechanical behavior of FDM specimens, is essential (Liu et al. 2021). More accurate simulations that account for all the intricacies of the FDM process must exist in order to select the best printing parameters.

Studies on simulation and experimental validation of AM components have been conducted. Tensile tests were carried out by Hambali et al. (2010) to examine the deformation behavior of sample FDM components produced in various orientations, and they contrasted the results with computer FEA analysis. They came to the conclusion that linear and static simulations weren't appropriate for the verification. Consequently, FEA that is nonlinear and dynamic may be better suitable for these kinds of investigations. The stress-deformation failure differences between a model made as a solid part and the other model made, taking into consideration each layer as solid and union forces between layers, were numerically simulated by Martinez et al. (2013). They came to the conclusion that the layer-based model achieved higher elastic values than the solid case under the identical charge settings, indicating that the second case is more constrictive than the first. They emphasized how the stresses from a failing ply could be dispersed among and carried by the remaining plies.

In their study, Garg et al. (Garg 2017) performed simulation and modeling of realistic models that take into account layers of various thicknesses and rasters positioned at various angles while retaining the interlayer and intralayer bonded region. By comparing them to observations, they empirically validated their findings. Additionally, experiments using the fractogram are done to determine how the FDM samples fail. They emphasized that the stress was more concentrated in the 0.178 and 0.254 mm layer thickness specimens towards the end of the gauge length, where the area begins to decrease, compared to solid simulation. They also came to the conclusion that at 90° raster angles, less stress is produced since layers are deposited perpendicular to the loading direction, and specimen failure is primarily dependent on adhesion between succeeding layers.

In order to experimentally validate the simulation results, four distinct infill patterns were printed and put to the test for tensile strength by Baikerikar and Turner (2017) using FEA simulations of FDM dog bone parts with bulk material parameters and derived material properties. In comparison to the experimental data, normal stress levels can be up to 70% off due to bulk isotropic features. The normal stress values obtained by FEA using derived material parameters were up to 45% off.

Implementing Industry 4.0 poses challenges for simulation and validation. However, new computing technologies and commercial software can help to close the gap. Additionally, there are some additional intricate aspects that will soon be taken into account throughout the water treatment process, such as the concentration of developing pollutants, oxygen reactive species, and other biological agents that are sensitive to a variety of conditions.

9.7 WATER TREATMENT IN THE INDUSTRY

Water treatment not only can be implemented by government facilities and public enterprises but also may be beneficial to the private sector. Examples include textile, food, and many other industries, as well as residential towers and metropolitan centers, as many new multifamily housing units have begun to offer their residents purified tab water. This gives the residents a lot of value because, first, tap water is cheaper than bottled water; secondly, this purified tab water has better taste and is cleaner than tap water; third, it doesn't pollute with trash as bottled water does; and finally, it is not needed to buy them in the supermarket and transport them all the way to housing.

Examples of water treatment in the industry reach the chiller system, which gives the company the opportunity to remove the heat power in excess, which is part of the cooling cycle, the air-conditioning systems, among other systems that allow to reduce the electrical cost of having just cooling systems based on refrigerants. The water treatment in this service is of great value to the user as the greatest cost of electricity in a company may come from air-conditioning and colling systems. Water treatment in companies may prevent scaling of pipes, valves, and pumping systems. Therefore, treatments addressed to remove salts of magnesium and calcium are widely used. In some cases, water in food must reach a high purity, and even salts are usually added to distilled water to create some formulations in pharma and food. Industry 4.0 already implemented in a manufacturing process, no matter the type of industry, should be as extensive as the control of the source allows, enhancing and optimizing the process a CPS can allow it.

Nonetheless, even if these creative solutions appears to be just a great value proposition, it comes with some limitations. Some of them includes the implementation of the system, which is highly expensive, the high maintenance cost, and that these processes can only be operated and fixed by an expert. Commonly, the main drawback that digitalization of water treatment has caused in many locations is that the system presents a problem and shuts itself down. However, this could be solved by applying sensors and neural networks to prevent and predict malfunctions, having digital twins and Internet of Things; therefore, the maintenance expert can solve the problem from his site and even make it so simple that even a beginner technician can fix the malfunction.

Even if the examples given previously may not seem like they occur in most apartment building or industries, there are some other things, like pools, heating systems, and even water systems for fire mitigation, which can be optimized via Industry 4.0 standards, which will guarantee better use of the resources and even better quality of services.

9.8 CONCLUSION

Implementation of Industry 4.0 principles to water treatment can be profitable, can be optimum, and will lead to sustainable use of water. Therefore, it is of the outmost importance for governments, industries, and utilities to invest in the water treatment to upgrade such systems, from using sensorization up to creating DT and a cyber-physical system. Although implementation will take time, for example, months and even years, the Industry 4.0 approach will provide an easy and accurate way to optimize water treatment and distribution, in quality and quantity, and even in cost revenue. Although some principles of Industry 4.0 are currently applied in companies that produce pipes, valves, and components for the water sector, there are plenty of opportunities in the creation of sensors through AM, FDM, and some other components to prevent leaks. In general, Industry 4.0 could induce a revolution in components of the water sector, but more importantly in the sustainability of the water sector and in the blue economy.

REFERENCES

Aguaded-Ramírez, E. 2017. Smart city and intercultural education. Procedia-Social and Behavioral Sciences, 237, 326–333.

Ahmed, A., Azam, A., Bhutta, M. M. A., Khan, F. A., Aslam, R., & Tahir, Z. 2021. Discovering the technology evolution pathways for 3D printing (3DP) using bibliometric investigation and emerging applications of 3DP during COVID-19. Cleaner Environmental Systems, 3, 100042.

Alcácer, V., & Cruz-Machado, V. 2019. Scanning the Industry 4.0: A literature review on technologies for manufacturing systems. Engineering Science and Technology an International Journal, 22(3), 899–919.

Aramand, M. 2008. Software products and services are high tech? New product development strategy for software products and services. Technovation, 28(3), 154–160.

Asif, M. B., & Zhang, Z. 2021. Ceramic membrane technology for water and wastewater treatment: A critical review of performance, full-scale applications, membrane fouling and prospects. Chemical Engineering Journal, 418, 129481.

Baikerikar, P. J., & Turner, C. J. 2017. Comparison of as-built FEA simulations and experimental results for additively manufactured dogbone geometries. In International Design Engineering Technical Conferences and Computers and Information in Engineering Conference 58110, V001T02A021. New York: American Society of Mechanical Engineers.

Baş, H., Elevli, S., & Yapıcı, F. 2019. Fault tree analysis for fused filament fabrication type three-dimensional printers. Journal of Failure Analysis and Prevention, 19, 1389–1400.

Beuken, R.H.S., Lavooij, C.S.W., Bosch, A., & Schaap, P.G., 2006. Low leakage in the Netherlands confirmed. In: Proceedings of the 8th Annual Water Distribution Systems Analysis Symposium. Cincinnati, USA.

Beuren, F.H., Ferreira, M.G.G., & Miguel, P. A. C. 2013. Product-service systems: A literature review on integrated products and services. Journal of cleaner production, 47, 222–231.

Boschert, S., & Rosen, R. 2016. Digital twin—the simulation aspect. In Mechatronic Futures (pp. 59–74). Springer, Cham.

Brentani, U.D. 1991. Success factors in developing new business services. European Journal of Marketing, 25(2), 33–59.

Buchanan, C., & Gardner, L. 2019. Metal 3D printing in construction: A review of methods, research, applications, opportunities and challenges. Engineering Structures, 180, 332–348.

Camero, A., & Alba, E. 2019. Smart City and information technology: A review. Cities, 93, 84–94.

Campbell, R S. 1983. Patent trends as a technological forecasting tool. World Patent Information, 5(3), 137–143.

Carroll, A.B. 1999. Corporate social responsibility: Evolution of a definitional construct. Business & Society, 38(3), 268–295.

Canedo, A. 2016. Industrial IoT lifecycle via digital twins. Proceedings of the IEEE 2016 International Conference on Hardware/Software Codesign and System Synthesis (CODES+ISSS), Washington, DC, 1–10.

Cerdas, F., Juraschek, M., Thiede, S., & Herrmann, C. 2017. Life cycle assessment of 3D printed products in a distributed manufacturing system. Journal of Industrial Ecology, 21(S1), S80–S93.

Chen, Z., Li, Z., Li, J., Liu, C., Lao, C., Fu, Y., . . . & He, Y. 2019. 3D printing of ceramics: A review. Journal of the European Ceramic Society, 39(4), 661–687.

Cho, J., & Lee, J. 2013. Development of a new technology product evaluation model for assessing commercialization opportunities using Delphi method and fuzzy AHP approach. Expert Systems with Applications, 40(13), 5314–5330.

Choudhari, C. M., & Patil, V. D. 2016. Product development and its comparative analysis by SLA, SLS and FDM rapid prototyping processes. In IOP Conference Series: Materials Science and Engineering (Vol. 149, No. 1, p. 012009). Philadelphia: IOP Publishing.

Contreras-Naranjo, J. E., Perez-Gonzalez, V. H., Mata-Gómez, M. A., & Aguilar, O. 2021.3D-printed hybrid-carbon-based electrodes for electroanalytical sensing applications. Electrochemistry Communications, 130, 107098.

Corradini, F., & Silvestri, M. 2021. A digital twin based self-calibration tool for fault prediction of fdm additive manufacturing systems. Annals of Daaam & Proceedings, 10(2), 607–616.

Deng, J., Zheng, Q., Liu, G., Bai, J., Tian, K., Sun, C., . . . & Liu, Y. 2021. A digital twin approach for self-optimization of mobile networks. Proceedings of the IEEE Wireless Communications and Networking Conference Workshops (WCNCW), Washington, DC, 1–6.

Domingo-Espin, M., Puigoriol-Forcada, J. M., Garcia-Granada, A. A., Llumà, J., Borros, S., & Reyes, G. 2015. Mechanical property characterization and simulation of fused deposition modeling Polycarbonate parts. Materials & Design, 83, 670–677.

Domoney, W. F., Ramli, N., Alarefi, S., & Walker, S. D. 2015. Smart city solutions to water management using self-powered, low-cost, water sensors and apache spark data aggregation. Proceedings of the IEEE 2015 3rd International Renewable and Sustainable Energy Conference (IRSEC), Washington, DC, 1–4.

Easingwood, C. J., & Storey, C. 1991. Success factors for new consumer financial services. International Journal of Bank Marketing, 9(1), 3–10.

Erdemir, A., Guess, T. M., Halloran, J., Tadepalli, S. C., & Morrison, T. M. 2012. Considerations for reporting finite element analysis studies in biomechanics. Journal of Biomechanics, 45(4), 625–633.

Errandonea, I., Beltrán, S., & Arrizabalaga, S. 2020. Digital Twin for maintenance: A literature review. Computers in Industry, 123, 103316.

Fernandez-Vicente, M., Calle, W., Ferrandiz, S., & Conejero, A. 2016. Effect of infill parameters on tensile mechanical behavior in desktop 3D printing. 3D Printing and Additive Manufacturing, 3(3), 183–192.

Gao, Y., Brennan, M., Joseph, P. F., Muggleton, J.M., & Hunaidi, O. 2005. On the selection of acoustic/vibration sensors for leak detection in plastic water pipes. Journal of Sound and Vibration, 283(3–5), 927–941.

Garg, A., & Bhattacharya, A. 2017. An insight to the failure of FDM parts under tensile loading: finite element analysis and experimental study. International Journal of Mechanical Sciences, 120, 225–236.

Hambali, R.H., Celik, H.K., Smith, P.C., Rennie, A.E. W., & Ucar, M. 2010. Effect of build orientation on FDM parts: A case study for validation of deformation behaviour by FEA. In: Ahmad Yusairi Bani Hashim (eds). International Conference on Design and Concurrent Engineering (iDECON 2010) (pp. 224–228). Penerbit Universiti (UTeM). ISBN 983-2948-84-1.

Hopkinson, N., & Dicknes, P. 2003. Analysis of rapid manufacturing—using layer manufacturing processes for production. Proceedings of the Institution of Mechanical Engineers, Part C: Journal of Mechanical Engineering Science, 217(1), 31–39.

Hozdić, E. 2015. Smart factory for industry 4.0: A review. International Journal of Modern Manufacturing Technologies, 7(1), 28–35.

Hu, L., Nguyen, N.T., Tao, W., Leu, M.C., Liu, X.F., Shahriar, M.R., & Al Sunny, S.N. 2018. Modeling of cloud-based digital twins for smart manufacturing with MT connect. Procedia Manufacturing, 26, 1193–1203.

Hunaidi, O., Wang, A., Bracken, M., Gambino, T., & Fricke, C. 2004. Acoustic methods for locating leaks in municipal water pipe networks. In International conference on water demand management (pp. 1–14). Jordan: Dead Sea.

Igunnu, E. T., & Chen, G. Z. 2012. Produced water treatment technologies. International Journal of Low-Carbon Technologies, 9(3), 157–177.

International Organization for Standardization (ISO). 2021. ISO 23247-1: Automation Systems and Integration—Digital Twin Framework for Manufacturing—Part 1: Overview and General Principles (ISO (DIS) 23247–1).

International Organization for Standardization (ISO). 2011. ISO 26000 — An emerging guidance on social responsibility. Business & the Environment, 22(1), 13–15.

Ivanov, S., Nikolskaya, K., Radchenko, G., Sokolinsky, L., & Zymbler, M. 2020. Digital twin of city: Concept overview. Proceedings of the IEEE 2020 Global Smart Industry Conference (GloSIC), Washington, DC, 178–186.

Jones, D., Snider, C., Nassehi, A., Yon, J., & Hicks, B. 2020. Characterising the Digital TwIn: A systematic literature review. CIRP Journal of Manufacturing Science and Technology, 29, 36–52.

Josifovska, K., Yigitbas, E., & Engels, G. 2019. Reference framework for digital twins within cyber-physical systems. In 2019 IEEE/ACM 5th International Workshop on Software Engineering for Smart Cyber-Physical Systems (SEsCPS), Washington, DC, 1, 25–31.

Kaushish, J.P. 2010. Manufacturing Processes. New Delhi, India: PHI Learning Pvt. Ltd.

Kidwell, J. 2017. Best practices and applications of 3D printing in the construction industry. Pomona, CA: California Polytechnic State University. https://digitalcommons.calpoly.edu/cgi/viewcontent.cgi?article=1090&context=cmsp. Retrieved from October 10.

Klocke, F., & Kuchle, A. 2009. Manufacturing Processes 4 (Vol. 2, p–433). Berlin: Springer. ISBN: 978-3-642-36772-4.

Kostenko, D., Kudryashov, N., Maystrishin, M., Onufriev, V., Potekhin, V., & Vasiliev, A. 2018. Digital twin applications: Diagnostics, optimisation and prediction. Proceedings of the 29th DAAAM International Symposium, 0574–0581.

Labuschagne, C., & Brent, A. C. 2005. Sustainable project life cycle management: the need to integrate life cycles in the manufacturing sector. International Journal of Project Management, 23(2), 159–168.

Lasi, H., Fettke, P., Kemper, H.G., Feld, T., & Hoffmann, M. 2014. Industry 4.0. Business & Information Systems Engineering, 6(4), 239–242.

Lee, E. A. 2008. Cyber physical systems: Design challenges. Proceedings of the 11th IEEE International Symposium on Object and Component-Oriented Real-Time Distributed Computing (ISORC), 363–369.

Lemu, H. G. 2012. Study of Capabilities and Limitations of 3D Printing Technology. In AIP Conference Proceedings 1431(1), 857–865. American Institute of Physics. New York: Melville.

Liu, M., Fang, S., Dong, H., & Xu, C. 2021. Review of digital twin about concepts, technologies, and industrial applications. Journal of Manufacturing Systems, 58, 346–361.

Liu, Z., & Kleiner, Y. 2013. State of the art review of inspection technologies for condition assessment of water pipes. Measurement, 46(1), 1–15.

Mabkhot, M. M., Al-Ahmari, A. M., Salah, B., & Alkhalefah, H. 2018. Requirements of the smart factory system: A survey and perspective. Machines, 6(2), 23.

Mamlook, R., & Al-Jayyousi, O. 2003. Fuzzy sets analysis for leak detection in infrastructure systems: A proposed methodology. Clean Technologies and Environmental Policy, 6(1), 26–31.

Martínez, J., Diéguez, J. L., Ares, E., Pereira, A., Hernández, P., & Pérez, J. A. 2013. Comparative between FEM models for FDM parts and their approach to a real mechanical behaviour. Procedia Engineering, 63, 878–884.

Mourtzis, D. 2016. Challenges and future perspectives for the life cycle of manufacturing networks in the mass customisation era. Logistics Research, 9(1), 1–20.

Nam, T., & Pardo, T. A. 2011. Conceptualizing Smart City with Dimensions OF Technology, People, and Institutions. Proceedings of the 12th Annual International Digital Government Research Conference: Digital Government Innovation in Challenging Times, 282–291.

Oztemel, E., & Gursev, S. 2018. Literature review of Industry 4.0 and related technologies. Journal of Intelligent Manufacturing, 1–56.

Paul, S. 2021. Finite element analysis in fused deposition modeling research: A literature review. Measurement, 178, 109320.

Pine, B. J. 1993. Mass customizing products and services. Planning Review 21(4), 6–55.

Pendergast, M. M., & Hoek, E. M. 2011. A review of water treatment membrane nanotechnologies. Energy & Environmental Science, 4(6), 1946–1971.

Pour-Ghaz, M., Kim, J., Nadukuru, S. S., O'Connor, S. M., Michalowski, R. L., Bradshaw, A. S., . . . & Weiss, W. J. 2011. Using electrical, magnetic and acoustic sensors to detect damage in segmental concrete pipes subjected to permanent ground displacement. Cement and Concrete Composites, 33(7), 749–762.

Qattawi, A., & Ablat, M. A. 2017. Design consideration for additive manufacturing: Fused deposition modelling. Open Journal of Applied Sciences, 7(6), 291–318.

Rajput, R. K. 2007. A textbook of manufacturing technology: Manufacturing processes. New Delhi, India: Laxmi Publications. ISBN: 978-013-180-2441.

Salonitis, K., & Al Zarban, S. 2015. Redesign optimization for manufacturing using additive layer techniques. Procedia Cirp, 36, 193–198.

Samuel, A. U., Oyawale, F., & Fayomi, O. S. I. 2019. Effects of waste management in beverage industries: A perspective. In Journal of Physics: Conference Series (Vol. 1378, No. 2, p. 022048). Philadelphia: IOP Publishing.

Semeraro, C., Lezoche, M., Panetto, H., & Dassisti, M. 2021. Digital twin paradigm: A systematic literature review. Computers in Industry, 130, 103469.

Shafto, M., Conroy, M., Doyle, R., Glaessgen, E., Kemp, C., LeMoigne, J., & Wang, L. 2012. NASA technology roadmap: Modeling, simulation, information technology & processing roadmap technology area 11. National Aeronautics and Space Administration, 32, 1–38.

Shahanas, K. M., & Sivakumar, P. B. 2016. Framework for a smart water management system in the context of smart city initiatives in India. Procedia Computer Science, 92, 142–147.

Singh, T., Kumar, S., & Sehgal, S. 2020. 3D printing of engineering materials: A state of the art review. Materials Today: Proceedings, 28, 1927–1931.

Sun, Q., Rizvi, G. M., Bellehumeur, C. T., & Gu, P. 2008. Effect of processing conditions on the bonding quality of FDM polymer filaments. Rapid Prototyping Journal, 14(2), 72–80.

Tijing, L. D., Dizon, J. R. C., Ibrahim, I., Nisay, A. R. N., Shon, H. K., & Advincula, R. C. 2020. 3D printing for membrane separation, desalination and water treatment. Applied Materials Today, 18, 100486.

UNESCO. 2019. UNESCO and NETEXPLO: Smart cities: Shaping the society of 2030. United Nations Educational, Scientific and Cultural Organization (UNESCO), Paris, France 2019. (ISBN 978-92-3-100317-2)

UNICEF. 2016. Strategy for Water, Sanitation and Hygiene 2016–2030. New York.

Wang, S., Wan, J., Li, D., & Zhang, C. 2016. Implementing smart factory of industrie 4.0: An outlook. International Journal of Distributed Sensor Networks, 12(1), 3159805.

Wang, Z. 2020. Review of real-time three-dimensional shape measurement techniques. Measurement, 156, 107624.

Wang, Z., Song, H., Watkins, D. W., Ong, K. G., Xue, P., Yang, Q., & Shi, X. 2015. Cyber-physical systems for water sustainability: Challenges and opportunities. IEEE Communications Magazine, 53(5), 216–222.

Warke, V., Kumar, S., Bongale, A., & Kotecha, K. 2021. Sustainable development of smart manufacturing driven by the digital twin framework: A statistical analysis. Sustainability, 13(18), 10139.

Yossef, M., & Chen, A. 2015. Applicability and limitations of 3D printing for civil structures. Proceedings of the 2015 Conference on Autonomous and Robotic Construction of Infrastructure (pp. 237–246). Ames IA: Center for Earthworks Engineering Research Institute for Transportation Iowa State University Research Park.

Yu, M., Zhang, W., & Meier, H. 2008. Modularization based design for innovative product-related industrial service. Proceedings of the IEEE International Conference on Service Operations and Logistics, and Informatics 1, 48–53. Washington, DC.

Zhang, J., Yao, X., Zhou, J., Jiang, J., & Chen, X. 2017. Self-organizing manufacturing: Current status and prospect for Industry 4.0. Proceedings of the IEEE 5th International Conference on Enterprise Systems (ES), 319–326. Washington, DC.

10 Computational Modeling and Assessment of Sustainable Materials Efficiency to Remove Heavy Metals from Drinking Water

Wessam S. Omara, S. Abdalla, and Sherif Kandil

CONTENTS

DOI: 10.1201/9781003260455-10

10.1   INTRODUCTION

Pollution is a common problem that made countries suffer for decades. Although the majority of the earth's surface is drinking water, the available sources of clean drinkable water are limited. This is due to the contamination of aquatic systems by different pollutants (Wołowiec et al. 2019). Heavy metals (HMs) arising from natural and human sources are the main causes of drinking water pollution. Heavy metals are a constant menace to various environments because they are continuously released during diverse human activities, both industrial (mineral extraction, metallurgical processes, smelting, electroplating, nanoparticle preparation, and agricultural activities) besides household (sewage, waste, and metal erosion); all within the scope of fast industrial development and urbanization (Tang et al. 2020). Heavy metals are included as persistent pollutants because they cannot be destroyed or decomposed (Farcasanu et al. 2018). The natural sources of HMs are mainly the products of rock weathering processes, atmospheric sediments, and rain (Herath et al. 2022). Anthropogenic sources include industrial wastes (Huang et al. 2014), fertilizers (Vakili et al. 2019)(Ding et al. 2021), and sewages (Wołowiec et al. 2019), where each of these sources can release heavy metal ions as by-products of these activities, which affect human health directly or indirectly (Aslam et al. 2021). Materials in raw form when used in many industries often contain one or more metal ions in their composition. As a result of the release of many chemicals and contaminants in the drinking water such as heavy metals and aldehydes that reaches the human being through drinking or irrigation in agriculture or both, it could be dangerous and causes human disease, when the heavy metals and their compounds enter into the human body (Omara et al. 2015; Balali-Mood et al. 2021).

The fame of heavy metals as toxins goes beyond their well-known reputation as the initiator of many industries (Vakili et al. 2019). In general, HMs are a set of trace elements that include metals, such as arsenic As, cadmium Cd, chromium Cr, cobalt Co, copper Cu, iron Fe, lead Pb, manganese Mn, mercury Hg, nickel Ni, tin Sn and zinc Zn. For example, arsenic (As)—"the poison of the king", is one of the most dangerous elements of heavy metals; it is a major semi-metallic component of more than 240 different types of minerals and is mainly associated with many elements such as sulfides with copper, nickel, lead, gold, cobalt, or other metals. The danger of arsenic is due to its release into the environment due to many factors. Although it is widely used (but not limited to, metallurgy, catalysis, electronics, glasswares, pesticides industries, wood protection materials, and growth promoter in some animals, and as medication to treat some diseases such as sleeping sickness, as well as syphilis), its side effects are very serious in some concentrations. As well as lead (Pb)(Balali-Mood et al. 2021), cobalt (Co), cadmium (Cd) (Wołowiec et al. 2019), and mercury (Hg), these metals have the same toxic effect and follow the same pattern in their toxicity. Table 10.1 shows the maximum allowable amount of exposure to some heavy metals according to World Health Organization WHO and United State Environmental Protection Agency US-EPA (Gaur et al. 2021; Nik Abdul Ghani et al. 2021). Mercury (Hg) differs in nature from other metals when it is pure mercury, as it is found in liquid form, while other metals are solid in their alloys (Lapo et al. 2018).

TABLE 10.1

Standard Limit of Five Selected Heavy Metals in Drinking Water

HMs Contaminants	Permissible Limit of Heavy Metals by USEPA (mg/L) (Gaur et al. 2021)	WHO Guideline Value (mg/L) (Nik Abdul Ghani et al. 2021)
Pb	0.05	0.01
Cd	0.005	0.003
As	0.05	0.01
Cu	0.05	2
Hg	0.002	0.006

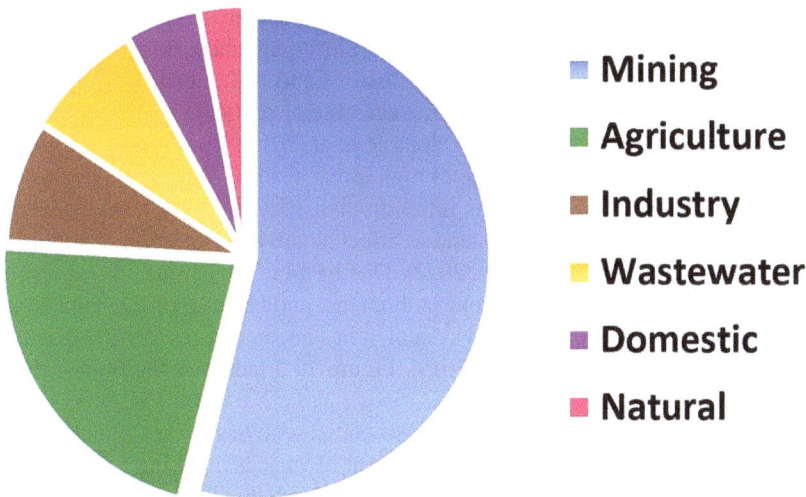

FIGURE 10.1 Percentages of heavy metals released in drinking water from different sources.

Figure 10.1 shows the percentages of heavy metals emitted from different industrial sources, although some metal ions are essential to human nutrition, such as copper Cu(II), which is a very important element for making the diet balanced, where it is found in nutritional supplements. Nevertheless, the increased levels of copper in the human body may cause poisoning, resulting in serious body disorders, such as hypoglycemia and dyslexia (Singhon 2014). Many industries produce Cu^{2+} ions during their manufacturing processes, which can find their way into the human food loop and accumulate in the human body (Zhang et al. 2019). The maximum allowable content of Cu^{2+} ions in foods is between 1.2 and 4.2 mg per day, as defined by the European Food Safety Authority (EFSA) (Zhang et al. 2022). The presence of Cu^{2+} in waste drinking water is primarily toxic to aquatic plants, even at relatively low concentrations. According to safe drinking water standards, the permissible limit for copper in drinking water is 1.3 mg/l. The Environmental Protection Agency has also set a limit on the use of copper in drinking water at the same concentration.

Long-term intake of copper by humans leads to undesirable changes in human organs, such as necrotic processes in the liver and kidneys, and irritation of the mucous membrane, which spreads widely, leading to damage to the capillaries. Depression, weakness, lethargy, and loss of appetite are among the most prominent symptoms of copper poisoning, as well as gastrointestinal damage and lung cancer (Pehlivan et al. 2012). Copper is, by nature, a minted metal and has been used by the human community for over 6,000 years. Copper reaches different sources of drinking water by mining activity, the use of various copper vessels, pipelines, taps, and the like. Low copper ion concentration in drinking water leads to health benefits. In some cases, copper accumulation in the body also leads to childhood cirrhosis, neurodegenerative diseases such as Wilson and Zimmer diseases, aging, and schizophrenia (Asokan et al. 2021).

Lead (Pb) also enters the ecosystem from anthropogenetic activities—for instance, fossil fuel combustion, landfill waste incinerators, some fertilizers, pesticides industries, and vehicle exhaust emissions. Lead (Pb) compounds exist in the form of complexes of sulfates, sulfides, and carbonates. Lead is the leading environmental pollutant, as it increasingly threatens the ecosystem, particularly in the vicinity of large industrial facilities and cities (Odobašić et al. 2019). Many works of literature have indicated that long-term uptake of high levels of Pb(II) in drinking water leads to critical health disorders, such as renal disease, nausea, coma, convulsions, and cancer, and these serious effects extend to affect metabolism as well as mental activities (Xing et al. 2018; Zhang et al. 2022). Due to these dangerous consequences of lead (II) environmental pollution and on humans, and because of the abundance of this heavy metal in the geochemical system, it has become necessary to remove lead (II) from effluents (Yuvaraja et al. 2019). Figure 10.2 indicates the organs that are affected by the five main heavy metals.

Form this concern, based on hazards associated with heavy metal contamination of drinking water, it has led to the development of various drinking water treatment technologies (Vakili et al. 2019). Recently, the technology of microfluid has become increasingly applied and further developed, with the aim of introducing precise control of final particle morphology and monodispersity, with controllable sizes and synthesis of different structures (Wang et al. 2019).

10.2   HEAVY METAL REMOVAL TECHNOLOGY

The major challenging techniques for heavy metal removal are categorized into chemical, physical, and biological categories (Surucu 2022). The sorbents or filter particles synthesized by advanced microfluidic technologies have shown tremendous potential in waste drinking water treatment. Conventional treatment processes have been developed and practiced, including filtration (Khandegar et al. 2021), ion exchange (Sabourian et al. 2016), carbonate or hydroxide precipitation, and a variety of absorbent materials, such as peat or Chelex 100 resin and amberlite, along with their inherent advantages and disadvantages in cleaning drinking water contaminated with heavy metals (Anas et al. 2019). Figure 10.3 highlights the three major methods approved in heavy metals removal from drinking water. Many adsorbents demonstrate excellent adsorption performance in removing metal ions, such

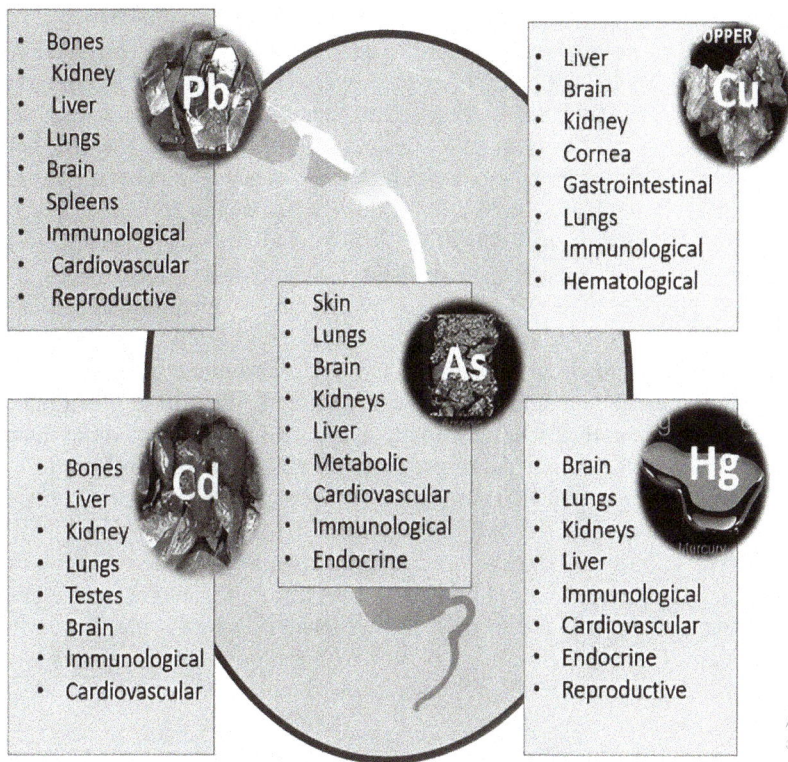

FIGURE 10.2   Organs affected by the five main heavy metals.

FIGURE 10.3   The three basic methods approved for heavy metals removal from drinking water.

as biochar, clay, carbon, and nano metal oxide. In addition, some cationic framework materials are used to remove harmful auxins, such as anionic exchange resins, cationic MOFs, MOFs, cationic polymeric organic networks, and multilayer double hydroxides. These substances play a vital role in drinking water treatment by physical adsorption of heavy metals (Zhao et al. 2022). Despite the efficiency of both the chemical and physical methods, they also have some drawbacks (Yang et al. 2022). Current technologies applied in drinking water treatment include UV ozone drinking water disinfection (Tahir et al. 2019), solar drinking water disinfection, reverse osmosis, etc.; these are very expensive, not easy to carry on, and occupy work site space (Asokan et al. 2021). There are many common methods, such as solvent extraction, active carbon adsorption, biodegradation, and extensive chemical oxidation, although they are effective, often costly, time-consuming, or both. During the past decade, research and studies have been directed to the use of green materials that are environmentally friendly in many applications. Moreover, these materials can be obtained at the nanoscale, where their structures range from 1 to 100 nm in size and are therefore a potential competitor in this regard due to their larger surface areas and superior surface reactivity (Václavíková et al. 2009).

For instance, filtration and precipitation processes are considered cost-effective techniques, but their efficiency is limited to removing extremely low concentrations of metal traces. Likewise, the success of ion exchange, which is able to reduce the concentration of metal ions to parts per level (to remove metal from waste drinking water), is reduced due to the existence of enormous amounts of monovalent and divalent cations as a compotator (Bassi et al. 2000). Also, composites have shown effective removal on both the large and growing scale but require complex preparation methods, prohibitive cost, and limited application ranges (Yan et al. 2022). New electrochemical techniques have acquired a vital role as one of the means to specifically remove mercury present in drinking water samples, where mercury-free drinking water samples were obtained after using mercury electrodeposition on the proposed electrode (Surucu 2022). Compared with previous chemical and physical techniques, biological drinking water treatment has received much attention and has been used in actual situations. Biological methods such as activated sludge and biofilm adsorption have better feasibility. It is also environmentally friendly and can interact with polluted drinking water with high efficiency without causing secondary pollution. Moreover, it is beneficial to the environment (Yan et al. 2022).

10.3 INTERNET OF THINGS AND COMPUTATIONAL MODELING

Machine learning and the Internet of Things are currently essential tools for commercializing advanced technology (Wen et al. 2020). Molecular modeling is a branch of computational science that uses theories and equations (Zhang 2015) to predict reaction possibilities or reaction products (de Brito Lira et al. 2021). In other words, molecular modeling at a different level of theory is a class of computational work that shows the possibility of describing the physical and chemical behavior of many systems and molecules covering many application areas. Modeling can provide data for systems and molecules where experimental techniques are limited and/or unavailable (Ezzat et al. 2020). From simplicity to complexity, there are a wide scale of functions and software that govern the accuracy and validity of these modulation data (de Brito Lira et al. 2021). The process model can only approximate the real process described mathematically. It simplifies the design and analysis of the entire system, reducing the number of experimental tests required, and accordingly, the higher the accuracy of the input data, the lower the error rate. These calculations enable one to evaluate and improve various operating conditions and design criteria in addition to defining the processes to be placed in the plant and disembarking, defining controls, designing and selecting a control law and strategy for the new process, in order to reach the training of plant operating personnel in a more effective manner. These calculations led to the use of molecular modeling and simulation to select and screen the functional materials most suitable for a particular application (Hughes and Walsh 2015). Thanks to continuous improvement and development in this field, researchers can save time, effort, and cost of laboratory work to solve many environmental, industrial, medical, and economic problems. Figure 10.4 presents the main framework elements for modeling and simulation. The first step depends on clearly defining the goals and limits of the system for certain model. The essential components of the system to be modeled, the pattern of interactions between them, and the level of detail and data required should be identified. It should be noted that the level of model complexity controls the difficulty and sometimes even the impossibility to solve. Depending on the comprehensive inspection of certain system

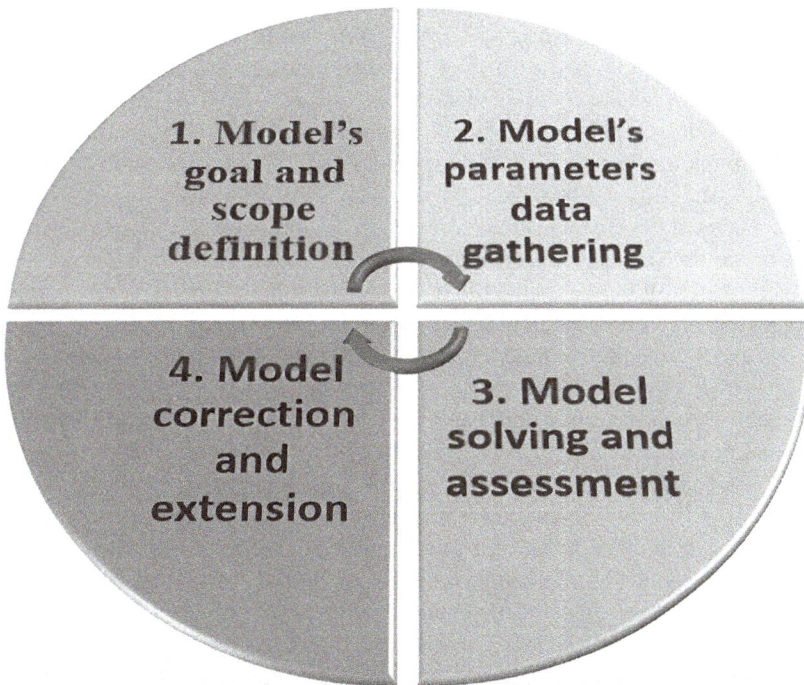

FIGURE 10.4   The main framework elements of modeling and simulation.
Source: Václavíková et al. (2009).

behavior, a process diagram is created with precision, and ground laws are employed to determine which phases should be applicable then, therefore, should be taken into account. This is a qualitative step in its nature and includes identifying relevant empirical data, such as information about relatively same legislation and regulations. In order to get a better model definition, the practitioner can consider 4Ws questions— that usm to ask over what, why, where, and when. Examples can include: What can be the objectives and real benefits of the model? Is it to be used to illustrate empirical patterns or to anticipate and design goals? Should you be looking at a steady state, a series of reactions governed by variable factors, or perhaps a transient interaction? How do the trade-offs balance the level of detail of the model, the information available about the interaction conditions, and the effort and time required to discover and evaluate the answer? An exact solution to these as well as many additional questions may assist to determine the hypotheses to be done the final difficulty of certain models, and also whether it is possible to obtain numerical values from the model's equations (Václavíková et al. 2009). The following step mainly depends on collecting all necessary physical and chemical data, including thermodynamic properties, conservation, and rate laws. Sometimes, experiments are required to obtain more information about the model, especially in more complex systems, where many different physical phenomena may occur. Depending on the obtained information, the law equations for the conservation of momentum, mass, energy, etc. need a formulation to explain the behavior of this system during the interaction. Based on those algebraic assumptions or differential equations obtained, it is more convenient in that case to write auxiliary equations additionally to conservation in form without dimension to define the main data governing the behavior of the system, and to allow the application of the results of the same model to other systems of different sizes but the same dimensionless information values. Lastly, it is a must to do a qualitative analysis of the model equations to confirm, including all the relevant aspects. The third step starts when the mathematical terms are used to completely define a system. A model is used to solve equations to evaluate the response of the system to either single or multiple independent variables. A differentiated set of analytic and numerical patterns is valid to be applied to a broad spectrum of models. Many computer packages are currently available in ready-to-use forms for modeling the whole process, such as computational fluid dynamics (CFDs) software. After getting a numerical or analytic solution model, it is important to evaluate whether the process is applicable and well describes both performance real behavior and the process/system under investigation. This is because in most cases, the equations of the model are highly nonlinear and may contain many solutions, so their validity must be qualitatively analyzed. Only then can the results be compared with either experimental or operational data in order to determine its suitability to represent the system of certain study. The final step is based on all previous analysis; sometimes it needs to either modify or expand the model, depending on its capability of describing this system under investigation and achieving the desired objectives. Of course, model extensions can be looked for if the obtained results are insufficient, and this process can be repeated continuously until the appropriate level of accuracy is reached. Final changes may require increasing the model complexity or readjusting it to reach a better taking into account of the factors that control the system behavior (Václavíková et al. 2009). As mentioned before, the complexity of the model depends largely on the competence of the practitioner, the data available, the computational power, and finally, the purposes of the study. Some researchers start in this field with a simplified model of the process, based on the fewest number of available information, then gradually, complexity is increased because knowledge about it increases gradually. Additionally, experiments to study the effect of working conditions (e.g., determination of temperature degree and its effect on mixture formation), which are usually time-consuming and expensive. Thus, modeling and simulation processes are excellent alternatives that may give a cost-effective solution to process design by allowing one to swiftly examine the influence of numerous design elements. The paradigm previously described is used in this study to develop an absorption system (Václavíková et al. 2009). In other words, the complexity of the model depends largely on the competence of the practitioner, the data available, the computational power, and finally, the purposes of the study. When modeling an adsorption system, the ultimate goal is to predict and characterize the performance and behavior of an adsorption unit. It is also critical to predicting the penetration curves of the molecule of interest. Meanwhile, these data are necessary for designing units and for determining production and renewal cycles. From what we mentioned previously, it is quite clear that drinking water contamination with heavy metals presents a serious challenge, and the danger is also exceeded when drinking water treatment technologies rely on high-cost and complex technologies (Shah et al., 2021), which cannot last forever due to the use of more resources, materials, and energies. Therefore, artificial intelligence, sustainable materials, and green technology are now the smartest tool for facing current and future problems (Akha et al. 2022).

10.3.1 Density Function Theory (DFT) of Molecular Modeling

In an amazing way, the number of uses of density functional formalism (DFF) in research related to materials sciences and chemistry has rapidly increased in the last 30 years. The sheer volume of research articles demonstrates that directed calculations are a tremendous success story, and many new researchers in the area are shocked to hear that extensive use of density functional approaches, notably in chemistry, did not begin until around 1990 (Jones, 2015). Density functional formalism (DFF) demonstrates that knowledge of the electron density distribution may be used to predict the ground state (GS) in addition to other features of an electron system in an external field $n(\mathbf{r})$ alone (Jones 2015). It should be noted that in 2010, the number of publications on DFT exceeded 1,500 articles compared to that in 1990, which was limited to a few hundred (Jones 2015). Structures are often optimized for minimum energy using density functional theory (DFT) (Ribeiro et al. 2019). Aside from ab initio molecular dynamics, DFT is always a great tool for calculating atoms, quantum states, molecules, and solids. It was first envisaged in its early simplified and approximated form by both Thomas and Fermi shortly after the establishment of quantum mechanics in 1927. In the mid-1960s, Hohenberg, Cohn, and Sham managed a logically rigorous functional theory of density quantum ground state based primarily on quantum mechanics; on the other hand, based on this construction, an explicit approximation theory known as the local density approximation emerged, which was confirmed for quantum ground state calculations of multiparticle systems to be superior to both the theories of Thomas Fermi and Hartry Fock. Since that time, density functional theory has developed more widely, and the result has been an abundance of computational research in various fields of physics, such as particle physics and solid physics. As a result of its success, there has always been a propensity to encourage the extension of the domains of application of density functional theory, and in these advances, emphasis has been focused on some topics that have remained rather vague in the basic theory from time to time (Eschrig 2003; Koskinen and Mäkinen 2009). As mentioned before, DFT is a powerful tool for studying and confirming the structure and chemical compositions of many compounds during their reactions (Ibrahim et al. 2020). Because frequency calculations are employed and no fictional frequency should be identified, this theory is a solid sign of assuring that the optimized structures were at their lowest energy (Ribeiro et al. 2019).

Therefore, the basic formalism of DFT is based on turbulent expansions so that the total energy of the electronic system N is inside DFT. The HK theory proves that N inside DFT is a functional function of electron density. However, in order to precisely estimate the system, various external factors are implicitly expected. The following are not all-inclusive:

- Both form and volume of unit cell volume and form.
- The chemical nature number and atomic locations in the unit cell.
- The characteristic wave vector, polarization, and intensity of the outer electromagnetic field.

This data is deemed external since it is separate from the system's electrical structure. In other words, the ground state of the system is reached for a certain set of external information by adjusting the electron density until the energy is decreased to a minimum. Let us, for example, observe this with $\{\lambda\}$ the set dependence on the electron density n and a parametrical dependence on $\{\lambda\}$. For any diverse set of $\{\lambda\}$, the basic theoretical framework can include the electron density that decreases $E\lambda[n]$. In a formal way, it is possible to consider the outer potential V that is based on the parameter λ and define the variations of V as due to the variations of λ around the value $\lambda = 0$, which is supposed to give the equilibrium system. The assumption of the dependency of $V\lambda$ with λ is defined through all orders (Finocchi, F. 2011):

$$V\lambda = V^{(0)} + \delta\lambda V^{(1)} + (\delta\lambda)^2 V^{(2)} + \tag{1}$$

Where $V(n) = \partial nV/\delta\lambda n$.

The previous equation differs from a Taylor expansion of V only for the absence of the 1/n factors. The corresponding ground state energy and electron density can be extended in a similar way:

$$n\lambda = n^{(0)} + \delta\lambda n^{(1)} + (\delta\lambda)^2 n^{(2)} + \tag{2}$$

$$E\lambda = E^{(0)} + \delta\lambda E^{(1)} + (\delta\lambda)^2 E^{(2)} + \tag{3}$$

This chapter will help the readers understand the vital role of molecular modeling in the heavy metal removal technology based on selecting the suitable sorbents. Also, researchers will be able to take up the impact of further research required parameters for pollutant removal using chitosan CS. Due to its biocompatibility, biodegradability, and antibacterial properties, chitosan is the most flexible natural and sustainable biopolymer for a wide range of applications. It exhibits a successful biological activity on pathogenic grammes, which is unique in a linear biocompatible and biodegradable polysaccharide natural polymer derived by the chitin deacetylating method, which is generally present in the hard exterior skeleton of shellfish, mollusks, and insects (Omer et al. 2022). Chitosan with a lower molecular weight typically has greater drinking water dissolvability and antibacterial capabilities than chitosan with a higher molecular weight; Cs's chemical alterations are the best alternatives for improving its qualities to be used in a wide range of applications. Cs has piqued the interest of researchers in recent decades since it is considered a suitable adsorbent in which a polymer film is capable of heavy metal adsorption from waste drinking water (Ezzat et al. 2020). This chapter also aspires that the present theoretical work is to elucidate the response mechanism and selectivity of chitosan CS at Pb (II) and Cu (II) through quantum mechanical calculations. Because chitosan is one of the most important natural sustainable biosorbents, it has a competitive advantage in drinking water treatment and heavy metal removal technologies, based on the nature of waste drinking water to be treated, industrial use, release standards, cost-effectiveness, regulatory restrictions, and even access to long-term environmental effects. Thus, it was necessary for this chapter to address in a simplified and focused manner and describe the modeling of the adsorption operations that are most practically applied for drinking water purification using chitosan and its derivatives and compounds in all different shapes and sizes.

10.4   SUSTAINABLE SORBENT MATERIALS FOR HEAVY METAL REMOVAL FROM DRINKING WATER

Concerns about environmental safety have led to a paradigm shift in a number of economic sectors, especially with regard to the manufacture of materials produced from sources of fossil origin, as is the case for plastics (Vasconcelos and Eleutério 2022). However, plastic products offer a solution to many problems of packaging, preservation, hygiene, filtration, and sterilization (Vasconcelos and Eleutério 2022; Gonçalves et al. 2022). Excessive accumulation of plastic waste increases pollution problems in both the environment and in drinking water. Moreover, plastics made from petroleum sources, an increasingly scarce natural resource, are nonbiodegradable and remain intact for a long time in landfills; even if burned, it releases gases that are toxic to both the environment and ground drinking water. As a result, by lowering our reliance on fossil fuels, we are contributing to the planet's sustainability and thus reducing greenhouse gas emissions. The use of renewable sources as raw materials in heavy metal removal techniques is receiving great attention in many industrial fields due to environmental and economic issues, especially global warming, and waste disposal management, as mentioned earlier.

The sorption is well described with the equation of Lagergren's first order (Pournara et al. 2019), shown next:

$$q_t = q_e \left[1 - exp\left(-K_L t \right) \right] \tag{4}$$

Where q_e is the amount (mg g^{-1}) of the metal ion sobbed in equilibrium, and KL the Lagergren or first order rate constant.

In terms of sustainability, renewable sources, as a result of their presence or production in large quantities, often attract great consideration, as sustainable materials are often obtained from natural or recycled resources and need less energy to generate. It uses few nonrenewable resources and has minimum environmental side effects. Many sorbents in use today are not sustainable in terms of energy usage and greenhouse gas emissions (Asdrubali, Schiavoni and Horoshenkov 2012). Sustainable materials include the four major classes of materials, such as carbon (Titirici et al. 2015), metals (Novikau and Lujaniene 2022; Es-sahbany et al. 2022), ceramics (Gonçalves et al. 2022; (Kausar et al. 2022; Dehghani et al. 2019), and polymers (Maliki et al. 2022; Syeda and Yap 2022).

10.4.1 CARBON

Carbon-based materials are the most diversified structures employed in current renewable energy (production and storage) and environmental sciences (e.g., refining/processing) (Pal et al. 2022). Carbonaceous compounds are also widely applied in drinking water treatment, separation of gas (e.g., carbon dioxide entrapment), storage, and as additives in soil (Pal et al. 2022; Nejadshafiee and Islami 2020). Carbon is a naturally occurring, sustainable substance found in nature in large quantities and has a great role in cost-effective drinking water purification technologies, especially when it is manufactured and used in nano and colloidal forms (Pal et al. 2022). As previously stated, many nanostructured carbon materials such as carbon nanotubes (CNT), graphene, carbon fibers, graphene nanosheets, graphene quantum dots, and others are widely recognized for their exceptional adsorption capabilities (Yadav et al. 2019; Gontrani et at. 2021). And since carbon is found in nature in various forms, such as coal, graphite, and diamond, this has become one of the most important motives for using carbon in many applications. For example, many studies have been extensively conducted on biochar because it is the main source of activated carbon and it is the main player in the absorption and removal of pollutants such as heavy metal ions (Zn, Cu, Ni, Co, Sb, Cr, Cd, etc.) from drinking water (Basiuk et al. 2014). Several researchers have studied the effect of pH on the demineralization ability of carbonaceous materials, where the pH of the carbonic colloids aqueous dispersion was 2.8 and the concentration of carbon nanoparticles was 250 ppm (Václavíková et al. 2009; Sharma et al. 2018). Among these carbon materials, carbon nanotubes are one of the preferred materials because they have a specific surface area of 952 m^{2}g^{-1} and a high aspect ratio, which ranges from 500 to 100,000. Yadav et al. (2019) reported an aspect ratio of 878.57 for MWCNT at 750°C. Manganese oxide–coated carbon nanotubes (MnO$_2$/CNT) were prepared and used to treat lead

(II) from drinking water. The Pb (II) adsorption capacity on MnO_2/CNT nanocomposite decreased with decreasing the pH of the solution (Yadav et al. 2019).

10.4.2 CLAY

Clay minerals, readily available naturally, are inexpensive materials and provide a cost-effective alternative to conventional processing. The clay's porous structure and high surface area provide fluid absorption and heavy metal absorption benefits (El-Maghrabi and Mikhail 2014). Natural minerals, including clay minerals and zeolites, from different regions of the world have been studied for the removal processes of heavy metal ions (adsorption and ion-exchange processes). Clay minerals, such as kaolinite, illite, or montmorillonite from Algeria, Brazil, China, Thailand, Tunisia, Pakistan, and Serbia, proved efficient for removing Cd^{2+}, Cr^{3+}, Cr^{6+}, Co^{2+}, Cu^{2+}, Fe^{3+}, Mn^{2+}, Ni^{2+}, Pb^{2+}, or Zn^{2+} from waste drinking waters. Many researchers studied natural minerals, including clay minerals and zeolite, from different parts of the world to remove heavy metal ions (adsorption and ion exchange processes) from waste drinking water (Bedelean et al. 2009). At present, the crude and modified clay/clay minerals are still receiving great research interest as potential adsorbents, especially from the point of view of adsorbent surface pollutant interactions, considering the existence of competition between those pollutants for adsorption sites (Novikau and Lujaniene 2022). Like carbon nanotubes, metallic clay is also applied as an adsorbent because of its good adsorption capacity for various ionic metals, such as arsenic and lead. The montmorillonite and kaolinite clay minerals could be modified with H_2SO_4 to remove copper, kaolin-supported bacterial biofilms of Bacillus species to absorb chromium, kaolinite columns using MnO_2 for cadmium, etc. The efficacy of metal ions removal was found to improve with increasing temperature, and vice versa (Yadav et al. 2019). On the other hand, the demineralization capacity increases with decreasing temperature in the case of Ti-pillared montmorillonite (Babaryk et al. 2021). In recent years, nanocomposites have gained increasing interest, as these composites are formed by integrating nanostructured materials within the interlayer space of clay metal. Because of their distinct physical and chemical features, as well as their potential applications in a variety of disciplines, researchers have focused their efforts to exploit these complexes in drinking water treatment and heavy metal removal technology (Yadav et al. 2019). In this regard, and as mentioned recently, natural clays in particular are known as promising absorbent materials for environmental protection because of their distinctive structural properties. The most important characteristic of clay minerals is the presence of tetrahedral and octahedral coordinated atoms representing tetrahedral and octahedral silicate sheets, respectively, with a plate thickness of usually less than 2 mm. The tetrahedral sheets are formed from SiO_4 tetrahedral structures, where they share three basic oxygen atoms, which exhibit a hexagonal pattern. Each tetrahedron's fourth tetrahedral or apical oxygen is perpendicular to the basal lamina (Novikau and Lujaniene 2022). The clay's modified composition allows access to low-cost compounds with high capacities of adsorption, ability to withstand harsh conditions, biosuitability, and the possibility of reuse with effective contaminant absorption even when an abundance of competing ions is present (Novikau and Lujaniene 2022). High pH is a major cause to marginalize the absorption of metal ions on all studied clays and clay minerals, especially when it is endothermic in all cases, except for Pd (II) and Ni (II). Equilibrium was reached in all cases by one hour. Depending on the pH and concentration of heavy metals, various types were formed—for example, at low pH, Cr and Pb oxides and hydroxides were obtained, whereas at high pH, Zn, Ni, Hg, Ag, Cd, and Mn oxides and hydroxides were obtained. It is worth noting that precipitation reaction was prevalent at high metal concentrations and at pH less than 6, while the adsorption process was prevalent at pH > 6. However, at pH > 6, the adsorption of Cu (II) and Cr (VI) is prevalent, and Pb (II) is impossible due to their presence. Weak solubility in the alkaline medium is because of precipitation, which was dominant in these cases. In highly alkaline media, the metal hydroxide forms are affected by the silanol and aluminol sites in montmorillonite to form inner-domain complexes (Novikau and Lujaniene 2022). Since the adsorption technology has been commonly applied for heavy metal removal, due to its efficacy and ease of use, recently, AHK natural Hade Court clay has been prepared, characterized, and applied in the adsorption process for the purification of solutions contaminated with heavy metals (Cu^{2+}, Ni^{2+}, Co^{2+}, and Pb^{2+}) (Es-sahbany et al. 2022). The slurry used was first characterized by XRD technique and then applied in a study on the adsorption behavior. The goal in this condition was to investigate the adsorption capacity and removal efficiency of Ni^{2+}, Cu^{2+}, Co^{2+}, and Pb^{2+} ions as a function of the following main parameters: the mass of the clay used contact duration, pH, and initial concentration. Moreover, the evaluation of Freundlich and Langmuir adsorption isotherms is achieved (Es-sahbany et al. 2022).

10.4.3 NATURAL BIOSORBENT SUSTAINABLE MATERIALS

The effective use of materials, and the manipulation of natural materials or waste, is one of the greatest goals of materials science researchers in terms of sustainable development (Dotto et al. 2019).

10.4.3.1 Biosorbent

Biological adsorption is the most unique and effective method for binding and removing ionic contaminants of heavy metals. It is considered an effective method as an ion exchange treatment even at low mineral concentrations. This method has environmentally friendly properties and is more affordable compared to traditional techniques. Plant origin is a source of biosustainable materials as different parts of plants that include bark, stem, leaf, root, flower, fruit biomass, peel, hull, skin, bark, bran, and stone can represent agricultural waste-based sorbents (AWBs). Common examples of sorbents include cellulose, hemicellulose, and lignin from AWBs that are high in hydroxyl groups (Sharma et al. 2018). As a result, these agricultural products have a good affinity to adsorb heavy metal ions (Sharma et al. 2018). Similarly, if we look at the animal kingdom, especially the marine, the volume of the global market's waste of seafood crustaceans, especially prawns, crabs, lobsters, and crabs, has recently reached several million tons annually, 50% of which is disposed of as crustacean waste. The shells contain significant amounts of both calcium carbonate and

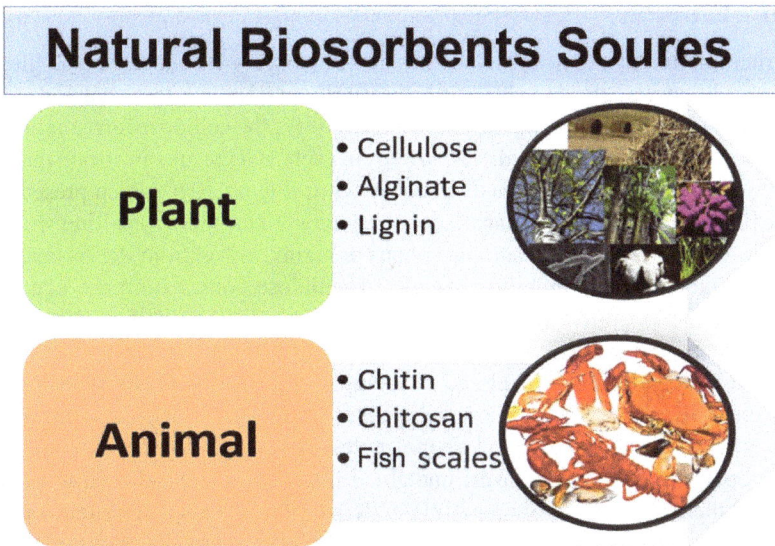

FIGURE 10.5   The natural biosorbents sources.

a chitin biopolymer (15–20% w/w), from which the N in biochitin can be deacetyl-ated, using concentrated NaOH, to produce chitosan polysaccharide, a copolymer of N- glucosamine and glucosamine. Chitin is the second most abundant biologi-cal polysaccharide polymer in the world. Because it originates as an essential sub-stance in the external protective skeleton of many living systems, chitin and chitosan each have numerous notable desirable qualities. Based on their chemical structures, these properties are currently used in many research and technological applications, where various research initiatives have been launched in recent years to explore and find possible uses for chitin and chitosan, and the level of interest in their potential applications has been reflected in the number of publications produced. Such appli-cations are involving textiles, paper and pulp, cosmetics, medicine, biotechnology, industries, food agriculture, chemical production, and environmental applications (Gerente et al. 2007). Figure 10.5 shows the sources of natural biosorbents.

10.5   THE MOLECULAR STRUCTURE OF MAJOR BIOPOLYMERS

As previously mentioned, biopolymers are a class of polymers that are obtained from biological sources; they are generally expected to interact significantly with one another if they include particular groups that generate reactive hydrogen bonds (H bonds) as well nonspecific interactions, such as van der Waals forces. The cel-lulose, chitin, and chitosan contain repeating units that are structurally linked and include functional groups (CH_3-CONH-, NH_2 and OH) which can interact strongly; as a result, as proved experimentally, relatively homogeneous nanocomposites (NC) emerge.

10.5.1 CELLULOSE

The structural unit of cellulose molecule is made up of (1→4) connected β-D-glucose units, also known as the anhydrous glucose unit (AGU). The torsional link forms every second of the AGU along the chain axis by 180°; the cellulose repeat unit is the AGU. The existence of triple hydroxyl groups in each AGU results in strong intramolecular and intermolecular H bonding, as shown in Figure 10.6, which presents the chemical structure and molecular model of cellulose. The semicrystalline structure of cellulose, with crystalline and amorphous domains, arises from the formation of these H bonds. The amphiphilic property of cellulose comes from the equatorial direction of the glucopyranose ring that has a hydrophilic character, although the hydrophobicity of the ring is in the axial direction.

Several publications have explored the comparative value of hydrophobic/solvent effects for the properties of cellulose, its drinking water insolubility, and nonorganic dipolar solvents. In short, the cellulose molecules are bound with each other by stacking glucopyranose rings in a layer; undefined interactions are believed to hold all layers together (Kostag and Seoud 2021). In terms of materials and sustainability, it should be mentioned that cellulose is the most sustainable material in the world, followed by chitin. Also, wastepaper is biomass based on cellulose. As a large amount of waste, paper is inevitably produced by manufacturers all over the world. Trash papers that contain cellulose are ranged from 40 to 55%, lignin is from 18 to 30%, and hemicellulose is from 25 to 40%, which is a potentially sustainable material that can be reused as cellulose nanocrystals and is environmentally friendly (Kaur et al. 2022). When applied and used as an alternative, low-cost sorbent materials for metal removal from drinking water and effluent treatment are effective (Kaur et al. 2022).

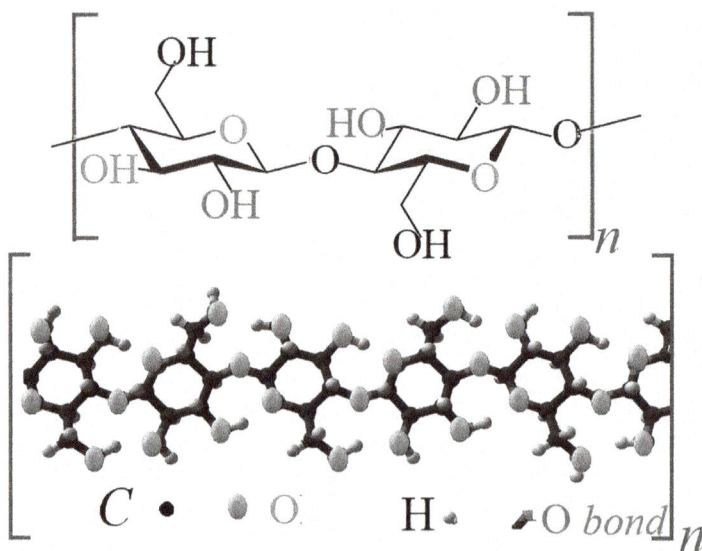

FIGURE 10.6   Chemical structure and molecular model of cellulose.

It is also used in the preparation of desalination membranes in drinking water treatment technology (Ali et al. 2020, 2021).

10.5.2 Chitin

Chitin is the precursor of chitosan. In 1811, Henri Braconnot discovered chitin by its extraction from mushrooms. Later in the 1820s, chitin was obtained from insect skin, where the "modified chitin" was obtained by C. Roughet, who treated chitin with an alkaline sodium hydroxide solution in 1859, though the treatment of chitin with potassium hydroxide as a different alkaline solution at 180°C and the production of chitosan was done in 1894 by F. Hoppe-Seyler. In 1902, Frankel and Kelly investigated the chemical structure of both chitin and chitosan, while in the 1950s, x-ray tests were used by numerous researchers, including H. Dory, to discover their chemical structures (Gedam et al., 2018).

The chemical structure of chitin consists of ((1→4) – 2-acetamido-2-deoxy- β-d-glucan), which forms a chain of N-acetylglucosamine, which is composed of glucosamine bound with an acetyl group. Figure 10.7 shows the molecular model and chemical structure of chitin. As mentioned earlier, its presence in nature follows cellulose as the second most abundant biopolymer. It is the building material that provides strength to the cell walls of fungi and exoskeletons of all crustaceans. The Greek word "chiton," which means "mail coat," is the origin of the name. Crab, krill, lobsters, and shrimp represent the major chitin commercial sources, with crabs estimated to produce 405,400 tons in 2018, lobster 150,000 tons in 2017, and shrimp 4.8 million tons, resulting in the production of more than 1 million tons of raw materials for conversion into chitin.

In crustaceans, chitin is strongly bound to molecules of lipids, proteins, and minerals in addition to pigments. The extraction process of industrial chitin is based on three essential steps: First one is protein removal. This can be done using aqueous

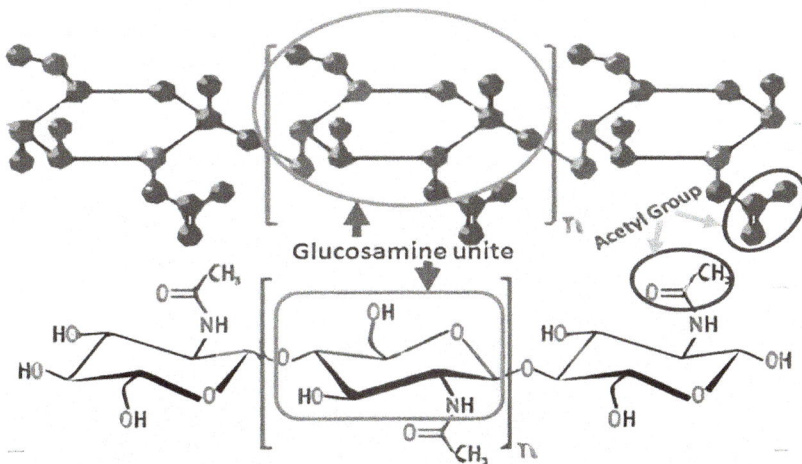

FIGURE 10.7   Molecular model and chemical structure of chitin.

mixtures of hydrogen carbonate, carbonate, sodium hydroxide, sulfite, sulfide, hydrogen sulfite, in addition to trisodium phosphate, carbonate, potassium hydroxide, and calcium hydroxide. The next step is acid demineralization to remove calcium carbonate, for example, the use of aqueous solutions of hydrochloric acid, and acids traces, such as sulfuric, nitric, acetic, formic. Hydrochloric acid is the most used. And finally, decolorizing to remove dyes. For example, this is done with acetone, with ethanol, or using hydrogen peroxide for bleaching process. The previous two steps also can be performed using some biological methods that have a less-hazardous impact on the environment, for example, protease-producing bacteria, followed by lactic acid treatment.

10.5.3 CHITOSAN

Chitosan is a natural hydrophobic biopolymer usually obtained by deacetylation of chitin using either biological or chemical methods, also by biopolymer and its derivatives concomitant depolymerization. Figure 10.8 shows the conversion (chitin → chitosan), which will be discussed next.

The insolubility of chitosan in drinking water, alkaline media, or organic solvents is due to existence of strong H-bonds between chitosan molecules. However, the protonation of the amine group makes it soluble in acidic solutions. Chitin and chitosan can be distinguished by their deacetylation (DA) degree. Chitosan is obtained when DA is above 50 mol %, which is soluble in dilute acidic solutions (Zargar et al. 2015). In other words, the degree of deacetylation (DD) of chitin and chitosan additionally to their solubility in 0.1 μM of dilute acids is used to distinguish between chitin and chitosan. If the degree of chitin deacetylation reduces 50%, it becomes

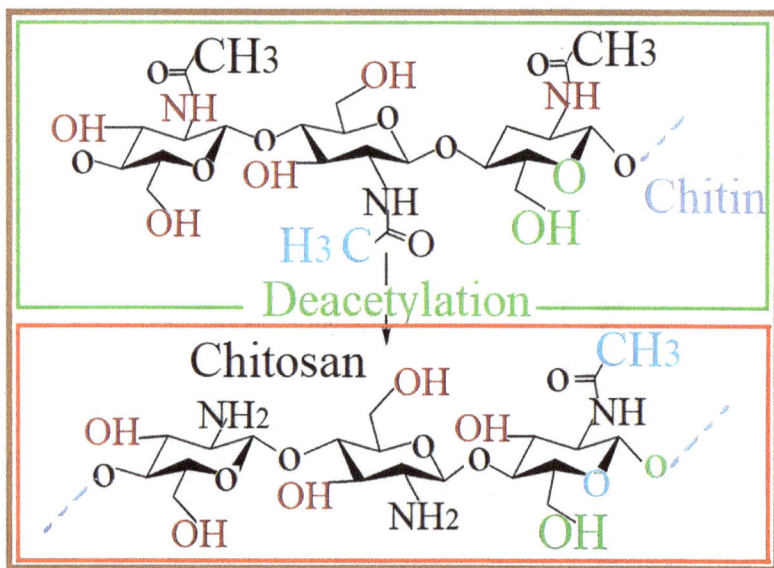

FIGURE 10.8 Chemical structures of both chitin and chitosan.

soluble in aqueous acidic medium, such as acetic acid, propionic acid, etc. (Gedam et al., 2018). When chitin is suspended in concentrated hot alkali for many hours, the process is called heterocyclic N-deacetylated (Zargar et al. 2015). While the process of placing chitin in a concentrated base solution is called homogeneous deacetylation, the polymer is biodegraded by rapid cooling below 5°C with crushed ice. In heterocyclic, DA is lower. When the DA is ranged from 48% to 55%, the product has a solubility due to the uniform distribution of acetyl groups along the chains (Gedam et al. 2018). Alternatively, chitin is dissolved and homogeneously demethylated in ionic liquids (ILs)—for example, tetraalkylammonium hydroxides (Kostag and Seoud 2021). Figure 10.9 Shows the chemical structure and molecular modeling of chitosan (Ezzat et al. 2020). One of the most environmentally friendly methods is the enzymatic treatment to convert chitin → chitosan into chitin deacetylases. This process is nondegrading and controlled to produce well-defined structures of chitosan by N-acetamido bonds active hydrolysis of chitin (Gerente et al. 2007; Le et al. 2019). In solid state, chitosan is a semicrystalline polymer. Each chitosan chain takes the form of an extended two-pronged helix, appearing as a zigzag structure. As in chitin, the "upper chains" are represented by the chitosan chains on the c-axis, while the "lower chains" are such as those in the unit cell. This indicates that the chitosan chains are packed in a nonparallel manner. The upper and lower chains are H-linked along the b-axis, resulting in plate stacking along the a-axis. The leaves

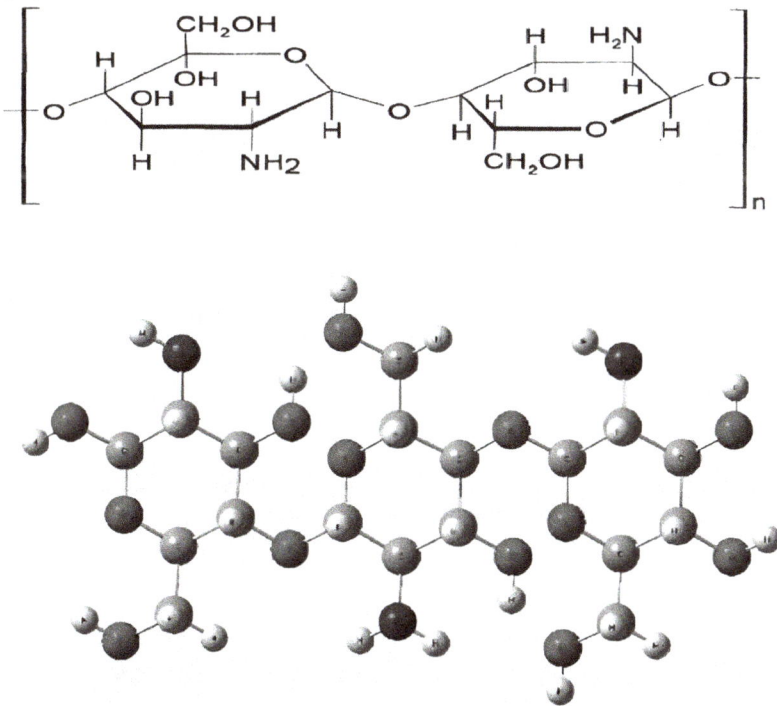

FIGURE 10.9 Chemical structure and molecular modeling of chitosan.

are bound to H by intercalated drinking water molecules. Commercially, due to their high nitrogen content (7.21%) compared to synthetically substituted cellulose (1.25%), chitin and chitosan are of great importance. Moreover, they have excellent properties, such as biodegradability and biocompatibility. Chitosan has been found to have high selective properties regarding the adsorption of metal ions, in which it can absorb transition metal ions and prevent alkali metal adsorption from aqueous solution (Gedam et al., 2018).

This has many forms and is considered the most abundant in chitosan samples. A somewhat-similar structure results from heating a sample of chitosan in an annealing process, in which the chains are packed in a parallel position. This is due to the absence of drinking water, and the chains appear directly linked in the form of H (Kostag and Seoud 2021).

From the presentation of the aforementioned properties of chitosan, it can be concluded that chitosan is a sustainable substance as it is found naturally in a renewable form, besides being biodegradable and nontoxic (Khandegar et al. 2021; Ikram et al. 2021). The existence of numerous hydroxyl and amino groups in the chitosan structure that suits chemical modifications is considered another important reason for using chitosan. As shown in Figure 10.9, the reactive groups present in chitosan are primary amino group in carbon atom number 2 and primary and secondary hydroxyl groups in carbon atoms number 6 and 3, respectively. The formation of bonds of the glycoside and the acetamide group also can present functional groups (Hassan et al. 2017). These functional groups allow for a large number of modifications, reactions, and applications, producing polymers with novel properties and behaviors (Aranaz et al. 2021). In contrast, sometimes active sites in polymer chains become blocked when preparing composite materials, as in the case of the polyaniline-gold nanocomposite, where the functional amine group bonded with capped Au nanoparticle, and this binding prevented further entrapment of the desired contaminant in this study, which was aldehyde (Omara and Elfeky 2016). Many facts of the wide versatility of chitosan make this material particularly interesting for the synthesis of suspensions, compounds, functional, or hybrid (nano) materials for various environmentally friendly applications. The intriguing polymorphic behavior exhibited by chitosan, along with its molar mass and degree of deacetylation, primarily determines its mechanical, physical, chemical, and biological properties (Maliki et al. 2022). There are different forms of chitosan-based materials (Gedam et al., 2018) that are present in Figure 10.10.

10.6   APPLICATIONS OF MODELING AND SIMULATION OF SOME SUSTAINABLE MATERIALS IN HEAVY METAL REMOVAL FROM DRINKING WATER

We mentioned earlier that molecular modeling is useful in analyzing processes for designing new uptake processes or improving understanding of existing operations. Based on their derivations, models of the process may be the product of one of three types: (1) theory, developed from fundamental principles of physics and chemistry; (2) empirical, obtained by performing an analysis of statistical or mathematical

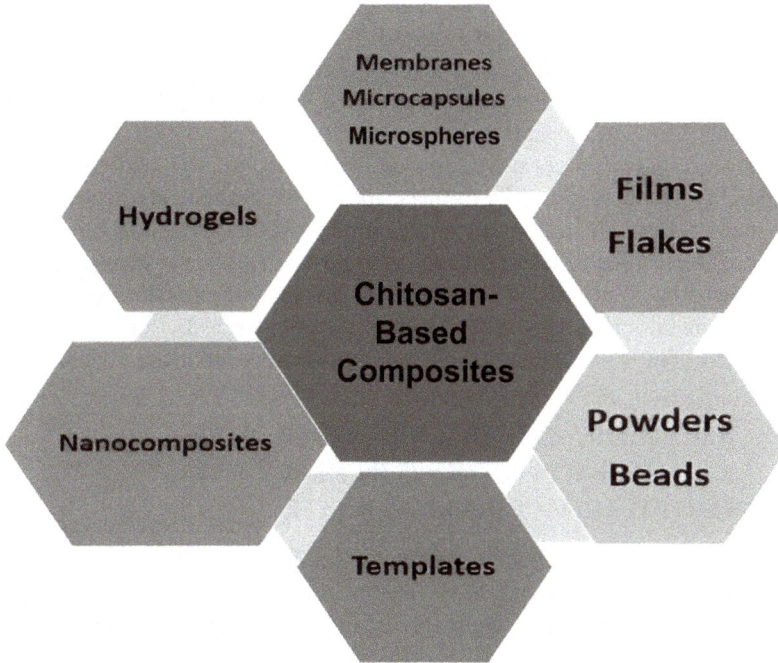

FIGURE 10.10   Different forms of chitosan-based materials.

calculations of the operating data of the system; or (3) a quasiempirical compromise based on the previous two types (Václavíková et al. 2009). At best, the process model in V approximates a mathematical description of the real process. It also supports easy overall system design and analysis, which limits the number of laboratory experiments required (*Colloidal and Particles* 2014). In this way, the researcher can better evaluate and modify the different operating conditions and design criteria and thus determine the sequence of operations that must be established from the beginning to the end of the work, arranging the control settings and setting a law and control strategy for the new process, and the matter may develop until the training of the plant operating staff and implement a more effective plan. A famous example of this interdisciplinary amalgamation is molecular modeling research directed at chitosan-based materials as a smart, cost-effective, and sustainable natural sorbent for the removal of heavy metals from aqueous solutions. The experimental results confirmed the proposed theoretical and computational data (Zarei et al. 2018). This became clear evidence that highlights the importance of integrating computational and experimental sciences in order to better understand and solve many problems. Performing experiments in combination with theoretical analysis is an excellent way to enhance our knowledge and improve processes aiming to remove heavy metals from drinking water (Hasanzade and Raissi 2019).

Density functional theory (DFT), which was confirmed before as a powerful tool to describe the physical and chemical interaction properties of metal ions with nanocellulose (Zarei et al. 2018; Kaur et al. 2022), graphene oxide GO, and chitosan CS (Wang et al. 2019; Portnyagin et al. 2015). Besides the DFT calculation, x-ray absorption spectroscopy (EXAFS) is a very good way to study the reaction mechanism. The exact structures and types of adsorbed heavy metal ions can be achieved by analyzing the EXAFS spectra (Wang et al. 2015).

Gerente's research group, their literature published in 2007, could demonstrate the capability of chitosan to adsorb metal ions from solution. The study's isothermal review is offered first, followed by a summary of the adsorption capabilities of several metal ions. The time dependency of chitosan metal ion systems is examined next, followed by an evaluation of the significant mass transport or kinetic mechanism in the metal ion removal rate. The rate control step is a critical parameter in designing the commercial metal ion adsorption systems (Gerente et al. 2007).

10.7   CASE STUDIES OF CHITOSAN MOLECULAR MODELING APPLICATIONS IN HEAVY METAL REMOVAL FROM DRINKING WATER

Chitosan was selected as a sustainable and natural biocompatible biosorbent with superior properties for removing heavy metals in aqueous media as a case study to illustrate its competitive removal properties and its high affinity towards effective removal of copper and lead ions from drinking water, where metal uptake capacity for the metal ions was calculated using equation 5:

$$q_e = \frac{C_0 - C_{eq}}{M} \tag{5}$$

Where q_e is the metal uptake at equilibrium (mg/g), C_0 is the initial concentration of metal ion (mg/L), C_{eq} is the equilibrium concentration of metal ion (mg/L), V is volume of solution (L), and M is the initial concentration of biosorbent in the solution (g/L) (Naeimi et al. 2022).

Also, a lot of experimental and theoretical research have been performed on chitosan and its derivatives to study, improve, and confirm their ability to build a permanent drinking water treatment system. Density function theory (DFT) was also used to confirm the synthesis of chitosan-based structures and to identify the active binding sites in each structure based on equations calculations (De Silva and Rajapakse 2015). According to the literature review, calculation methods based on hybrid functions and equivalence base groups are one of the most extensively used approaches in theoretical analyses of the composition of metal ions with chitosan derivatives (Portnyagin et al. 2015). Density functional theory was used to investigate the vibrational spectrum of D-glucosamine using the hybrid B3LYP functional group and 6–31 g**. The interaction of Pb (II) with chitosan and its derivatives was studied using the functional B3LYP and the LanL2DZ base group. This was done to optimize the computational method choice required for performing initial calculations for many groups using heavier 6–31 g* base for nickel and 6–311 g* base for other elements and wB97X-D dispersion corrected functional dispersion, where DFT is currently the most suitable method for studying interactions noncovalent (Portnyagin et al. 2015).

A. S. Portnyagin and others in 2015 investigated using (DFT) the Ni^{2+} complexes with chitosan and its N-heterocyclic derivatives. Calculation of 11 typical structures corresponding to the types of "bridge" and "necklace" complexes was done. Their findings indicated that the creation of "necklace" complexes is more appropriate in most circumstances. Nitrogen atoms in chitosan and its heterocyclic derivatives were shown to play an important role in Ni^{2+} binding, and the degree of charge transfer from the bonding to the central ion in the complexes was found to relate to complex stability. A class of stability of complexes depending on the type of functional substitution in large chitosan was in close agreement with the data obtained experimentally from the isotherms of nickel ion adsorption on chitosan and its derivatives (Portnyagin et al. 2015). They concluded that DFT simulations revealed the potential of forming "bridge" and "necklace" complexes between Ni^{2+} ion, chitosan, and IMC in the same proportion. Due to the existence of sterile obstacles as well as a restricted number of closely spaced saccharides within the polymer network, "suspended" structures may provide a more advantageous situation in the case of crosslinked polymers. The most favorable complexes of Ni^{2+} with 2-PEC, on the other hand, are represented in the form of an "asymmetric bridge" structure, indicating that the adsorption capacity of 2-PEC may be lower than the number of functional groups when adsorption is carried out from highly concentrated Ni^{2+} ion solutions. It is worth noting that the probability of Ni^{2+} interactions with nitrogen atoms in the heterocycle derivative is extremely rare for the "necklace" model, and for this, most likely, only "asymmetric bridge" complexes will be formed. Most likely, in most Ni^{2+} formed complexes, Ni^{2+} ions are most often associated with unsubstituted chitosan rings, but from it, a small fraction of "bridge"-type complexes with pyridyl-substituted rings can also be formed. The researchers were able to write a general class of complex stability based on the type of functional substitution in large chitosan. The results they obtained were in agreement with their experimental data on Ni^{2+} sorption by chitosan and its N-heterocyclic derivatives (Portnyagin et al. 2015).

Bhabesh Chandra Deka and Pradip Kr. Bhattacharyya (2015) described a density functional study that was conducted in order to investigate the interaction between chitosan, the gene carrier, and the nucleobases, where their study focused on the pivotal role played by the hydrogen bond in the formation and the link between chitosan and nucleobase. And they were able to find the inclusion of the aqueous medium to reach the greatest impact on the interaction between the two. The thermochemical study revealed that there is a substantial thermodynamic driving force in the gas phase to create the approximation. In the aqueous phase, however, no such driving force was determined. Additionally, apart from adenine, the aromaticity of the nucleobases was enhanced by rounding formation with chitosan. One of the most important findings is that the protons of the nuclear bases lead to the disintegration of two components of the approaches (Deka and Bhattacharyya 2015).

This research group reported a bioadsorbent biologically absorbent chitosan CS, a chitin polysaccharide derivative consisting of D-glucosamine and N-acetyl-D-glucosamine randomly distributed as an effective agent for purifying drinkable drinking water, where the percentage of elimination of Pb (II), Cd (II), and Cr (VI) of drinking water using CS under optimal conditions was as follows: 64%, 94%, and 70%. However, many attempts have been done to investigate the mechanism of chitosan-heavy metal interaction. Hence, their study aimed to simulate chitosan's infrared spectrum through the calculations of DFT, also the determination of vibrational bands related to potential sites of metal binding that can be used as a tool to study the mechanism of interaction between heavy metals and chitosan (De Silva and Rajapakse 2015). Based on theoretical analysis of the infrared spectrum of chitosan, the researchers were able to clearly distinguish the vibrational bands and accompanying vibrational patterns in the vicinity of the sites most consistent with metallic bonding and thus were able to correctly set the FTIR (experimental) spectrum of chitosan. In addition, they were able to determine the assignment structure of the formed chitosan metal complex resulting from the presence of a large shift in the infrared bands near the specific coordination sites for the formation of bonds with the metal. Through their study, the method they used became a diagnostic tool to determine the metal binding preferences of chitosan-based biosorbents as one of the sustainable biological compounds (De Silva and Rajapakse 2015).

Basila Hassan et al in 2015 used density functional theory (DFT) to study the interaction of Pb (II) ions with chitosan biopolymer in question. They used Schiff bases and N-alkylated/aryl chitosan derivatives as absorbent materials for Pb ions besides studying them at the monomer level. They were also able to perform a natural bond orbital (NBO) for chitosan analysis and its derivatives to reach an understanding of the interactions of the donor and acceptor. Thus, they were able to map the molecular electrostatic potential (MEP) of the absorbent materials with the color code. They then calculated the global reaction coefficients for the sorbents based on the frontier molecular orbital (FMO) energies. Also, they tested complex's structure formed between Pb (II) ion with chitosan and its derivatives independently by DFT at the B3LYP/LanL2DZ level, then discussed the stability of pools based on Eads values. It was then observed that the reduced chitosan derivative carboxydehyde pyridine (RPC) has the ability to form a more stable complex with Pb (II) ions than other derivatives (Hassan et al. 2015). In more details, the Pb (II) adsorption by chitosan monomer and its derivatives was studied at the B3LYP/LanL2DZ theoretical level. This research also investigated the entire potential adsorbent sites for lead (II) adsorbents. They were also able to find chitosan coordinating the formation of Pb (II) ions at different sites based on the position of the Pb (II) placed on it. They presented the potential geometries for the Pb–chitosan reaction in seven optimized modes. Their data indicated that the oxygen of the heterocyclic ring could also be involved in the metal reaction as well as the OH substitutions in chitosan (Hassan et al. 2015). They concluded that a theoretical study was carried out using the DFT method to describe the adsorption behavior of chitosan and its derivatives for lead (II) ions. Modeling and characterization of Schiff bases and N-alkylated/arylated chitosan's derivatives were done. MEP maps showed the surface properties. Global reaction variables provided the reactivity of chitosan and its derivatives, whereas the derivatives containing aromatic rings and electron-rich substituents showed better absorption and more reactivity than chitosan. The structure of complexes consisting of sorbents and Pb (II) ions was studied and compared. Finally, to understand complexes stability, the calculation of Eads values for the complexes was done. It is noteworthy that among the derivatives, RPC forms stronger complexes with the lead ion.

Basila Hassan and others in 2017 conducted a theoretical study based on the DFT of toxic metal ion adsorption of Hg (II) using chitosan monomer as well as two of its derivatives, where they prepared both citralidine and salicylidin chitosan. Then they analyzed the effect of structural characteristics on the stability of the complexes under study using the Gaussian 03 software package. They were also able to study all possible matches of these adsorbents using the simplest universal geometric shapes. They also studied all the adsorption sites by placing the metal ion in an intermediate position between the atoms and determining the stable match of the adsorbed metal ion compound. They found that the interaction of Hg (II) and the adsorbents is electrical, where the bonding of the metal ion to the oxygen atoms is weaker than to the nitrogen atom in all cases based on the enhancement of the nitrogen charge density upon the formation of the Schiff base. Derivatives on chitosan monomer are stable in acidic medium. The ΔE value of the complexes is in the order SC—Hg (II) > chitosan—Hg (II) > CC—Hg (II), which leads to the fact that the stability of the complexes increases with the increase of the energy gap. Their studies also revealed that the aromatic Schiff base derivatives of chitosan are better at taking up Hg (II) than the aliphatic derivatives (Hassan et al. 2017). The most noticeable part is the stability of the Hg–polymer complex, and from their results it was concluded that CC and SC were more convenient for Hg (II) adsorption. They were able to report the formation of Schiff base derivatives to enhance the chitosan's property. This was because through their study the aliphatic and aromatic bases were modeled. From their calculations using density functional theory performed on the reaction of Hg (II) with chitosan monomer and two of its derivatives, CC and SC, the results confirmed that the aromatic chitosan derivative of Schiff base is better at taking up Hg (II) than the aliphatic derivative. The CC-Hg (II) compound is less stable than the Hg (II) chitosan compound. They also found that the interaction between Hg (II) and the adsorption process was electrical. Likewise, in all cases, the strength of the bond between the metal ion and the nitrogen atom was stronger than the strength of the bond with the oxygen atoms. Also, they were able to improve the nitrogen charge density when forming a Schiff base. But the metal ion–CC compound was less stable than the metal ion–chitosan compound. Derivatives based on chitosan monomer are distinguished by their stability in acidic media, where they observed a lower ΔE value for the derivatives than that of the monomer of chitosan that was indicating an improvement in conductivity; the complex SC shows the lowest value. The E value of the complexes is in the order SC—Hg (II) > chitosan—Hg (II) > CC—Hg (II), indicating that the increase in the energy gap led to increasing the stability of the complexes. This study considers the monomer of chitosan and its derivatives as a model for a good adsorption of mercury (II) (Hassan et al. 2017).

Gendam and her team (Gedam et al, 2018) described chitosan (deacetylated chitIn: β (1→4) D glucosamine); it was described as a desirable wading sorbent that is low-cost, biodegradable, and biocompatible to dilute few heavy metals from drinking water. Where they modified the flexible skeleton of chitosan by inserting a few organic/inorganic cracks to obtain biocompounds to adsorb several diverse pollutants such as heavy metals, for example, as they mentioned in their paper, where they discussed the batch adsorption of Pb (II) ions from aqueous solution by graphite-saturated chitosan compound (GDCC) as an absorbent compound. They found that the maximum absorption capacity of Pb (II) ions was 6.711 mg/g (from Langmuir) at optimum pH 6, with a dose of 1 g/L in 120 minutes. They recommended the use of the biosorption mechanism within the scope of drinking water cleaning and treatment procedures (Gedam et al, 2018). The adsorption mechanism of graphite-doped chitosan composite (GDCC) biosorbents possess is mainly dependent on various functional groups, like hydroxyl carboxyl, amino, phosphate, and so on, as we mentioned before. These functional groups have the ability to provide active binding sites for the absorption of heavy metals in different ways. Considering that the biological sorption mechanism is very complex due to the structural diversity of biological sorbents, one of the most important factors affecting effective bioabsorption on the surface of biosorbent materials is the availability of a number of active binding sites, and as a result of the convergence of pollutants to the surface of the bioabsorbent material loaded with a number of diverse functional groups that can interact with the donor and the recipient with the heavy metal or its ions. The process of adsorption is a combination of ion exchange, complexing, precipitation, and other processes that are highly influenced by the pH of the solution. Similarly, in order to comprehend the mechanism of adsorption, the pH of the zero-point charge (pHpzc) of the adsorbents must be determined; pHpzc is of critical importance in the field of ecology, because it determines how easily toxic ions are absorbed, so that the relationship between initial pH values and initial pH (pH) can be plotted. In addition to at the point of intersection of the resulting curve, where the difference between pH = 0 is denoted as the pH of the zero-point charge. Positive adsorption is preferred at pH > pHpzc, while anionic adsorption is preferred at pH < pHpzc. This is since at lower pH values, the concentration of hydronium ions that have a competitive relationship with the cationic pollutants of the adsorption sites on the surface of the adsorbent increases, whereas when the pH increases, the concentration of hydroxide ions increases and the adsorbent surface becomes negative, thus increasing the potential for attraction between the adsorbent surface with cationic pollutants. Hence, pH > pH pzc becomes suitable for cationic absorption. In the case of anionic adsorption, the pH of the solution is required to be lower than pzc for the surface of adsorbent to become positively charged to support the adsorption capacity of anionic pollutants on the adsorbent surface (Gedam et al. 2018). As we explained earlier, the amino group of chitosan plays a major role in the binding and removal of metals, as is the case with Pb (II) ions, through the adsorption property, where it can form a coordination site for metal ions. In the acidic medium, the amino group can acquire protons by reacting with H^+ ions (Gedam et al. 2018).

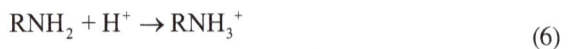

$$RNH_2 + H^+ \rightarrow RNH_3^+$$

(6)

PROTONATED CHITOSAN

Where the authors of this research dealt with the method of reducing drinking water pollution with heavy metal ions Pb (II) using chitosan-doped graphite compound

(GDCC). The maximum capacity of Pb (II) ions absorption was 6.711 mg/g (based on Langmuir) at optimum pH 6, with a dose of 1 g/L in 120 minutes. The most important thing for choosing these materials is to pay attention to increasing the efficiency of absorption quality, providing a safe, easy-to-maintain, and easy-to-use method, reducing residual mass production, lowering the cost of capital, and not being toxic (Gedam et al., 2018).

In 2019, Camilla L. Vieira et al. used the reaction between lauroyl chloride and chitosan to synthesize the chitosan-N-lauroyl (CL) composite; two proportions were prepared to obtain CL1P (1:1) and CL2P (2:1) adsorbents. Then these were applied to studies in Pb (II) and Cu (II) aqueous solutions on the adsorption capacity of metal ions and the interaction between the metal/adsorbent. Afterward, they were able to evaluate the effect of pH, temperature of the reaction medium, selectivity, kinetics, and rate of adsorption. Then, verifying the inter-ion complexity feasibility and the theoretically adsorbed by DFT approach, they were able to calculate the maximum theoretical capacities of absorption (q_{max}) for CL1P and CL2P for Cu(II)/Pb(II) ions 36.4/41.0 and 48.5/37.6 mg.g^{-1}, respectively. The laboratory q_{max} values were lower than the theoretical values, ranging from 31.5 to 33.0 mg/g. They described the process of ion adsorption as endothermic, spontaneous, and connected with pseudo-second order kinetics at the optimum pH value (5.5). They used the calculations of quantum mechanical to support their results, and they identified the active interaction sites for the chitosan ion and indicated the role of functional groups, such as hydroxyl, carbonyl, and amide, in stabilizing the complexes. Given the shortage of studies evaluating the performance of polar chitosan-based adsorbents to remove heavy metals at that time, their findings are promising, as they allowed a more reliable explanation of the relationship between polar groups in case pH is controlled, which was able to help re-envision and structure polymeric adsorbents (Vieira et al. 2019).

In the first step, they improved the structures of CL, CL-Pb, and CL-Cu at the level of B3LYP/6–311 G(d)/LANL2DZ. Additionally, they showed through data analysis a number of effective changes in the geometric parameters of both composites (CL-Pb and CL-Cu) as a result of the presence of metal ions. The installation was possible for both assemblies due to the installation of the seven-member ring. Comparing CL-Pb and CL-Cu with chitosan structure complexity revealed effective differences in the parameters of geometry. This is supported by the binding site of metal ions in chitosan. Their computational data show no significant role of the lauroyl group to the complexing process for both composites. They demonstrated the possibility of accessing additional details about the geometric parameters of all structures through Cartesian coordinates by performing quantum chemical calculations, and based on this, they determined the bond order parameters to evaluate the results related to the interactions between CL, Cu (II), and Pb (II) ions. An important step in studying such long-term interactions is to confirm that these interactions occur. In this regard, fulfilling Poppelier's criteria is considered the most appropriate way to attribute the existence of an interaction, defined in the *Quantum Theory of the Formulation of Atoms into Molecules* (QTAIM). The criteria are (1) the bond critical point (BCP) existence between the acceptor and donor atoms, and (2) electron density values [ρ (r)] that must ranged from 0.002 to 0.035 a.u. They listed the parameters of CL,

Pb (II), and Cu (II) ions in a table to compare the stability and strength of these interactions. Depending on their data, they found that only CL-Cu only was able to accumulate at two defined critical bonding points. In systems with higher polarity, the Mayer bond order and fuzzy magnitudes parameters differ. The parameters of fuzzy take longer to calculate, but the sensitivity of the base set is significantly reduced. In most cases, there is an overlap between the ambiguous bond arrangement and the experimental bond arrangement, and from the obtained data, the value of the interaction can be determined to a large extent. The cooperative nature of this type of interaction leads to great complexity between the minerals and the chitosan samples under study. This is because the presence of interactions between CL, Pb (II), and Cu (II) ions was determined, and the geometrical analyses and bond ordering allow evaluating the stability of these interactions, which depends on the thermodynamic theoretical information. The thermodynamic data obtained suggested that the process of sorption was spontaneous and endothermic, with a degree of randomness at interface of the surface solution. The absorbent material did not show selectivity for the metal ions under investigation. Astana, through theoretical calculations in APFD/6–311+G(d,p)/Def2-TZVP//B3LYP/6–311 G(d)/LANL2DZ stable interaction in hydroxyl, amide, and carbonyl groups in CL with ions [Pb (II) and Cu (II)] (Vieira et al. 2019).

Also, in 2019, Igor Hernandes Santos Ribeiro et al. used DFT to understand the process of adsorption of copper (II), cadmium (II), mercury (II), lead (II), chromium (III), and zinc (II) ions using a vanillin-derived methacrylate monomer (VMA). Various analyses were performed, and all are essential to expect the complexes formation, such as adsorption energy, molecular electrostatic potential (MEP), boundary orbitals, hardness, harmonic analysis, and ductility. By means of the molecular electrostatic potentials and the frontier molecular orbital (FMOs), they were able to find the best places for bonding and adsorption, so they placed each metal ion in the study near the nitrogen and oxygen atoms of the imine and carboxyl groups of the vanillin monomer, respectively. They proved that the strength of the bonding of metal ions with an atom of nitrogen is larger than with oxygen atoms, and this was due to the nitrogen charge density, which increases in the Schiff base formation near the aromatic ring. They showed that the monomer has a high adsorption capacity for Pb (II), Cu (II), and Cr (III) ions due to the high energy values used. The QTAIM analysis was studied to understand the interaction nature between the monomer of vanillin and the metal ions, which has been described in almost all cases as a covalent moiety. Thus, they were able to prove that the monomer derived from vanillin is characterized by good drinking water stability, and based on the results they reached, they proved that this derived compound represents a good material for effluents and poisoning cases treatments (Ribeiro et al. 2019).

In 2020, Hend and others reported the use of the density functional theory (DFT) to explore chitosan chemical structure and studied its competitive effectiveness in removing heavy metals from polluted drinking water. In this paper, they were able to make an effective attempt to develop a cost-effective and reliable natural polymer structure based on Cs with high ability to bind and sequester some heavy metals that can cause drinking water contamination. The authors of this paper studied the relationship between heavy metal, Cs, and their stability by applying density functional

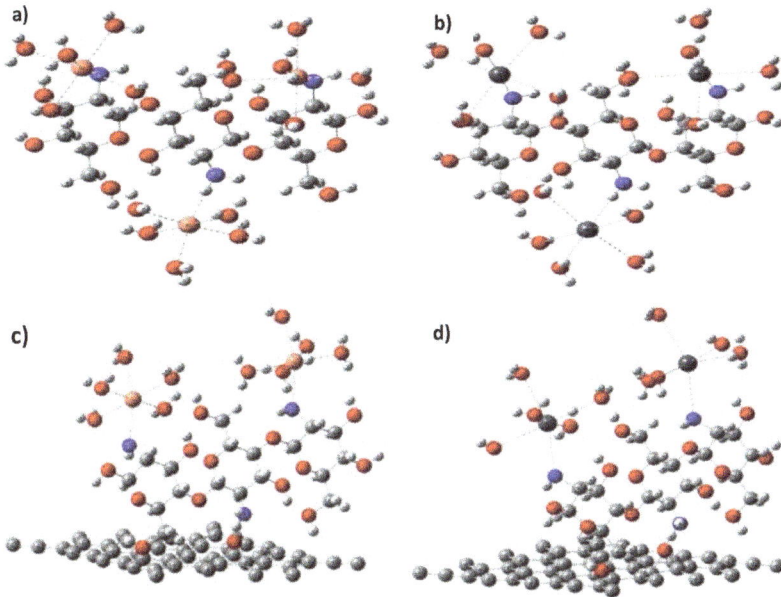

FIGURE 10.11   B3LYP/LANL2DZ calculated optimized structures for (a) chitosan + Cu.; (b) chitosan + Pb; (Ezzat et al., 2020), chitosan/graphene oxide with hydrated metals models at basis set B3LYP/LANL2DZ: (c) for Cs/GO + Cu, (d) Cs/GO + Pb, respectively (Menazea et al., 2020).

theory (DFT) using B3LYP with the LANL2DZ base set. Their results showed the observation that the band gap energy values for Cs, Cs-Fe, Cs-Ni, Cs-Cu, Cs-As, Cs-Cd, and Cs-Pb were 2.571, 0.6147, 0.7176, 0.3396, 0.6648, 1.5007, and 0.4128 eV. Figure 10.11a and b shows the B3LYP/LANL2DZ calculated optimized structures of both CS-Cu and CS-Pb. Straight later showed that the Cs structure has high selectivity and binding affinity with the hydrated metals Cu and Pb, respectively, instead of Fe, Ni, As, and Cd. This research was also able to reveal the examined relationship between Cu and Pb with Cs, which indicated the possibility of using Cs in sensing applications to detect and remove copper and lead from waste to drinking water (Ezzat et al. 2020). In the same year, those researchers studied the potential interactions between the previously studied heavy metals and chitosan/graphene oxide (Cs/GO), where they were able to apply DFT at the level of B3LYP with a LANL2DZ basis set to study the interaction potential. The aforementioned heavy metals were selected due to their hazardous effect when present in industrial waste drinking water. The interactions between heavy metals and Cs that were previously investigated were then compared with the Cs/GO complex. To study the interaction between heavy metals and Cs in comparison with both the effect of the nitrogen-containing groups on the surface of Cs and oxygen-containing groups on the surface of GO, the proposed model was applied. Figure 10.11c and d shows the B3LYP/LANL2DZ calculated optimized structures of both (CS/GO)-Cu and (CS/GO)-Pb independently. The interaction between the Cs/GO complex and the hydrated metal showed better metal selectivity than the Cs complex alone. They praised the possibility of applying their results in the field of environmental pollution reduction by removing heavy metals from industrial waste drinking water (Menazea et al. 2020).

In 2021, Manny Anthony and others used for the first time the synthesis of chitosan–manganese magnetic ferrite at low combustion of temperature without using cross-linkers and its removal of copper (II) and lead (II) from single-component metal solutions, where they used the nonlinear Langmuir model to describe the isothermal data better, while the pseudo-nonlinear second order model described the kinetic data better, and through their study they were able to infer the use of Cu (II) or Pb (II) monolayer chemisorption and adsorption as a step to determine the average, respectively. Their results showed that the adsorption capacities using magnetic chitosan and manganese ferrite increased for both metals continuously than those found by manganese ferrite, which indicated that chitosan was able to enhance the performance of magnetic adsorbents. The maximum absorption capacity of magnetic ferrite for chitosan and manganese for copper (II) and lead (II) was 14.86 and 15.36 mg/g, while the amplitudes of manganese ferrite were 2.59 and 13.52 mg/g, respectively. In addition, through their results, they showed that the adsorbents had a higher affinity and bonding absorption for Pb (II) than Cu (II) due to the ability of the former to form the inner field complexes with Mn ferrite and magnetic Mn chitosan. Finally, thermodynamic studies revealed that the adsorption of either Pb (II) or Cu (II) by magnetic manganese chitosan ferrite was spontaneous and endothermic. They have carried out a characterization of the prepared adsorbents in order to know the information about their surface functional locations, elemental composition, morphology, and particle size by energy-dispersive spectroscopy scanning electron microscopy, dynamic light scattering technique, and Fourier transform infrared spectroscopy. Based on their results, they relied upon documenting their theoretical study (Taguba et al. 2021).

Also, Minghu Zhao's group in 2021 was able to recover gold from waste drinking water because of its great interest in the past years due to its scarcity and high economic value globally. They were able to successfully synthesize an innovative chitosan-based adsorbent (CS-GTU) using formaldehyde as a crosslinking compound between chitosan and guanylthiourea, and it was applied to the selective adsorption of Au (III) ion from an aqueous medium. Through batch experiments, the maximum CS-GTU adsorption capacity for Au (III) reached 695.63 mg/g at pH 5.0, and the adsorption process followed the pseudo-second-order kinetic model and Langmuir isothermal models, and they indicated that the monolayer chemical adsorption might have occurred on absorbent surfaces. Adsorption was described as a spontaneous enthalpy-catalyzed chemical process based on thermodynamic analysis. In addition, the adsorbent showed a remarkable selectivity towards removing Au (III) from solutions in the presence of different metals, and their results showed during five repeated experiments that the adsorption of CS-GTU can be regenerated efficiently. In addition to this, they successfully used the Thomas model to simulate the experimental penetration curves, which can match the experimental data with the correlated curve ($R2 > 0.9$) well for sulfur/nitrogen. They considered that CS-GTU beads are a suitable and effective absorbent material towards absorbing gold ions in aqueous solutions (Zhao et al. 2021).

Bingjie Wang and others in 2021 highlighted the importance of using ion-printed polymers with special recognition cavities. Therefore, they concluded that checking the composition and improving the performance of functional materials using the trial-and-error design method is a waste of time and money. Therefore, their study was based on the design of high-performance chitosan microspheres (ICSMs) by calculating density functional theory (DFT) and synthesized by facile microfluidic technology. The synthesized ICSMs showed a very uniform profile (Dav = 420.6 µM, CV = 3.6%) and a very high adsorption capacity (q_{max} = 107.12 mg g). The adsorption isotherms were best fitted with the Langmuir model, while the kinetic data followed the pseudo-second order model, and from that they inferred the dominant role of chemical tests. In addition, ICSM showed acceptable stability and reusability (95.34 mg g, after five cycles). Also, they studied the mechanism of selective adsorption quantitatively by means of electronegativity, electrophilic affinity index, adsorption energy (Ea), and bond length. They expected their study to lay the foundation for the design and synthesis of high-performance biosorbent materials for future drinking water treatment (Wang et al. 2021).

Recently, in 2022, Ahmed M. Omar and others were able to collect much research based on the use of one of the well-known and cost-effective techniques for waste drinking water treatment based on pollutant adsorption using natural polymers such as chitosan and extraction by environmentally friendly and low-cost methods. Despite this, Cs suffers from multiple challenges, such as low absorbency, low reusability, and low surface area. Therefore, they engaged a review writer to obtain an overview of the latest developments in chitosan-based adsorbents that established better adsorption processes for several hazardous heavy metals, including Cu (II), Au (III), Cd (II), Pb (II), Cr (VI), and As (V) ions. In addition, they discussed the ability of chitosan-based adsorbents to remove contaminant anions, including phosphates and nitrates. In addition, they explained the potential mechanisms of adsorption of these pollutants on chitosan adsorbents and deduced optimal conditions for adsorption processes considering the currently reported studies. Moreover, they emphasized major research weaknesses and future directions that may inspire researchers to discover better solutions to drinking water treatment problems (Omer et al. 2022).

Jian Chen and others in 2022 were able to easily prepare the porous chitosan/ Zr-MOF (UiO-66) (CSF @UiO-66) chitosan/Zr-MOF (CSF @UiO- 66) composite foam by grafting UiO-66 into polymerized chitosan by adding a binding agent, such as epichlorohydrin. Their results showed that UiO-66 flexible foams with high loads had superior solid–liquid separation performance and low density (15.0 ~ 23.4 mg/ cm3). The hierarchical pores and abundant functional groups contributed to the effective removal of ketoprofen (KPF) from drinking water by the composite foams, which resulted in the highest capacity of adsorption of 209.7 mg/g at pH 4, which outperformed most of the adsorbents indicated in their previous studies. Langmuir isothermal model and the pseudo-second order kinetic model were used to describe KPF adsorption, as shown by batch adsorption. Notably, composite foams have shown promising performance even in the presence of some ions and good applicability for engineering applications. In addition, the calculations of DFT revealed that the mechanism of adsorption between KPF and composite foams can be attributed to hydrogen bonding and π-interaction. Basically, they considered this work as an innovative strategy to prepare composite foam filled with high loads of MOFs to achieve an effective removal of KPF adsorption in liquids, which can overcome the negative effects of agglomeration due to small particle size and poor stability of MOFs (Chen et al. 2022).

Dinu et al. (2022) presented the evaluation of sorption studies of a solution of heavy metal ions containing five-component Zn^{2+}, Pb^{2+}, Cd^{2+}, Ni^{2+}, and Co^{2+} at equimolar concentrations in fixed-base columns using some chitosan-functionalized clinoptilolite from APCA (CS-CPL) cryogel sorbent materials compared to unmodified composites. They followed the general sorption tendency of APCA-operated sorbent composites of the sequence $Co^{2+} < Zn^{2+} < Cd^{2+} \leq Pb^{2+} < Ni^{2+}$, which indicated that Co^{2+} ions have less affinity for the functional groups of the sorbent, while Ni^2 ions have greater strength and selectivity. They were able to gain more perceptions into using microbeads composite in a continuous flow setting by describing the kinetic data by Thomas and Yoon Nelson models. They estimated a theoretical maximum HMI adsorption capacity of 145.55 mg/g and a 50% penetration time of 121.5 minutes for the absorbent column of the CSEDTA-CPL cryogel; both values were significantly greater than those achieved by the column filled with pure CSCPL adsorbents. Moreover, they successfully achieved uptake of HMIs from synthesized microbeads under dynamic conditions by using 0.1 M HCl aqueous solution. In addition, they analyzed theoretically the APCA structures associated with the complex adsorbents and their spatiotemporal structures within complex syntheses with systematically transition metals. Starting with EDTA, the most stable conformer, coordination groups with HMIs can be determined by energy consumption of 1 kcal/mol, which is sufficient to convert the spatial structure into a form suitable for HMI chelation (Dinu et al. 2022).

10.8 CONCLUSION

The ability of sustainable materials to remove heavy metals from drinking water was discussed. The role of Internet of Things was elaborated, especially using molecular modeling in assessing the performance of chitosan as one of the most effective sustainable sorbents for heavy metal removal. In this aspect, molecular modeling represents an effective tool for the prediction of reactions and products. It provides a guideline for much experimental research to solve many problems related to drinking water security. The accuracy of data obtained based on employing sustainable materials such as chitosan and its derivatives can help in developing a cost-effective drinking water treatment system for heavy metal removal from drinkable water.

REFERENCES

Akha, N.Z., Salehi, S. and Anbia, M., 2022. Removal of arsenic by metal organic framework/chitosan/carbon nanocomposites: Modeling, optimization, and adsorption studies. International Journal of Biological Macromolecules, 208, pp. 794–808.

Ali, A.S.M., Fadl, E.A., Soliman, M.M. and Kandil, S.H., 2020. Optimization of the evaporation step in cellulose acetate membranes preparation by dry—wet phase inversion technique for drinking-water desalination applications. Desalinat Drinking-water Treat, 174, pp. 63–70.

Ali, A.S.M., Soliman, M.M., Kandil, S.H. and Khalil, M., 2021. Emerging mixed matrix membranes based on zeolite nanoparticles and cellulose acetate for drinking-water desalination. Cellulose, 28(10), pp. 6417–6426.

Anas, N.A.A., Fen, Y.W., Omar, N.A.S., Daniyal, W.M.E.M.M., Ramdzan, N.S.M. and Saleviter, S., 2019. Development of graphene quantum dots-based optical sensor for toxic metal ion detection. Sensors, 19(18), p. 3850.

Aranaz, I., Alcántara, A.R., Civera, M.C., Arias, C., Elorza, B., Heras Caballero, A. and Acosta, N., 2021. Chitosan: An overview of its properties and applications. Polymers, 13(19), p. 3256.

Asdrubali, F., Schiavoni, S. and Horoshenkov, K.V., 2012. A review of sustainable materials for acoustic applications. Building Acoustics, 19(4), pp. 283–311.

Aslam, M.M.A., Kuo, H.W., Den, W., Usman, M., Sultan, M. and Ashraf, H., 2021. Functionalized carbon nanotubes (Cnts) for drinking-water and waste drinking-water treatment: Preparation to application. Sustainability, 13(10), p. 5717.

Asokan, K., Vivekanand, P.A. and Muniraj, S., 2021. An eco-friendly method to remove copper ion from drinking-water by using homemade bio-adsorbent in tip-tea-bag. Materials Today: Proceedings, 36, pp. 883–885.

Babaryk, A.A., Adawy, A., García, I., Trobajo, C., Amghouz, Z., Colodrero, R.M., Cabeza, A., Olivera-Pastor, P., Bazaga-García, M. and dos Santos-Gómez, L., 2021. Structural and proton conductivity studies of fibrous π-Ti 2 O (PO 4) 2· 2H 2 O: application in chitosan-based composite membranes. Dalton Transactions, 50(22), pp. 7667–7677.

Balali-Mood, M., Naseri, K., Tahergorabi, Z., Khazdair, M.R. and Sadeghi, M., 2021. Toxic mechanisms of five heavy metals: Mercury, lead, chromium, cadmium, and arsenic. Frontiers in Pharmacology, 12, p. 643972.

Basiuk, E.V., Martinez-Herrera, M., Alvarez-Zauco, E., Henao-Holguín, L.V., Puente-Lee, I. and Basiuk, V.A., 2014. Noncovalent functionalization of graphene with a Ni (II) tetraaza [14] annulene complex. Dalton Transactions, 43(20), pp. 7413–7428.

BASSI, R., PRASHER, S.O. and Simpson, B.K., 2000. Removal of selected metal ions from aqueous solutions using chitosan flakes. Separation Science and Technology, 35(4), pp. 547–560.

Bedelean, H., Maicaneanu, A., Burca, S. and Stanca, M., 2009. Removal of heavy metal ions from waste drinking-waters using natural clays. Clay Minerals, 44(4), pp. 487–495.

Chen, J., Ouyang, J., Chen, W., Zheng, Z., Yang, Z., Liu, Z. and Zhou, L., 2022. Fabrication and adsorption mechanism of chitosan/Zr-MOF (UiO-66) composite foams for efficient removal of ketoprofen from aqueous solution. Chemical Engineering Journal, 431, p. 134045.

de Brito Lira, J.O., Riella, H.G., Padoin, N. and Soares, C., 2021. An overview of photoreactors and computational modeling for the intensification of photocatalytic processes in the gas-phase: State-of-art. Journal of Environmental Chemical Engineering, 9(2), p. 105068.

De Silva, S.M. and Rajapakse, C.S.K., 2015. Use of computational method to identify metal binding sites of chitosan as a tool to investigate the interaction mechanism of chitosan and heavy metals. Technical Session—Mineral Resources Technology. Proceedings of the Research Symposium of Uva Wellassa University, January 29–30, 2015, p. 68.

Dehghani, F., Rezaie, B., Sachan, A. and Ghosh, T., 2019. Effects of Grinding on Particle Shape: Silica and Magnetite. SME Annual Conference & Expo and CMA 121st National Western Mining Conference: Smart Mining: Resources for a Connected World Denver, Colorado, USA, February 24–27, 2019, ISBN: 978-1-5108-8466-3, pp. 438–441.

Deka, B.C. and Bhattacharyya, P.K., 2015. Understanding chitosan as a gene carrier: A DFT study. Computational and Theoretical Chemistry, 1051, pp. 35–41.

Ding, Q., Li, C., Wang, H., Xu, C. and Kuang, H., 2021. Electrochemical detection of heavy metal ions in drinking-water. Chemical Communications, 57(59), pp. 7215–7231.

Dinu, M.V., Humelnicu, I., Ghiorghita, C.A. and Humelnicu, D., 2022. Aminopolycarboxylic acids-functionalized chitosan-based composite cryogels as valuable heavy metal ions sorbents: Fixed-Bed column studies and theoretical analysis. Gels, 8(4), p. 221.

Dotto, J., Fagundes-Klen, M.R., Veit, M.T., Palacio, S.M. and Bergamasco, R., 2019. Performance of different coagulants in the coagulation/flocculation process of textile waste drinking-water. Journal of Cleaner Production, 208, pp. 656–665.

El-Maghrabi, H.H. and Mikhail, S., 2014. Removal of heavy metals via adsorption using natural clay material. Journal of Environment and Earth Science, 4(19), pp. 38–46.

Eschrig, H., 2003. The Fundamentals of Density Functional Theory (Vol. 2). Leipzig: Edition Am Gutenbergplatz.

Es-sahbany, H., El Yacoubi, A., El Hachimi, M.L., Boulouiz, A., El Idrissi, B.C. and El Youbi, M.S., 2022. Low-cost and eco-friendly Moroccan natural clay to remove many bivalent heavy metal ions: $Cu2+$, $Co2+$, $Pb2+$, and $Ni2+$. Materials Today: Proceedings, 58, pp. 1162–1168.

Ezzat, H., Menazea, A.A., Omara, W., Basyouni, O.H., Helmy, S.A., Mohamed, A.A., Tawfik, W. and Ibrahim, M., 2020. DFT: B3LYP/LANL2DZ study for the removal of Fe, Ni, Cu, As, Cd and Pb with Chitosan. Biointerface Research in Applied Chemistry, 10, pp. 7002–7010.

Farcasanu, I.C., Popa, C.V. and Ruta, L.L., 2018. Calcium and cell response to heavy metals: Can yeast provide an answer? Calcium and Signal Transduction, 1, pp. 23–41.

Finocchi, F., 2011. Density Functional Theory for Beginners: Basic Principles and Practical Approaches. Institut des NanoSciences de Paris (INSP) CNRS and University Pierre et Marie Curie, Paris.

Gaur, V.K., Sharma, P., Gaur, P., Varjani, S., Ngo, H.H., Guo, W., Chaturvedi, P. and Singhania, R.R., 2021. Sustainable mitigation of heavy metals from effluents: Toxicity and fate with recent technological advancements. Bioengineered, 12(1), pp. 7297–7313.

Gedam, A.H., Narnaware, P.K. and Kinhikar, V., 2018. Blended composites of chitosan: Adsorption profile for mitigation of toxic pb (ii) ions from water. Chitin-Chitosan-Myriad Function Science Technol, 6, pp. 99–118.

Gerente, C., Lee, V.K.C., Cloirec, P.L. and McKay, G., 2007. Application of chitosan for the removal of metals from waste drinking-waters by adsorption—mechanisms and models review. Critical Reviews in Environmental Science and Technology, 37(1), pp. 41–127.

Gonçalves, F.A., Santos, M., Cernadas, T., Ferreira, P. and Alves, P., 2022. Advances in the development of biobased epoxy resins: Insight into more sustainable materials and future applications. International Materials Reviews, 67(2), pp. 119–149.

Gontrani, L., Pulci, O., Carbone, M., Pizzoferrato, R. and Prosposito, P., 2021. Detection of Heavy Metals in Water Using Graphene Quantum Dots: An Experimental and Theoretical Study Molecules, 26(18), p. 5519.

Hasanzade, Z. and Raissi, H., 2019. Assessment of the chitosan-functionalized graphene oxide as a carrier for loading thioguanine, an antitumor drug and effect of urea on adsorption process: Combination of DFT computational and molecular dynamics simulation studies. Journal of Biomolecular Structure and Dynamics, 37(10), pp. 2487–2497.

Hassan, B., Muraleedharan, K. and Mujeeb, V.A., 2015. Density functional theory studies of Pb (II) interaction with chitosan and its derivatives. International Journal of Biological Macromolecules, 74, pp. 483–488.

Hassan, B., Rajan, V.K., Mujeeb, V.A. and Muraleedharan, K., 2017. A DFT based analysis of adsorption of $Hg2+$ ion on chitosan monomer and its citralidene and salicylidene derivatives: prior to the removal of Hg toxicity. International Journal of Biological Macromolecules, 99, pp. 549–554.

Herath, I.K., Wu, S., Ma, M. and Ping, H., 2022. Heavy metal toxicity, ecological risk assessment, and pollution sources in a hydropower reservoir. Environmental Science and Pollution Research, 29(22), pp. 32929–32946.

Huang, Y., Li, J., Chen, X. and Wang, X., 2014. Applications of conjugated polymer based composites in waste drinking-water purification. RSC Advances, 4(107), pp. 62160–62178.

Hughes, Z.E. and Walsh, T.R., 2015. Computational chemistry for graphene-based energy applications: progress and challenges. Nanoscale, 7(16), pp. 6883–6908.

Ibrahim, M.S., El-Mageed, A. and El-Salam, A., 2020. Density functional theory calculations on the grafting copolymerization of 2-substituted aniline onto chitosan. Polymer Bulletin, 77(12), pp. 6391–6407.

Ikram, R., Mohamed Jan, B., Abdul Qadir, M., Sidek, A., Stylianakis, M.M. and Kenanakis, G., 2021. Recent Advances in Chitin and Chitosan/Graphene-Based Bio-Nanocomposites for Energetic Applications. Polymers, 13(19), p. 3266.

Jones, R.O., 2015. Density functional theory: Its origins, rise to prominence, and future. Reviews of Modern Physics, 87(3), p. 897.

Kaur, J., Sengupta, P. and Mukhopadhyay, S., 2022. Critical Review of Bioadsorption on Modified Cellulose and Removal of Divalent Heavy Metals (Cd, Pb, and Cu). Industrial & Engineering Chemistry Research.

Kausar, A., Rehman, S.U., Khalid, F., Bonilla-Petriciolet, A., Mendoza-Castillo, D.I., Bhatti, H.N., Ibrahim, S.M. and Iqbal, M., 2022. Cellulose, clay and sodium alginate composites for the removal of methylene blue dye: Experimental and DFT studies. International Journal of Biological Macromolecules, 209, pp. 576–585.

Khandegar, V., Kaur, P.J. and Chanana, P., 2021. Chitosan and Graphene Oxide-based Nanocomposites for Drinking-water Purification and Medical Applications: A Review. BioResources, 16(4), pp. 8525–8566.

Koskinen, P. and Mäkinen, V., 2009. Density-functional tight-binding for beginners. Computational Materials Science, 47(1), pp. 237–253.

Kostag, M. and El Seoud, O.A., 2021. Sustainable biomaterials based on cellulose, chitin and chitosan composites-A review. Carbohydrate Polymer Technologies and Applications, 2, p. 100079.

Lapo, B., Demey, H., Zapata, J., Romero, C. and Sastre, A.M., 2018. Sorption of Hg (II) and Pb (II) ions on chitosan-iron (III) from aqueous solutions: Single and binary systems. Polymers, 10(4), p. 367.

Le, T.T.N., Le, V.T., Dao, M.U., Nguyen, Q.V., Vu, T.T., Nguyen, M.H., Tran, D.L. and Le, H.S., 2019. Preparation of magnetic graphene oxide/chitosan composite beads for effective removal of heavy metals and dyes from aqueous solutions. Chemical Engineering Communications, 206(10), pp. 1337–1352.

Maliki, S., Sharma, G., Kumar, A., Moral-Zamorano, M., Moradi, O., Baselga, J., Stadler, F.J. and García-Peñas, A., 2022. Chitosan as a Tool for sustainable development: A mini review. Polymers, 14(7), p. 1475.

Menazea, A.A., Ezzat, H.A., Omara, W., Basyouni, O.H., Ibrahim, S.A., Mohamed, A.A., Tawfik, W. and Ibrahim, M.A., 2020. Chitosan/graphene oxide composite as an effective removal of Ni, Cu, As, Cd and Pb from waste drinking-water. Computational and Theoretical Chemistry, 1189, p. 112980.

Naeimi, A., Amini, M. and Okati, N., 2022. Removal of heavy metals from wastewaters using an effective and natural bionanopolymer based on Schiff base chitosan/graphene oxide. International Journal of Environmental Science and Technology, 19(3), pp. 1301–1312.

Nejadshafiee, V. and Islami, M.R., 2020. Intelligent-activated carbon prepared from pistachio shells precursor for effective adsorption of heavy metals from industrial waste of copper mine. Environmental Science and Pollution Research, 27(2), pp. 1625–1639.

Nik Abdul Ghani, N.R., Jami, M.S. and Alam, M.Z., 2021. The role of nanoadsorbents and nanocomposite adsorbents in the removal of heavy metals from waste drinking-water: A review and prospect. Pollution, 7(1), pp. 153–179.

Novikau, R. and Lujaniene, G., 2022. Adsorption behaviour of pollutants: Heavy metals, radionuclides, organic pollutants, on clays and their minerals (raw, modified and treated): A review. Journal of Environmental Management, 309, p. 114685.

Odobašić, A., Šestan, I. and Begić, S., 2019. Biosensors for determination of heavy metals in waters. Biosensors for Environmental Monitoring BoD–Books on Demand. pp. 139–151.

Omara, W. and A Elfeky, S., 2016. Effect of Gold/Amine Nanoparticles on Polyaniline Electrochemical Sensitivity to Formaldehyde. Recent Patents on Nanotechnology, 10(2), pp. 157–164.

Omara, W., Amin, R., Elhaes, H., Ibrahim, M. and A Elfeky, S., 2015. Preparation and characterization of novel polyaniline nanosensor for sensitive detection of formaldehyde. Recent Patents on Nanotechnology, 9(3), pp. 195–203.

Omer, A.M., Dey, R., Eltaweil, A.S., Abd El-Monaem, E.M. and Ziora, Z.M., 2022. Insights into recent advances of chitosan-based adsorbents for sustainable removal of heavy metals and anions. Arabian Journal of Chemistry, 15(2), p. 103543.

Pal, B., Matsoso, J.B., Parameswaran, A.K., Roy, P.K., Lukas, D., Luxa, J., Marvan, P., Azadmanjiri, J., Hrdlicka, Z., Jose, R. and Sofer, Z., 2022. Flexible, ultralight, and high-energy density electrochemical capacitors using sustainable materials. Electrochimica Acta, 415, p. 140239.

Pehlivan, E., Altun, T. and Parlayici, Ş., 2012. Modified barley straw as a potential biosorbent for removal of copper ions from aqueous solution. Food Chemistry, 135(4), pp. 2229–2234.

Portnyagin, A.S., Bratskaya, S.Y., Pestov, A.V. and Voit, A.V., 2015. Binding Ni (II) ions to chitosan and its N-heterocyclic derivatives: Density functional theory investigation. Computational and Theoretical Chemistry, 1069, pp. 4–10.

Pournara, A.D., Margariti, A., Tarlas, G.D., Kourtelaris, A., Petkov, V., Kokkinos, C., Economou, A., Papaefstathiou, G.S. and Manos, M.J., 2019. A Ca 2+ MOF combining highly efficient sorption and capability for voltammetric determination of heavy metal ions in aqueous media. Journal of Materials Chemistry A, 7(25), pp. 15432–15443.

Ribeiro, I.H.S., Reis, D.T. and Pereira, D.H., 2019. A DFT-based analysis of adsorption of Cd2+, Cr3+, Cu2+, Hg2+, Pb2+, and Zn2+, on vanillin monomer: A study of the removal of metal ions from effluents. Journal of Molecular Modeling, 25(9), pp. 1–9.

Sabourian, V., Ebrahimi, A., Naseri, F., Irani, M. and Rahimi, A., 2016. Fabrication of chitosan/silica nanofibrous adsorbent functionalized with amine groups for the removal of Ni (II), Cu (II) and Pb (II) from aqueous solutions: Batch and column studies. RSC Advances, 6(46), pp. 40354–40365.

Shah, J., Židonis, A. and Aggidis, G., 2021. State of the art of UV drinking-water treatment technologies and hydraulic design optimisation using computational modelling. Journal of Drinking-water Process Engineering, 41, p. 102099.

Sharma, R., Dhillon, A. and Kumar, D., 2018. Biosorbents from agricultural by-products: Updates after 2000s. In Bio-and Nanosorbents from Natural Resources (pp. 1–20). Springer, Cham.

Singhon, R., 2014. Adsorption of Cu (II) and Ni (II) Ions on Functionalized Colloidal Silica Particles Model Studies for Waste drinking-water Treatment (Doctoral Dissertation, Besançon). Universite Du Franche- comite, France.

Surucu, O., 2022. Electrochemical removal and simultaneous sensing of mercury with inductively coupled plasma-mass spectrometry from drinking-water. Materials Today Chemistry, 23, p. 100639.

Syeda, H.I. and Yap, P.S., 2022. A review on three-dimensional cellulose-based aerogels for the removal of heavy metals from drinking-water. Science of the Total Environment, 807, p. 150606.

Taguba, M.A.M., Ong, D.C., Ensano, B.M.B., Kan, C.C., Grisdanurak, N., Yee, J.J. and de Luna, M.D.G., 2021. Nonlinear Isotherm and Kinetic Modeling of Cu (II) and Pb (II) Uptake from Drinking-water by MnFe2O4/Chitosan Nanoadsorbents. Drinking-Water, 13(12), p. 1662.

Tahir, M.B., Kiran, H. and Iqbal, T., 2019. The detoxification of heavy metals from aqueous environment using nano-photocatalysis approach: A review. Environmental Science and Pollution Research, 26(11), pp. 10515–10528.

Tang, J., He, M., Luo, Q., Adeel, M. and Jiao, F., 2020. Heavy Metals in agricultural soils from a typical mining city in China: Spatial distribution, source apportionment, and health Risk assessment. Polish Journal of Environmental Studies, 29(2), 1379–1390.

Titirici, M.M., White, R.J., Brun, N., Budarin, V.L., Su, D.S., del Monte, F., Clark, J.H. and MacLachlan, M.J., 2015. Sustainable carbon materials. Chemical Society Reviews, 44(1), pp. 250–290.

Václavíková, M., Vitale, K., Gallios, G.P. and Ivanicová, L. eds., 2009. Drinking-Water Treatment Technologies for the Removal of High-Toxity Pollutants. Springer, The Netherlands.

Vakili, M., Deng, S., Cagnetta, G., Wang, W., Meng, P., Liu, D. and Yu, G., 2019. Regeneration of chitosan-based adsorbents used in heavy metal adsorption: A review. Separation and Purification Technology, 224, pp. 373–387.

Vasconcelos, H.C. and Eleutério, T., 2022. Sustainable Materials for Advanced Products. In Handbook of Sustainability Science in the Future: Policies, Technologies and Education by 2050 (pp. 1–17). Springer International Publishing, Cham.

Vieira, C.L., Neto, F.O.S., Carvalho-Silva, V.H. and Signini, R., 2019. Design of apolar chitosan-type adsorbent for removal of Cu (II) and Pb (II): An experimental and DFT viewpoint of the complexation process. Journal of Environmental Chemical Engineering, 7(3), p. 103070.

Wang, B., Bai, Z., Jiang, H., Prinsen, P., Luque, R., Zhao, S. and Xuan, J., 2019. Selective heavy metal removal and drinking-water purification by microfluidically-generated chitosan microspheres: Characteristics, modeling and application. Journal of Hazardous Materials, 364, pp. 192–205.

Wang, B., Xuan, J., Yang, X. and Bai, Z., 2021. Synergistic DFT-guided design and microfluidic synthesis of high-performance ion-imprinted biosorbents for selective heavy metal removal. Colloids and Surfaces A: Physicochemical and Engineering Aspects, 626, p. 127030.

Wang, X., Chen, Z. and Yang, S., 2015. Application of graphene oxides for the removal of Pb (II) ions from aqueous solutions: Experimental and DFT calculation. Journal of Molecular Liquids, 211, pp. 957–964.

Wen, F., Sun, Z., He, T., Shi, Q., Zhu, M., Zhang, Z., Li, L., Zhang, T. and Lee, C., 2020. Machine learning glove using self-powered conductive superhydrophobic triboelectric textile for gesture recognition in VR/AR applications. Advanced Science, 7(14), p. 2000261.

Wołowiec, M., Komorowska-Kaufman, M., Pruss, A., Rzepa, G. and Bajda, T., 2019. Removal of heavy metals and metalloids from water using drinking water treatment residuals as adsorbents: A review. Minerals, 9(8), p. 487.

Xing, C., Huang, W., Xie, Z., Zhao, J., Ma, D., Fan, T., Liang, W., Ge, Y., Dong, B., Li, J. and Zhang, H., 2018. Ultrasmall bismuth quantum dots: Facile liquid-phase exfoliation, characterization, and application in high-performance UV—Vis photodetector. ACS Photonics, 5(2), pp. 621–629.

Yadav, S., Goel, N., Kumar, V., and Singhal, S. 2019. Graphene oxide as proficient adsorbent for the removal of harmful pesticides: comprehensive experimental cum DFT investigations, *Anal. Chem. Lett.* 9(3), pp. 291–310.

Yan, C., Qu, Z., Wang, J., Cao, L. and Han, Q., 2022. Microalgal bioremediation of heavy metal pollution in drinking-water: Recent advances, challenges, and prospects. Chemosphere, 286, p. 131870.

Yang, Y., Zhang, Y., Zheng, H., Zhang, B., Zuo, Q. and Fan, K., 2022. Functionalized dual modification of covalent organic framework for efficient and rapid trace heavy metals removal from drinking-water. Chemosphere, 290, p. 133215.

Yuvaraja, G., Pang, Y., Chen, D.Y., Kong, L.J., Mehmood, S., Subbaiah, M.V., Rao, D.S., Pavuluri, C.M., Wen, J.C. and Reddy, G.M., 2019. Modification of chitosan macromolecule and its mechanism for the removal of Pb (II) ions from aqueous environment. International Journal of Biological Macromolecules, 136, pp. 177–188.

Zarei, S., Niad, M. and Raanaei, H., 2018. The removal of mercury ion pollution by using Fe3O4-nanocellulose: Synthesis, characterizations and DFT studies. Journal of Hazardous Materials, 344, pp. 258–273.

Zargar, V., Asghari, M. and Dashti, A., 2015. A review on chitin and chitosan polymers: Structure, chemistry, solubility, derivatives, and applications. ChemBioEng Reviews, 2(3), pp. 204–226.

Zhang, D., Li, N., Cao, S., Liu, X., Qiao, M., Zhang, P., Zhao, Q., Song, L. and Huang, X., 2019. A layered chitosan/graphene oxide sponge as reusable adsorbent for removal of heavy metal ions. Chemical Research in Chinese Universities, 35(3), pp. 463–470.

Zhang, J., 2015. Computational Modeling and Evaluation of Reverse Osmosis and Forward Osmosis Drinking-water Treatment Technologies Integrated with Cchp Waste Heat. Stanford University, United States.

Zhang, Z., Zhao, W., Hu, C., Guo, D. and Liu, Y., 2022. Colorimetric copper (II) ions detection in aqueous solution based on the system of 3' 3' 5' 5'-tetramethylbenzidine and AgNPs in the presence of Na2S2O3. Journal of Science: Advanced Materials and Devices, 7(2), p. 100420.

Zhao, M., Li, X., Huang, Z., Wang, S. and Zhang, L., 2021. Facile cross-link method to synthesize chitosan-based adsorbent with superior selectivity toward gold ions: Batch and column studies. International Journal of Biological Macromolecules, 172, pp. 210–222.

Zhao, X.J., Liu, S.H., Zhou, J.H., Tan, L.X. and Sun, J.K., 2022. Nanoporous cationic organic cages for trapping heavy metal oxyanions. ACS Applied Nano Materials, 5(1), pp. 890–898.

11 Circular Economy Models in Water and Wastewater Sectors

Alaa El Din Mahmoud and Eyas Mahmoud

CONTENTS

11.1 INTRODUCTION

Fresh water makes up 2.5% of the world's water, but we can access just 1% of it because most of it is trapped in glaciers and snowfields, in addition to the rapid growth of population. The United Nations declared that water usage is double the population growth's rate. By 2025, it is estimated that 1.8 billion people will live in drought areas (geographic 2021); 2 billion people lack safe drinking water, and 1.7 billion lack basic sanitation. Climate change and environmental pollution make the problem more complicated (Zarei 2020). It is expected that water stress level (projected ratio of water withdrawals to water supply) will be the highest by 2040, as shown in Figure 11.1.

Nowadays, some countries are about to reach 0% fresh water, such as Cape Town (a coastal country in South Africa), which suffered from water shortage and, consequently, drought in 2017. Due to a decrease in the amount of rainfall and overconsumption, the water level in dams decreased. Therefore, they applied the circular economy pillars to save water as much as they can, because they needed and appreciated every drop of water, and people started sharing tips on social media on how to make each drop count (Warner and Meissner 2021).

DOI: 10.1201/9781003260455-11

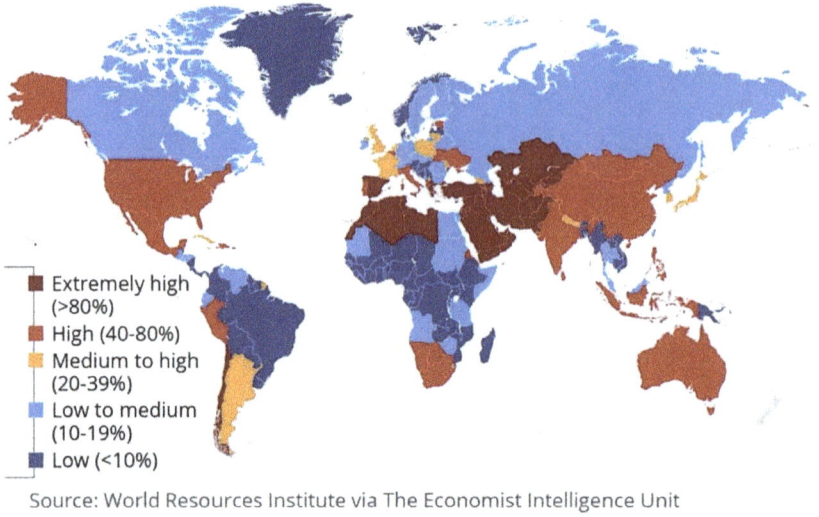

Source: World Resources Institute via The Economist Intelligence Unit

FIGURE 11.1 Water stress level around the world by 2040.

The circular economy is the first step to achieving sustainability. This concept was created by Leontief in 1928 and applied by many countries since the end of 1990s because of increasing environmental issues (Barreiro-Gen and Lozano 2020).

A circular economy (CE) aims to achieve a zero-waste system. The traditional economy produces waste because it is linear (take, make use, dispose). On the other hand, the raw material in the circular economy remains in the system as long as possible because it gets reused or recycled again into new products. This is so powerful in maintaining our natural resources (Neczaj and Grosser 2018).

Global policy makers recommend CE as an alternative model. CE represents an alternate to a traditional linear economy (LE) (see Table 11.1), in which resources are kept in use and the maximum value is extracted whenever possible, then the product and materials are recovered and regenerated at the end of their service lives (D'Amato et al. 2017). LE is the traditional way of production which generates large amount of waste and drains lots of resources and raw materials, while CE is a restorative and regenerative economic system that keeps using products and materials as long as they provide value, so it reduces the amount of waste and promotes the secondary material and reusing, recycling, upcycling, and other pillars (Hossain et al. 2020; Munaro et al. 2020).

Building upon CE as a sustainable model, which continuously recycles and restores the components and raw materials involved in the value chain, this model presents a complete challenge to solve, without which it will be a costly model (Ramadoss et al. 2018).

The model is globally adopted by corporations and governments due to the potential they possess in order to capture economic value while promoting sustainable environment. The idea of a circular economy aims to restore and regenerate. It begins with the principle of biological (regenerative) and technological nutrients from cradle

TABLE 11.1

Linear Economy versus Circular Economy

Linear Economy	Circular Economy
• A linear economy has typically followed the "take-to-dispose" strategy. This entails gathering and processing raw materials into usable items that are eventually discarded as garbage. In this economic system, value is created by manufacturing and selling as many items as possible. (Sørensen 2018). • Wastewater should be collected for treatment after consumption. Following that, depending on the economic water usage model, the treated water in the linear model is disposed of without being used again (Sgroi et al. 2018).	• Circular economy is a systemic approach to business, society, and the environment aimed at fostering business growth. Unlike the linear model of "taking waste," there is a reviving circular economy aimed at grading economic growth away from the consumption of finite resources (Gubeladze and Pavliashvili 2020). • Another definition by Kirchherr et al. (2017): "[A]n economic system that is based on business models which replace the 'end-of-life' concept with reducing, alternatively reusing, recycling, and recovering materials in production/distribution and consumption processes, thus operating at the micro level (products, consumers, companies), meso level (eco-industrial parks) and macro level (city, region, nation), with the aim to accomplish sustainable development, which implies creating environmental quality, economic prosperity, and social equity, to the benefit of current and future generations." • A circular economy encourages technological, organizational, and social innovation across and within value chains, in addition to waste prevention and reduction (Bicket et al. 2014).

to cradle (restorative). The model stresses that the system can be optimized by continuously maintaining the materials at their optimum use and value (Mahmoud 2020a).

Figure 11.2 illustrates the major elements of the CE concept as well as the possible benefits of its adoption. CE could minimize the use of raw materials and, in turn, minimize waste and pollution. CE increases competitiveness and opens new markets with employment opportunities and social benefits. Furthermore, the circular economy has pillars, principles, or frameworks to implement in any industry. They are rethinking about how to create sustainable economy, reduce water usage and pollution at the source, reuse wastewater, and find alternative sources, like recycling, by extracting water from wastewater for other uses, recovery of resources, such as nutrients and energy from water-based waste, and removal (reclamation) of the pollutants from water (Mahmoud et al. 2022a; Mishra et al. 2022; Smol et al. 2020).

Cape Town adapted to this situation by putting restrictions and implementing these principles—for example, the country reduced the demand of water from 1,000 ML to 500 ML a day and banned some activities, such as washing cars and filling swimming pools. Everyone was allowed just 50 liters a day (Warner and Meissner 2021). The citizens used recycled water (gray water) and made a campaign with the slogan "If it's yellow, let it mellow," promoting flushing the toilet only when necessary (Edmond n.d.).

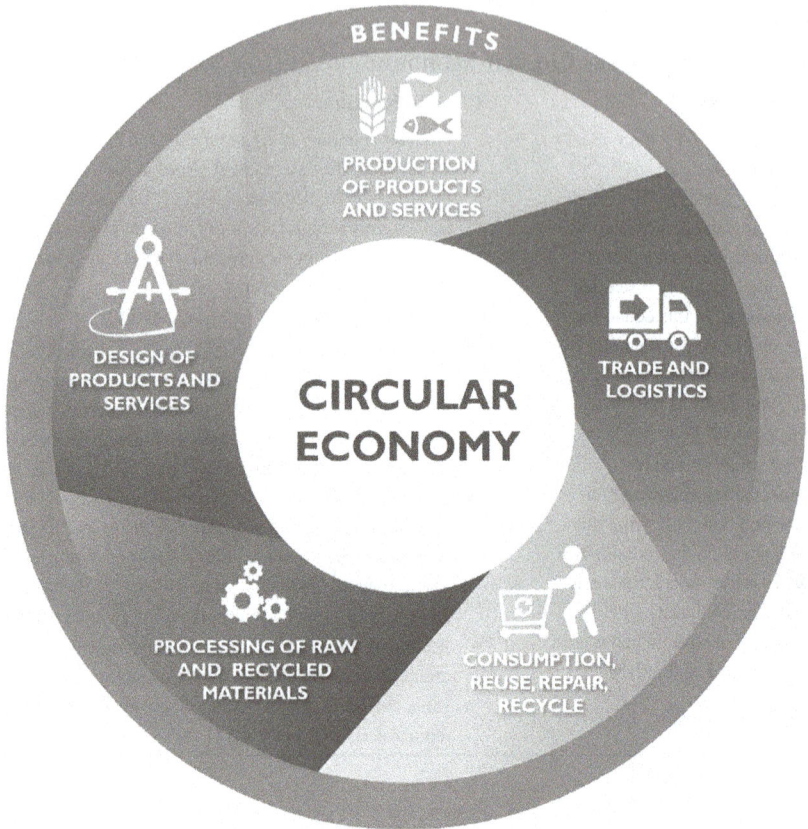

FIGURE 11.2　Circular economy and its benefits of adoption.

Source: Adapted from Berg et al. (2018).

The team of Levi Strauss & Co. in the Epping area of Cape Town tried to use other water sources, so they drilled a borehole that allowed them to draw brackish as a short-term phase. They also installed a water treatment plant so that treated saltwater or recycled water could be used (Millstein 2020). Due to sewage and wastewater from several industries spewed into rivers, fresh water turns into wastewater. According to the government's admission, 75% of the 910 municipality-run wastewater treatment works achieved 50% compliance to minimum compliance standards last year, causing health and environmental issues to the citizens and animals. The urban wastewater treatment plants are a significant part of circular sustainability because of the integration of energy production and resource recovery during clean water production (Neczaj and Grosser 2018).

The national Department of Water Affairs and Forestry (DWAF) has warned that Cape Town will face a water crisis by 2015, ten years before Day Zero (Warner and Meissner 2021). That's why we should take serious steps towards circular economy from now on to achieve sustainability.

Due to the integration of energy generation and resource recovery during clean water production, urban wastewater treatment plants (WWTPs) can be an important aspect of circular sustainability. WWTPs will become "ecologically sustainable" technology systems in the near future. However, over the last 25 years, a number of key factors have emphasized the importance of recovering the resources found in wastewater. Global nutrient needs, as well as the recovery of water and energy from wastewater, are the primary drivers of the wastewater industry's growth (Neczaj and Grosser 2018).

Artificial intelligence has demonstrated promises as a viable platform for accelerating circular economic transition. The Fourth Industrial Revolution is dominated by artificial intelligence (IA). AI promises to estimate that the potential value of AI, unchanged, to assist waste disposal in a circular food economy in 2030 will reach 127 billion USD per year (Tsegaye et al. 2021).

11.2 PRINCIPLES OF CIRCULAR ECONOMY (CE)

The CE principles are adopted to achieve the Sustainable Development Goals (SDGs) of the United Nations on water and sanitation by reducing raw material consumption, removing waste from material use (Zarei 2020).

The circular economy aims to achieve the "zero-waste" principle, which is based on three rules (Flores et al. 2018):

(1) All items with a lengthy life span (durables) must be reused rather than discarded or downcycled.
(2) All products with a short life span (consumables) should be used to their full potential before being securely returned to the biosphere.
(3) Natural resources, to the extent that they can be renewed, should be used in a sustainable manner.

The circular economy pillars or framework begins with "reduce," which is the most important one, because it saves the environment and prevents wastewater generation at the source through sustainable planning and design. Household water use is the largest growing sector that will be increased up to 80% over the next 25 years; with this, we need to adopt some changes in our routine to reduce water consumption. We can achieve this by raising people's awareness. This, in turn, reduces individual expenses and time used obtaining water. Additionally, this would improve the availability of water for other uses and WWTP (Smol et al. 2020).

We should do all these things and save water so that we don't experience Cape Town's situation, where they had to adopt these habits to live. They used gray water, built WWTP. For example, they prevent washing cars and filling pools with fresh water and other activities.

The next one is removing pollutants (bacteria, pathogens, impurities, toxic substances, etc.) from wastewater and saving disposal after treatment according to national laws and regulations (reclamation). We shouldn't put the pollutants back to clean water; this way, we waste the clean one, applying the national laws that obligate factors to clean the wastewater before discharge. Wastewater also contains phosphorus, nitrogen, and sulfur, which cause eutrophication. Therefore, there are three kind of treatment—biological, chemical, and physical (El Din Mahmoud and Fawzy 2016). We can reuse the treated, clean water in parks, industry, and agriculture. This would provide a dependable water supply during the irrigation period (Smol et al. 2020). As a result, the chance of crop failure and financial loss is reduced. Additionally, treated wastewater provides a controlled level of nutrients that can be used instead of fertilizers (Mahmoud and Fawzy 2021; Neczaj and Grosser 2018).

If we can't reduce or reuse water, we can use some technologies, such as nanofiltration (NF), reverse osmosis (RO), etc., to recycle wastewater. Wastewater is a significant source of some elements, such as phosphorus, so wastewater treatment helps us recover it. European countries fulfill 20–30% of their needs through recovered phosphorus instead of mining and importing. Before taking any step, we should rethink how we use resources in such a way to build a sustainable economy that eliminates waste and emissions. We can achieve sustainability and CE through collaboration among different sectors and municipalities (Smol et al. 2020).

11.3 CIRCULAR ECONOMY IN WATER AND WASTEWATER

The use of an economic system to tackle the global challenge of climate change and pollution may be useful. In particular, the application of a circular economy to water and wastewater sector, in which raw materials are used in closed loops, with attention paid to prevent the loss of their value as much as possible, may help address these problems. Integrating the circular economy in water and wastewater sectors involves combining these sectors with an economic system that reduces pollution, waste, and climate change (Fungaro et al. 2021; Sharma et al. 2021). The main objective of the water and wastewater sector is to provide safe drinking water for the public. When properly treated, it prevents disease and safeguards the environment (Gwenzi et al. 2022; Mahmoud et al. 2021b, 2022b).

The main issues associated with the water and wastewater sector are contamination with dyes (Mahmoud et al. 2018), heavy metals (Badr et al. 2020; Mahmoud et al. 2021a), toxic gases (Tiwari et al. 2019), even natural disasters, and cyberattacks (CISA, 2015). The consequences of such incidents could lead to large numbers of illnesses or casualties that could impact public health and economic vitality (Khan et al. 2021; Mansoor et al. 2021; Mousazadeh et al. 2021). Other sectors in the economy, such as food and agriculture, energy, and transportation, could suffer due to a denial of service from the water and wastewater sector (Dotaniya et al. 2022).

Applying the circular economy pillars in saving water is a significant way to achieve sustainability to overcome water scarcity and stress (Mahmoud and Kathi 2022). Even in indirect ways, there are economic measures that could reduce microplastic pollution in water (Khan et al. 2022). Therefore, it is vital to understand the benefit of CE principles in water dimensions use as a service, a source of energy, and a carrier (Figure 11.3).

Circular water management includes the 5Rs: reduce, reuse, recycle, restore, and recover water resources (Figure 11.4). Establishing wastewater treatment plants is a rational solution to get clean water, energy, and resource recovery because we can return the water back to water cycle and the environment, as we see in Figure 11.2.

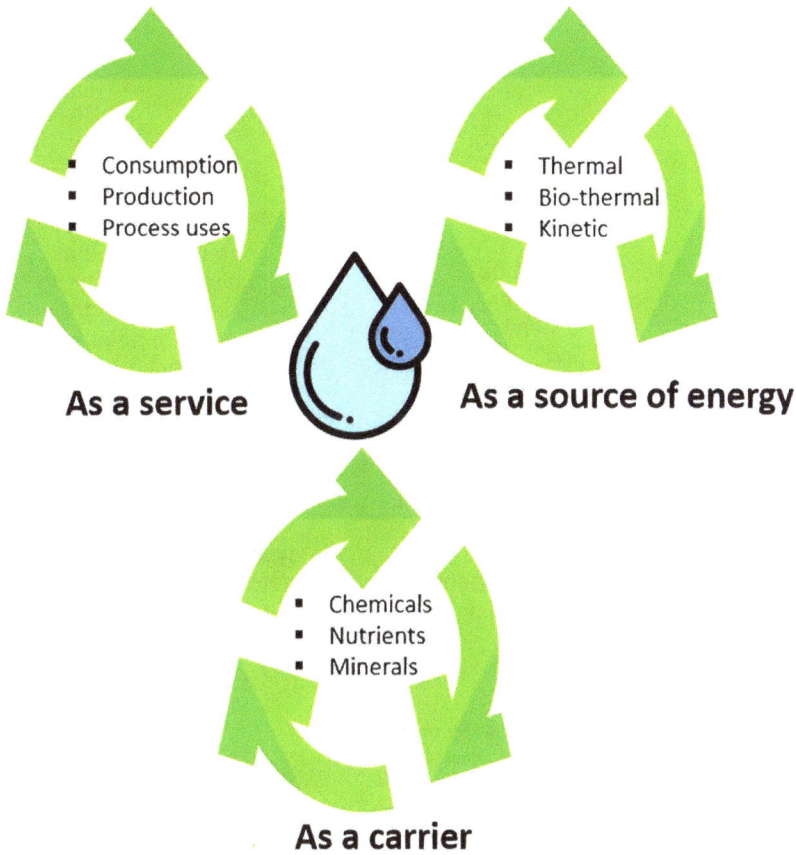

FIGURE 11.3 Benefit of CE principles in water dimensions use as a service, a source of energy, and a carrier.

Thus, the water management cycle begins with procurement of fresh water. It then proceeds with its treatment and distribution. Consumers then use the water, and it is collected. At this point, it is posttreated, reused, or returned to the environment (Mahmoud 2020b; Vambol et al. 2021; Zarei 2020).

Figure 11.5 illustrates the natural hydrological cycle, which begins with evaporating water from the oceans and surface water, which makes clouds, then different shapes of rain that return again to oceans and surface water through runoff on ground or filtration into the ground to make underground water. This water can be utilized in different activities that resulted in various pollutants from the industrial activities that could change the water's nature pH, temperature, physical, and chemical properties. Agriculture, commercial, institutional, and residential activities add organic and inorganic matters to water besides the fertilizers, detergents, and so on. Therefore, we need to apply the water treatment with the basic standards before running off it again to the main resources to save the ecological environment underwater, fish wreath, water quality and avoid acid rains, eutrophication, and so on.

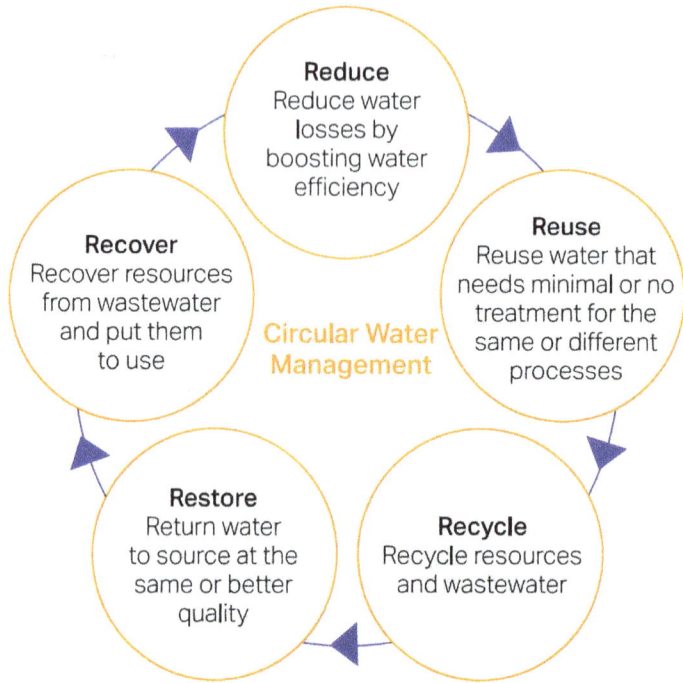

FIGURE 11.4 The 5Rs of circular water management.

Source: Adapted from WBCSD (2017).

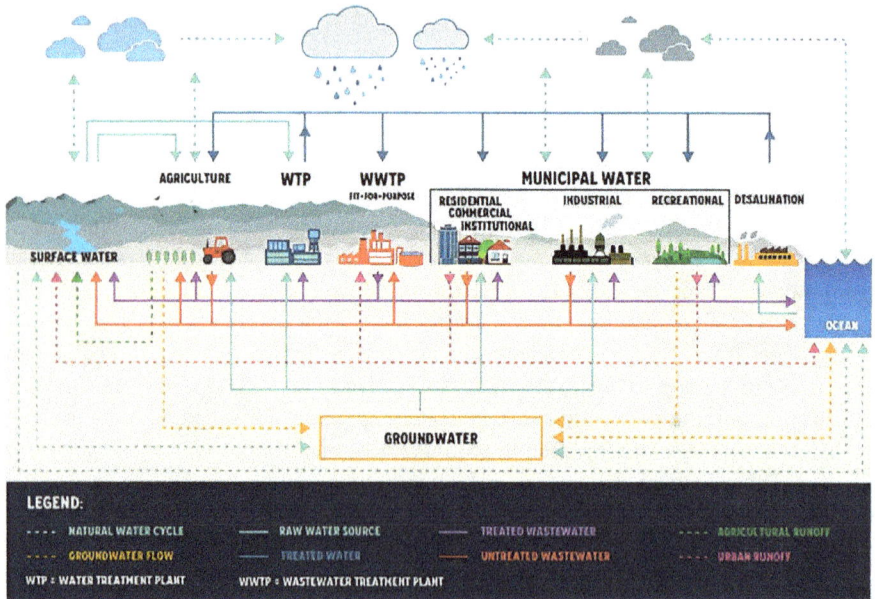

FIGURE 11.5 Wastewater in the water cycle.

Source: Adapted from WWAP (UN Water 2017).

The soil, atmosphere, and water bodies are connected, so any defect in one component has effects on the other ones and, of course, on plants, animals, and humans. Consequently, our economic, social, and environmental aspects will be affected. On the other hand, wastewater treatment plants can add value through recovery of the resources contained in wastewater, such as nutrients, energy, and water as an eco-friendly approach for wastewater resources management (Chaukura et al. 2022; Neczaj and Grosser 2018).

Due to the combination of energy generation and resource recovery during clean water production, urban wastewater treatment plants (WWTPs) may be an essential aspect of circular sustainability (Rashidi et al. 2015). WWTPs will become "ecologically sustainable" technology systems in the near future (Neczaj and Grosser 2018). However, during the last 25 years, a number of key factors have emphasized the need for recovering the resources found in wastewater. Global fertilizer demands, as well as the recovery of water and energy from wastewater, are the primary drivers of the wastewater industry's growth. Phosphate rock was designated as a key raw material by the European Commission in May 2014, highlighting the necessity of recovering it from renewable sources. Humans excrete 20% of the mineral phosphorus they ingest, according to Cordell et al. (2009). It is also projected that excreta streams might provide enough phosphorus to serve the mineral phosphorus market (including domestic animals). Potassium has not had the same economic influence on farming as phosphorus, but it may be fully serviced from waste streams, whereas the nitrogen market is at a high level (50%). The effluent of a WWTP can be recycled and used for agricultural and industrial processes, thereby reducing demand for potable water.

WWTPs are an important part of the smart city, which balances various social, economic, and ecological aspects (Neczaj and Grosser 2018). Figure 11.6 depicts the intelligent wastewater system concept. The system is considered to be intelligent

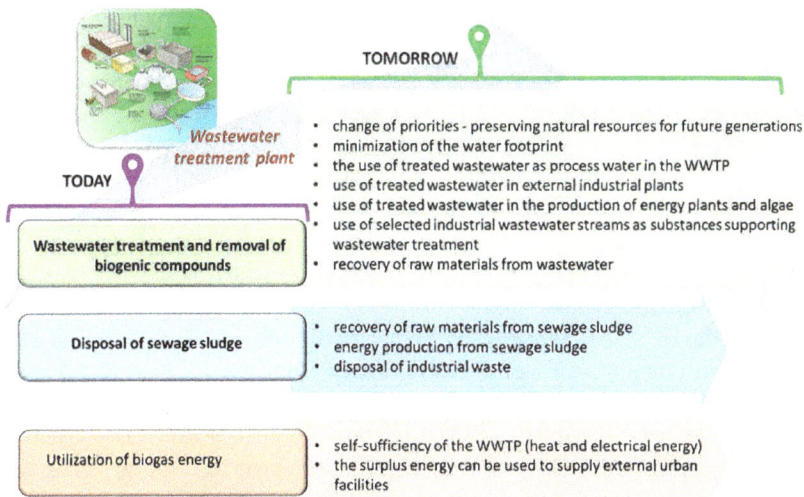

FIGURE 11.6 Wastewater treatment plant (WWTP) today and in the future.

Source: Adapted from Neczaj and Grosser (2018).

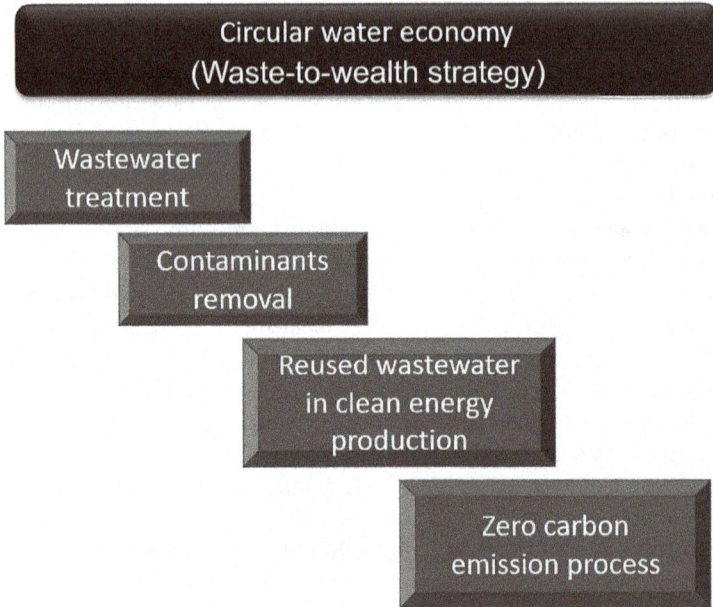

FIGURE 11.7   Waste-to-wealth strategy with a circular water economy.

because it not only efficiently treats wastewater but also allows for effluent reuse. Furthermore, energy is generated in the process and fertilizers are produced. Smart concepts are being implemented by various cities in the world in increasing numbers. For example, a WWTP in Borås, a city in Sweden, provides renewable fuel for a city power plant (Nhede 2016). This recycling approach seeks to harvest energy in the city's waste streams and minimize the city's dependence on fossil fuels.

Figure 11.7 provides a simple representation of the waste-to-wealth strategy with a circular water economy. Recently, the treated effluents could be utilized for hydrogen production that does not emit CO_2 when used like fossil fuel. Hydrogen produced could be used to meet the energy requirements of the wastewater transportation and treatment process (Malek et al. 2021). As a result, the process will be self-sustaining.

In summary, Delgado et al. (2021) revealed the framework of water in circular economy and resilience (WICER; Figure 11.8), which is really relevant to Sustainable Development Goals (SDGs). Figure 11.8 shows WICER connection to the achievement of SDG6, clean water and sanitation for all, and linked to other SDG goals and targets, such as SDG1.4 (achieving universal access to basic services), SDG3.9 (reducing water pollution–related deaths), SDG7.2 (increasing the share of renewable energy), SDG7.3 (improving energy efficiency), SDG8.4 (improving resource efficiency to decouple economic growth from environmental degradation), SDG9.1 (developing quality, reliable, sustainable, and adoption of clean and environmentally sound technologies), and SDG11 (making cities inclusive, safe, resilient, and sustainable).

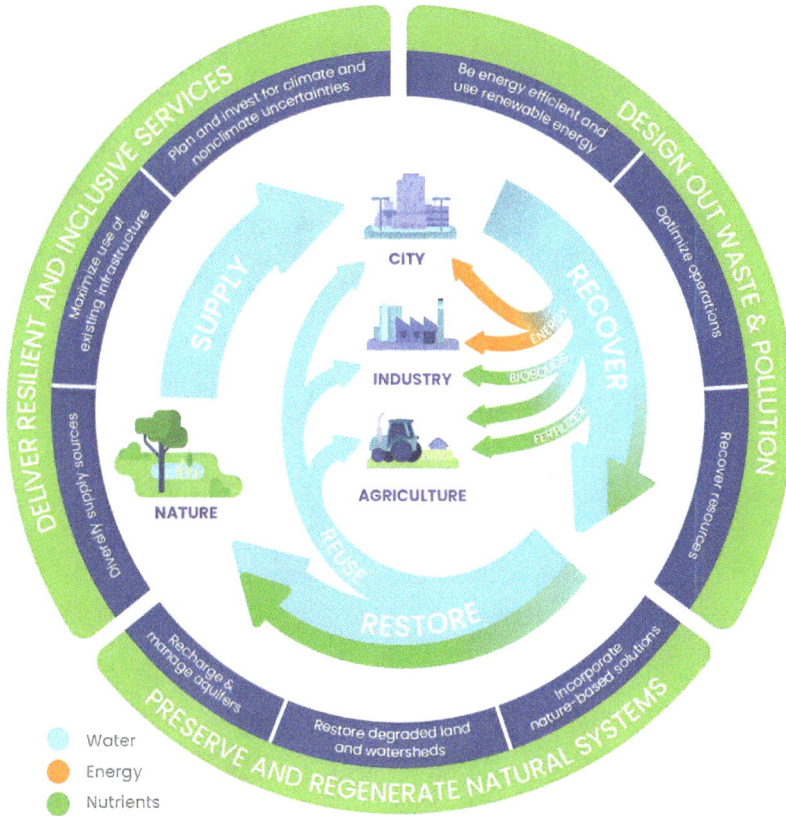

FIGURE 11.8 Framework of water in circular economy and resilience (WICER).

Source: Adapted from Delgado et al. (2021).

11.3.1 TACKLING FOOTPRINT OF WATER

Carbon footprint has become more popular as people have become more aware about climate change and its impacts. However, it is important to note that this is different from water footprint in that carbon footprint measures the total amount of greenhouse gases, which includes carbon dioxide and methane, that are released in a process.

The water footprint, whose standard was defined in 2009, indicates freshwater allotment resulting in water consumed or polluted over a given period of time (Ercin and Hoekstra 2012). It is measured in unit volumes of water per unit time (L day^{-1} or m^3 month^{-1}) or amount of product (L kg^{-1}, L Kcal^{-1}, L g^{-1} of protein) and can be defined for a particular process or certain group of consumers or producers. This allows one to gauge the amount of water that is used directly or indirectly in a particular location over a particular period of time. The water footprint is very specific to when and where water is consumed or polluted. This means that you cannot interchange one unit of water footprint with another for offsetting. To calculate water

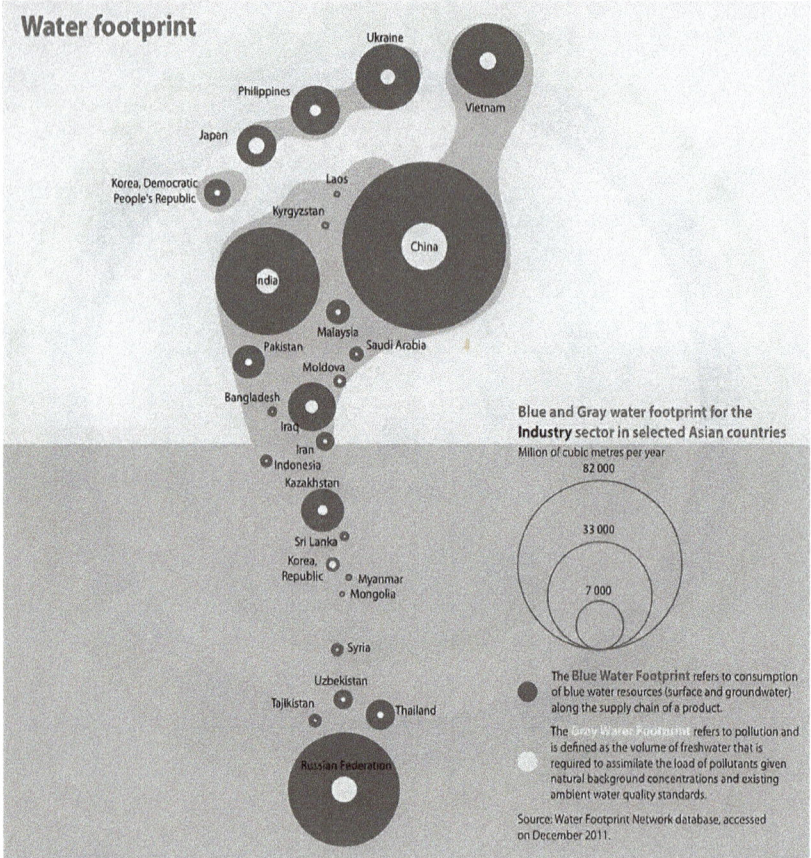

FIGURE 11.9 The blue and gray water footprint for the industry in selected Asian countries in millions of cubic meters per year.

footprint, bottom-up or top-down approaches are used, depending on what is to be analyzed. Bottom-up approaches are used for processes, products, supply chains, as well as sector, national, and global studies. A top-down approach can also be used for sector, national, and global studies.

The sustainability of the water footprint can also be assessed based on the catchment area, the availability of fresh water, and the waste assimilation capacity. There are also benchmarks that can be used to assess the sustainability of the water footprint. In order to tackle water footprint, one must consider the three components contributing to total water footprint. It consists of what is known as blue water footprint, green water footprint, and gray water footprint.

The usage of surface water and groundwater is linked to a blue water footprint. The use of rainwater stored as soil moisture is linked to green water footprint. The amount of fresh water necessary to assimilate the load of pollutants is calculated using ambient water quality criteria as the gray water footprint. The gray and blue water footprints of selected Asian countries in 2011 are shown in Figure 11.9. To decrease water footprint, one must consider what is meant by *consumption* in this context. It is the loss of water from the ground-surface water body because of evaporation, incorporation into a product, or transportation to another area or the sea.

Water footprint is considered to be a pressure indicator on the environment. It is important to note that it indicates how much water is consumed or contaminated and not the environmental impact, which includes the resulting changes in the water quality in rivers and aquifers or runoff. In addition, the impacts of these changes are not measured by the water footprint, including the loss of biodiversity and effects on human health. Despite not directly measuring the impact on the environment, water footprint is still a useful measure because it provides an indication of the pressure on the environment and the level of the environmental impact that could result. To tackle water footprint, the sources contributing to overexploitation have to be decreased using the method of calculation as a guide. This can be done by policy makers to address overexploitation. Reducing water footprint helps lessen water scarcity and deteriorating water quality.

11.3.2 Case Studies

A. China

Wastewater collection, treatment, and reuse were suggested by the municipal government of Lingyuan City as a way to alleviate the city's water scarcity and pollution issues while also promoting circular economy concepts, as indicated in Figure 11.10 (WorldBank 2021c). The city improved its wastewater treatment facility, raised the

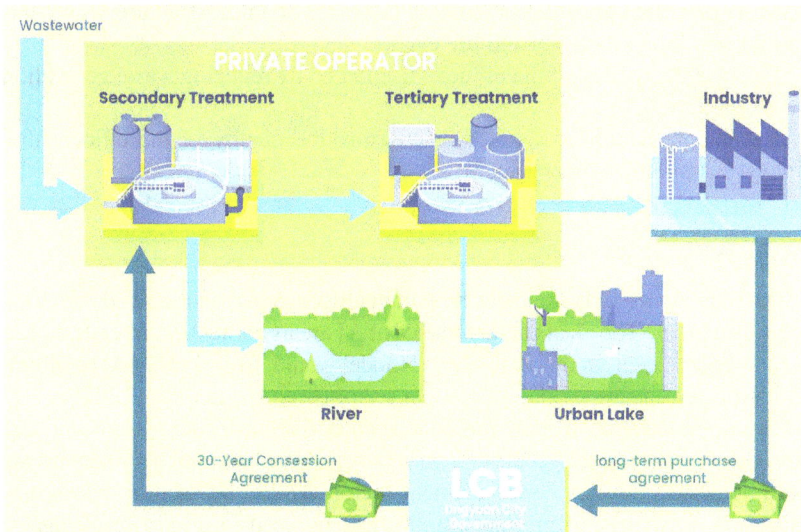

FIGURE 11.10 Scheme of reusing treated wastewater for industrial reasons and restoration of ecosystem in China.

Source: Adapted from WorldBank (2021c).

number of linked houses to 90%, and constructed separate stormwater and wastewater drainage systems. The city now sells a portion of its treated wastewater to industrial users in order to recoup operational costs and generate additional revenue. The remaining wastewater is utilized to replenish an urban lake in order to restore urban biodiversity and sustain the shallow aquifer surrounding the lake. The project has also resulted in the aquifer being replenished, as industries are no longer extracting water from it.

B. Indonesia

The municipal service of Malang could effectively reduce nonrevenue water (NRW) to 16% and was able to supply > 95% of the population as well as increased energy efficiency through mechanical and electrical equipment (WorldBank 2021b). The utility improved its operating efficiency by establishing district meter regions, pressure management with pressure-reducing valves, active leakage control, and new technology (control instruments and supervisory control and data collecting systems).

C. Brazil

The World Bank's 2030 Water Resources Group for Brazil launched an initiative in 2019 to improve the performance of four large wastewater treatment plants. Instead of investing in expanding or building new plants to enhance capacity, the company decided to save costs. The utility will be able to postpone investments in tertiary treatment and cut spending in physical growth as a result of plant optimization (WorldBank 2021a).

The program's goals are to:

- Maximize the utilization of existing materials and infrastructure, as well as improve asset efficiency and life cycles.
- Improve the quality of the effluent discharged by reducing waste generation and pollution.
- Regenerate natural systems by increasing the quality of the effluent discharged into the Upper Tietê River.

11.3.3 Application Models

Artificial intelligence (AI) simulates human intelligence in machines and take the corresponding actions. The ideal characteristics of AI are its ability to deliberate and act to achieve a specific goal. In this case, the AI model would involve using AI to minimize water footprint of the water and wastewater sector (Malviya and Jaspal 2021).

AI has some advantages that enable it to enter the water and wastewater sectors because of its efficiency, speed, and autonomy. It can be applied in various ways. For example, in the wastewater treatment sector, AI can be used to determine the efficiency of water treatment plants, and it can be used to eliminate nitrogen and sulfur in streams. In addition, it can be used to make predictions, including to predict the turbidity and hardness of waste streams. AI in wastewater treatment uses three main models which are based on artificial neural networks (ANN), fuzzy logic algorithms, and genetic algorithms. AI can also be used to determine the biological oxygen demand (BOD) and chemical oxygen demand (COD) in the wastewater sector, with determination coefficient values of 0.99. In turn, AI can assist in improving pollution removal, energy efficiency, and cost in the wastewater sector.

An ensemble approach based on AI was used to assess the performance of a wastewater treatment plant (Nourani et al. 2018). The performance of the plant was assessed based on the BOD in the effluent, COD, and total nitrogen. The assessment was done using three different AI models. These AI models were the feed-forward neural network (FFNN), adaptive neuro-fuzzy interference system (ANFIS), support vector machine (SVM), and the classical multilinear regression (MLR). The models also allowed for the prediction of the performance. ANFIS provided the best results. The neural network ensemble (NNE) reliably predicted the performance of wastewater treatment plant in Nicosia.

AI can also be applied for biological wastewater treatment (Manu and Thalla 2017). To anticipate nitrogen removal from wastewater, support vector machine (SVM) and ANFIS can be employed. This model has several inputs, which include COD, pH, Kjeldahn nitrogen, TS, ammonia nitrogen, free ammonia, and total solids (TS). Based on these inputs, the AI model allows one to predict the removal efficiency of Kjeldahl nitrogen. Real data recorded at a wastewater treatment plant in Mangalore was used to assess the model. The ANFIS model performance was assessed using root mean square error, Nash Sutcliff error (NSE), and correlation coefficients (CC). The SVM model provided better results and was able to make better predictions as compared to other AI models. The results from the SVM model indicate that water quality parameters are related for aerobic biological wastewater treatment.

AI can also be applied to manage rainwater (Marković 2018). Fuzzy cognitive maps were used to manage rainwater or gray water in order to make a process for water usage that was well-oriented and evaluated. AI was used to make a decision system to make the best use of wastewater. This cognitive process improved the economics of wastewater management and decreased water footprint.

AI can also be applied for water treatment in an adsorption process (Alam et al. 2022). AI models have been successfully applied to separate various pollutants from water streams by adsorption. However, it is important to note that there are some challenges associated with the application of AI for water treatment. One must be careful about how the data used for the model is selected, and in some cases, there is a lack of data available for the process of interest. Furthermore, the application of these models to real systems needs to be completed as AI is applied more and more in this field.

AI has also been applied for wastewater treatment done using membrane bioreactors (Kamali et al. 2021). For membrane bioreactors, ANN is the most widely implemented model. ANN was used to make sure that the membrane bioreactors were efficient, reliable, and cost-effective. It can be used to commercialize these technologies. In addition to predicting performance, AI techniques can also be applied for control systems for water and wastewater management (Wen and Vassiliadis 1998). This can be accomplished by using hybrid AI systems. This approach can be effective when you have a complex mixture and need to reduce effluent levels of certain pollutants below set values. The hybrid AI system allows to make decisions that apply to various conditions that may not be typical while considering the cost and energy required for the separation. This allows one to improve the efficiency and has in fact been developed by Gensym Corporation.

AI can also be applied for technoeconomic optimization of water management (Tariq et al. 2021). Oftentimes, the systems that are being optimized are hybrid energy systems. Different plants can be integrated using hybrid energy systems, such as wastewater treatment plants and potable water filtration plants. In the technoeconomic optimization, AI can be flexible enough to target the optimization of one technoeconomic objective or multiple technoeconomic objectives at the same time. For hybrid systems, a digital twin model, which is also known as the metamodel or surrogate model, is used. Also, for systems with multiple energy sources, the Hybrid Optimization of Multiple Energy Resources (HOMER) software is used. This may be followed up with statistical analysis of the sensitivity of the results. This information made available by AI helps to aid policy makers to propose the most effective policies. For the optimization scenarios, photovoltaics, wind turbines, diesel, and batteries may be considered. Furthermore, the analysis may be detailed enough to include carbon dioxide (CO_2) emissions and the cost of water. Based on the results from HOMER, electricity cost ranging from 0.205 to 0.229 US KWh^{-1} was found for photovoltaic, diesel, and battery systems.

11.4   CONCLUSION AND RECOMMENDATIONS

In this chapter, some strategies to address the concern regarding the availability of water, its contamination, and the increasing demand for water from a growing population were provided. This issue has garnered more attention as climate change has become more of an issue. CE models in the water and wastewater sectors may be effective to address these problems. The key is to reduce, reuse, recycle, reclaim, recover, and restore water. AI has proven to be useful in CE models in the water and wastewater sectors. Some operational benefits resulting from the implementation of CE models include increased efficiency and additional revenue streams. In addition to the operational benefits, the carbon footprint associated with water should be considered for these systems. Further development of CE applied to the water and wastewater sectors requires investment and a legal framework. Despite some of these challenges, CE applied to the water and wastewater sectors has opened the door to many opportunities.

11.5   ACKNOWLEDGMENTS

Dr. Alaa El Din Mahmoud acknowledges the support from Egypt's Academy of Scientific Research and Technology (ASRT) for SA-Egypt Joint Project.

REFERENCES

Alam, G., Ihsanullah, I., Naushad, M., Sillanpää, M., 2022. Applications of artificial intelligence in water treatment for optimization and automation of adsorption processes: Recent advances and prospects. Chemical Engineering Journal 427, 130011.

Badr, N.B.E., Al-Qahtani, K.M., Mahmoud, A.E.D., 2020. Factorial experimental design for optimizing selenium sorption on Cyperus laevigatus biomass and green-synthesized nano-silver. Alexandria Engineering Journal 59(6), 5219–5229.

Barreiro-Gen, M., Lozano, R., 2020. How circular is the circular economy? Analysing the implementation of circular economy in organisations. Business Strategy the Environment 29(8), 3484–3494.

Berg, A., Antikainen, R., Hartikainen, E., Kauppi, S., Kautto, P., Lazarevic, D., Piesik, S., Saikku, L., 2018. Circular Economy for Sustainable Development. Reports of the Finnish Environment Institute.

Bicket, M., Guilcher, S., Hestin, M., Hudson, C., Razzini, P., Tan, A., ten Brink, P., van Dijl, E., Vanner, R., Watkins, E., 2014. Scoping Study to Identify Potential Circular Economy Actions, Priority Sectors, Material Flows and Value Chains. Luxembourg Publications Office of the European Union, Luxembourg, Luxembourg. https://doi.org/10.2779/29525

Chaukura, N., Muzawazi, E.S., Katengeza, G., Mahmoud, A.E.D., 2022. Chapter 20 – Remediation Technologies for Contaminated Soil Systems, In: Gwenzi, W. (Ed.), Emerging Contaminants in the Terrestrial-Aquatic-Atmosphere Continuum. Elsevier, Amsterdam, The Netherlands pp. 353–365.

CISA, Water and Wastewater Systems Sector, 2015. www.cisa.gov/water-and-wastewater-systems-sector.

Cordell, D., Drangert, J.-O., White, S., 2009. The story of phosphorus: Global food security and food for thought. Global Environmental Change 19(2), 292–305.

D'Amato, D., Droste, N., Allen, B., Kettunen, M., Lähtinen, K., Korhonen, J., Leskinen, P., Matthies, B.D., Toppinen, A., 2017. Green, circular, bio economy: A comparative analysis of sustainability avenues. Journal of Cleaner Production 168, 716–734.

Delgado, A., Rodriguez, D.J., Amadei, C.A., Makino, M., 2021. Water in Circular Economy and Resilience. World Bank, Washington, DC.

Dotaniya, M.L., Meena, V.D., Saha, J.K., Dotaniya, C.K., Mahmoud, A.E.D., Meena, B.L., Meena, M.D., Sanwal, R.C., Meena, R.S., Doutaniya, R.K., Solanki, P., Lata, M., Rai, P.K., 2022. Reuse of poor-quality water for sustainable crop production in the changing scenario of climate. Environment, Development and Sustainability.

Edmond, C., n.d. Cape Town Almost Ran Out of Water. Here's How It Averted the Crisis. World Economic Forum. Available at: https://www.weforum.org/agenda/2019/08/cape-town-was-90-days-away-from-running-out-of-water-heres-how-it-averted-the-crisis/ (accessed 11 October 2022).

El Din Mahmoud, A., Fawzy, M., 2016. Bio-based Methods for Wastewater Treatment: Green Sorbents, In: Ansari, A.A., Gill, S.S., Gill, R., Lanza, G.R., Newman, L. (Eds.), Phytoremediation: Management of Environmental Contaminants, Volume 3. Springer International Publishing, Cham, pp. 209–238.

Ercin, E., Hoekstra, A.Y., 2012, Carbon and Water Footprints: Concepts, Methodologies and Policy Responses. World Water Assesment Programme (4). United Nations Educational, Scientific and Cultural Organization (UNESCO), Paris, France.

Flores, C.C., Bressers, H., Gutierrez, C., de Boer, C., 2018. Towards Circular Economy—A Wastewater Treatment Perspective, the Presa Guadalupe Case. Management Research Review. Emerald Group Holdings Ltd., Volume 41, No. 5, pp. 554–571.

Fungaro, D.A., Silva, K.C., Mahmoud, A.E.D., 2021. Aluminium tertiary industry waste and ashes samples for development of zeolitic material synthesis. Journal of Applied Materials Technology 2(2), 66–73.

Geographic, N., 2021. www.nationalgeographic.com/environment/article/freshwater-crisis.

Gubeladze, D., Pavliashvili, S., 2020. State assessment and future vision. International Journal of Innovative Technologies in Economy, RS Global Sp. z O.O., 5(32). Available at: https://doi.org/10.31435/RSGLOBAL_IJITE/30122020/7286

Gwenzi, W., Selvasembian, R., Offiong, N.-A.O., Mahmoud, A.E.D., Sanganyado, E., Mal, J., 2022. COVID-19 drugs in aquatic systems: A review. Environmental Chemistry Letters 20(2), 1275–1294.

Hossain, M.U., Ng, S.T., Antwi-Afari, P., Amor, B., 2020. Circular economy and the construction industry: Existing trends, challenges and prospective framework for sustainable construction. Renewable and Sustainable Energy Reviews. Pergamon 130, 109948.

Kamali, M., Appels, L., Yu, X., Aminabhavi, T.M., Dewil, R., 2021. Artificial intelligence as a sustainable tool in wastewater treatment using membrane bioreactors. Chemical Engineering Journal 417, 128070.

Khan, A.H., Abutaleb, A., Khan, N.A., El Din Mahmoud, A., Khursheed, A., Kumar, M., 2021. Co-occurring indicator pathogens for SARS-CoV-2: A review with emphasis on exposure rates and treatment technologies. Case Studies in Chemical and Environmental Engineering 4, 100113.

Khan, N.A., Khan, A.H., López-Maldonado, E.A., Alam, S.S., López López, J.R., Méndez Herrera, P.F., Mohamed, B.A., Mahmoud, A.E.D., Abutaleb, A., Singh, L., 2022. Microplastics: Occurrences, treatment methods, regulations and foreseen environmental impacts. Environmental Research 215, 114224.

Kirchherr, J., Reike, D., Hekkert, M., 2017. Conceptualizing the circular economy: An analysis of 114 definitions. Resources, Conservation and Recycling 127, 221–232.

Mahmoud, A.E.D., 2020a. Eco-friendly reduction of graphene oxide via agricultural byproducts or aquatic macrophytes. Materials Chemistry and Physics 253, 123336.

Mahmoud, A.E.D., 2020b. Graphene-based nanomaterials for the removal of organic pollutants: Insights into linear versus nonlinear mathematical models. Journal of Environmental Management 270, 110911.

Mahmoud, A.E.D., Al-Qahtani, K.M., Alflaij, S.O., Al-Qahtani, S.F., Alsamhan, F.A., 2021a. Green copper oxide nanoparticles for lead, nickel, and cadmium removal from contaminated water. Scientific Reports 11(1), 12547.

Mahmoud, A.E.D., Fawzy, M., 2021. Nanosensors and Nanobiosensors for Monitoring the Environmental Pollutants, In: Makhlouf, A.S.H., Ali, G.A.M. (Eds.), Waste Recycling Technologies for Nanomaterials Manufacturing. Springer International Publishing, Cham, pp. 229–246.

Mahmoud, A.E.D., Fawzy, M., Khairy, H., Sorour, A., 2022a. Environmental Bioremediation as an Eco-sustainable Approach for Pesticides: A Case Study of MENA Region, In: Siddiqui, S., Meghvansi, M.K., Chaudhary, K.K. (Eds.), Pesticides Bioremediation. Springer International Publishing, Cham, pp. 479–494.

Mahmoud, A.E.D., Fawzy, M., Abdel-Fatah, M.M.A., 2022b. Technical Aspects of Nanofiltration for Dyes Wastewater Treatment, In: Muthu, S.S., Khadir, A. (Eds.), Membrane Based Methods for Dye Containing Wastewater: Recent Advances. Springer Singapore, Singapore, pp. 23–35.

Mahmoud, A.E.D., Kathi, S., 2022. Chapter 7 — Assessment of Biochar Application in Decontamination of Water and Wastewater, In: Kathi, S., Devipriya, S., Thamaraiselvi, K. (Eds.), Cost Effective Technologies for Solid Waste and Wastewater Treatment. Elsevier, pp. 69–74.

Mahmoud, A.E.D., Stolle, A., Stelter, M., 2018. Sustainable synthesis of high-surface-area graphite oxide via dry ball milling. ACS Sustainable Chemistry & Engineering 6(5), 6358–6369.

Mahmoud, A.E.D., Umachandran, K., Sawicka, B., Mtewa, T.K., 2021b. 26 — Water Resources Security and Management for Sustainable Communities, In: Mtewa, A.G., Egbuna, C. (Eds.), Phytochemistry, the Military and Health. Elsevier, pp. 509–522.

Malek, A., Rao, G.R., Thomas, T., 2021. Waste-to-wealth approach in water economy: The case of beneficiation of mercury-contaminated water in hydrogen production. International Journal of Hydrogen Energy 46(52), 26677–26692.

Malviya, A., Jaspal, D., 2021. Artificial intelligence as an upcoming technology in wastewater treatment: a comprehensive review. Environmental Technology Reviews 10(1), 177–187.

Mansoor, S., Hameed, A., Anjum, R., Maqbool, I., Masoodi, M., Maqbool, K., Dar, Z.A., Hamadani, A., Mahmoud, A.E.D., 2021.21 — Obesity: CAUSES, Consequences, and Disease Risks for Service Personnel, In: Mtewa, A.G., Egbuna, C. (Eds.), Phytochemistry, the Military and Health. Elsevier, pp. 407–425.

Manu, D.S., Thalla, A.K., 2017. Artificial intelligence models for predicting the performance of biological wastewater treatment plant in the removal of Kjeldahl Nitrogen from wastewater. Applied Water Science 7(7), 3783–3791.

Marković, G., 2018. Wastewater management using artificial intelligence, E3S Web of Conferences. EDP Sciences, France p. 00050.

Millstein. M, Z.P., 2020. Unzipped Staff, 2020. How the Cape Town Water Crisis Spurred Progress. Levi Strauss & Co 27 Nov 2021.

Mishra, B., Tiwari, A., Mahmoud, A.E.D., 2022. Microalgal potential for sustainable aquaculture applications: Bioremediation, biocontrol, aquafeed. Clean Technologies and Environmental Policy. Springer, Amsterdam, The Netherlands pp. 1–13.

Mousazadeh, M., Naghdali, Z., Rahimian, N., Hashemi, M., Paital, B., Al-Qodah, Z., Mukhtar, A., Karri, R.R., Mahmoud, A.E.D., Sillanpää, M., Dehghani, M.H., Emamjomeh, M.M., 2021. Chapter 9 — Management of environmental health to prevent an outbreak of COVID-19: A review, In: Hadi Dehghani, M., Karri, R.R., Roy, S. (Eds.), Environmental and Health Management of Novel Coronavirus Disease (COVID-19). Academic Press, pp. 235–267.

Munaro, M.R., Tavares, S.F., Bragança, L., 2020. Towards circular and more sustainable buildings: A systematic literature review on the circular economy in the built environment. Journal of Cleaner Production 260, 121134.

Neczaj, E., Grosser, A., 2018. Circular economy in wastewater treatment plant—Challenges and barriers. *Proceedings* 2(11), 614.

Nhede, N., 2016. Swedish Utility Selects Veolia for Wastewater to Energy Project. www.smart-energy.com/news/swedish-utility-selects-veolia-for-rollout-of-wastewater-project/.

Nourani, V., Elkiran, G., Abba, S.I., 2018. Wastewater treatment plant performance analysis using artificial intelligence—an ensemble approach. Water Science and Technology 78(10), 2064–2076.

Ramadoss, T.S., Alam, H., Seeram, R., 2018. Artificial intelligence and Internet of Things enabled circular economy. The International Journal of Engineering Science 7(9), 55–63.

Rashidi, H., GhaffarianHoseini, A., GhaffarianHoseini, A., Nik Sulaiman, N.M., Tookey, J., Hashim, N.A., 2015. Application of wastewater treatment in sustainable design of green built environments: A review. Renewable and Sustainable Energy Reviews 49, 845–856.

Sgroi, M., Vagliasindi, F.G.A., Roccaro, P., 2018. Feasibility, sustainability and circular economy concepts in water reuse. Current Opinion in Environmental Science & Health 2, 20–25.

Sharma, H.B., Vanapalli, K.R., Samal, B., Cheela, V.R.S., Dubey, B.K., Bhattacharya, J., 2021. Circular economy approach in solid waste management system to achieve UN-SDGs: Solutions for post-COVID recovery. Science of the Total Environment 800, 149605.

Smol, M., Adam, C., Preisner, M., 2020. Circular economy model framework in the European water and wastewater sector. Journal of Material Cycles Waste Management 22, 682–697.

Sørensen, P.B., 2018. From the linear economy to the circular economy: A basic model. Finanz-Archiv: Zeitschrift für das Gesamte Finanzwesen 74(1), 71–87.

Tariq, R., Cetina-Quiñones, A.J., Cardoso-Fernández, V., Daniela-Abigail, H.-L., Soberanis, M.A.E., Bassam, A., De Lille, M.V., 2021. Artificial intelligence assisted technoeconomic optimization scenarios of hybrid energy systems for water management of an isolated community. Sustainable Energy Technologies and Assessments 48, 101561.

Tiwari, A., Alam, T., Kumar, A., Shukla, A., 2019. Control of odour, volatile organic compounds (VOCs) & toxic gases through biofiltration—an overview. International Journal of Tech. Innov. Mod. Engineering Science 5, 1–6.

Tsegaye, B., Jaiswal, S., Jaiswal, A.K., 2021. Food waste biorefinery: Pathway towards circular bioeconomy. Foods 10, 1174.

UN Water, 2017. Waste water: The untapped resources. Facts and figure. The United Nations World Water Development Report.

Vambol, S., Vambol, V., Mozaffari, N., Mahmoud, A.E.D., Ramsawak, N., Mozaffari, N., Ziarati, P., Khan, N.A., 2021. Comprehensive insights into sources of pharmaceutical wastewater in the biotic systems, Pharmaceutical Wastewater Treatment Technologies. IWA, p. 17.

Warner, J.F., Meissner, R., 2021. Cape Town's "Day Zero" water crisis: A manufactured media event? International Journal of Disaster Risk Reduction 64, 102481.

WBCSD, 2017. Business Guide to Circular Water Management: Spotlight on Reduce, Reuse and Recycle. Geneva.

Wen, C.-H., Vassiliadis, C.A., 1998. Applying hybrid artificial intelligence techniques in wastewater treatment. Engineering Applications of Artificial Intelligence 11(6), 685–705.

WorldBank, 2021a. Water in Circular Economy and Resilience (WICER): The Case of São Paolo, Brazil: Optimizing Wastewater Treatment Plants in the Metropolitan Area of São Paulo. World Bank, Washington, DC.

WorldBank, 2021b. Water in Circular Economy and Resilience (WICER): The Case of Indonesia: Promoting Nonrevenue Water Reduction and Energy Efficiency in Indonesia's Water Utilities. World Bank, Washington, DC.

WorldBank, 2021c. Water in Circular Economy and Resilience (WICER): The Case of Lingyuan City, China: Unconventional Water Resources in a WaterScarce City. World Bank, Washington, DC.

Zarei, M., 2020. Wastewater resources management for energy recovery from circular economy perspective. Water-Energy Nexus 3, 170–185.

12 Water Circular Economy Strategic Approaches
Morocco Case Study

Hanae Hamidi and Houda El Haddad

CONTENTS

12.1 INTRODUCTION

The world's population is growing by about 80 million people a year, which translates into an increase in freshwater demand of about 64 billion cubic meters a year (Hinrichsen et al. 1997). 90% of the 3 billion people estimated to be added by 2050 will be in developing countries, especially in regions that do not currently have sustainable access to security drinking water and adequate sanitation (United Nations 2007).

Access to water is unequal between countries. Figure 12.1 compares the percentages of surface freshwater resources stored in reservoirs, total renewable water resources, access to improved water supply services, and access to improved sanitation services (World Bank1995; Banque Mondiale 2007) in different regions over the period 1998–2002, with more than 660 million people without adequate sanitation living on less than US$2 a day and more than 385 million on less than US$1 a day (World Bank Group's Water Global Practice and Water and Sanitation Services 2015).

47% of the world's population by 2030 would be living in areas of high water stress (OECD 2008). The years between 2016 and 2020 showed that access to safe drinking water at home increased from 70% to 74%, for sanitation services from 47% to 54%, and for handwashing facilities with soap and water, from 67% to 71%. (United Nations News, 2021).

DOI: 10.1201/9781003260455-12

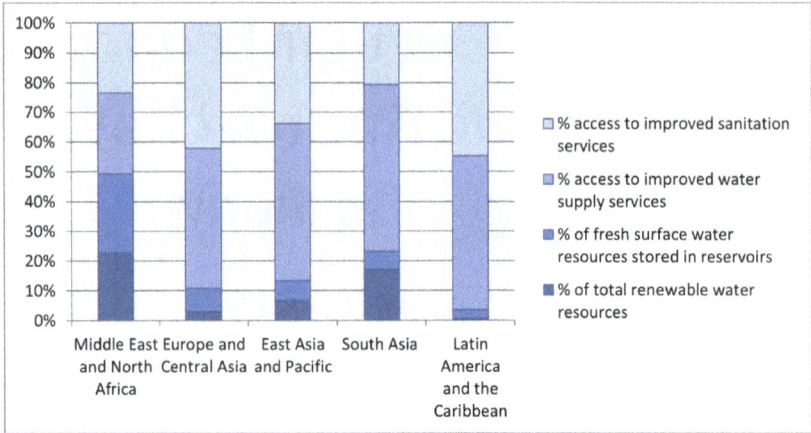

FIGURE 12.1 Percentages of fresh surface water resources, total renewable water resources, access to water supply, and sanitation services.

Source: Based on FAO AQUASTAT data, 1998–2002.

Globally, less than 20% of drainage basins show near-perfect water quality (United Nations 2021a), yet competition for water exists at all levels and should grow with water demands in almost all countries. Irrigated agriculture is the world's biggest water user, accounting for between 70% and 90% (in some regions) of water withdrawals. About 20% of the total water used in the world comes from underground sources, both renewable or nonrenewable. This share increases rapidly in arid areas (United Nations 2021b), where 85% of the world's population resides in the driest half of the Earth (Shawiza 2018).

Desalination is rarely used for agriculture (1%), but its use for high-value crops in greenhouses is gradually increasing. Desalination represents only 0.4% of water used in 2004 (about 14 cubic kilometers per year), but it is expected to double by 2025 (United Nations 2021c).

The International Energy Agency reports that the world will need 60% more energy in 2030 than in 2002. As renewable energy is insufficient to meet the huge increase in energy demand expected by 2030, fossil fuel extraction and nuclear power development will continue to increase, along with their effects on water supplies and the environment, as water is needed for the production of all types of energy. (Kirby 2004).

The Kingdom of Morocco, like most other countries in the world, has been putting enormous pressure on its natural resources for almost 50 years (Kuper et al. 2012), as shown in Figure 12.2 (World Bank 2007). Between 1998 and 2002, the percentage of water resources withdrawn by Morocco that belongs to the North African country is around 74%. Its production and consumption models are essentially linear compared to those of most other countries: extraction of resources—transformation—consumption—production of waste (Conseil Economique, Social et Environnemental 2019).

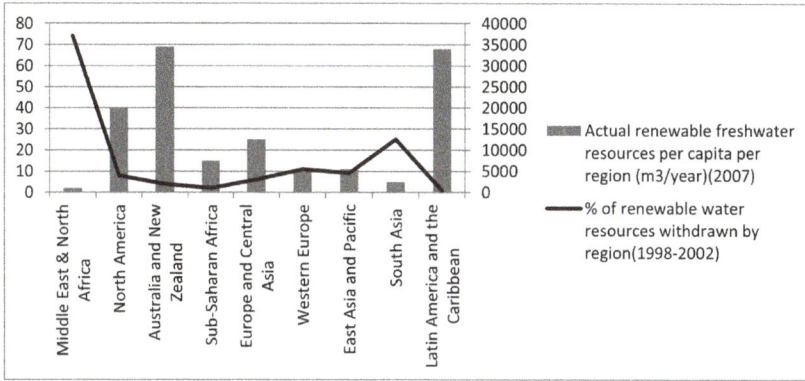

Chart showing two y-axes (left: 0–80, right: 0–40000) for regions: Middle East & North Africa, North America, Australia and New Zealand, Sub-Saharan Africa, Europe and Central Asia, Western Europe, East Asia and Pacific, South Asia, Latin America and the Caribbean.

Legend:
- Actual renewable freshwater resources per capita per region (m3/year)(2007)
- % of renewable water resources withdrawn by region(1998-2002)

FIGURE 12.2 Renewable freshwater resources withdrawn by regions and actual by inhabitants and regions.

Source: Based on FAO AQUASTAT data, 1998–2002; 2007.

At present, the exploitation of its groundwater is undergoing a "silent revolution" and is invisible (Llamas and Martinez-Santos 2005), as in Southern Europe, Asia, and the Americas. The country's water deficit is likely to be an urgent problem for decades to come and will impose significant constraints on its economic development and food security. The agricultural sector will be the first victim because, firstly, it is the main consumer of water in Morocco, consuming more than 85% of the available water. Secondly, about 40% of the country's workforce works in this sector (Bennouna 2020a).

In order to accompany the country's increasing demographic, irrigation (Rosegrant et al. 2002), and economic expansion, and to cope with climate change, which exacerbates the problem of water scarcity (Boukhari et al. 2019), Morocco has always been interested in setting up innovative strategies for the development and rational management of water resources in order to remedy the most urgent water problems, to balance water with these new needs (Molle and Berkoff 2006), and to make water a determining factor for sustainability (Ministry of Energy Transition and Sustainable Development 2009).

In this chapter we discuss the history of water policy in Morocco since 1914 and, after independence, the implementation of structural adjustment programs in the 1980s, and the operation of institutional innovations since the 1990s. And we cite the operation of institutional innovations since the 1990s. The first phase of the dam policy, from the 1960s to the 1980s, with the creation of reservoirs and water allocation systems, where supply exceeded demand, contributed to the end of water rationing and cutoffs and surviving the drought of the 1980s without enormous damage (Stour and Agoumi 2008). The second phase, from the 1980s to the 2000s, is marked by the stage of meeting water demand and then exceeding supply. In this chapter, we also mention the implementation, from 2001 onwards, of reforms in water management, which is judged to be globally insufficient through the management of the demand for this resource and the rationalization of its consumption. In addition, we mention the demographic, economic, and unfavorable climatic conditions that have put the country in a situation of water stress since 2010. In conjunction with this stress situation, the new water strategy focuses on solutions for the circular water economy, such as wastewater reuse, desalination, and groundwater preservation. On this basis, we conclude that Morocco's water policy, if it follows its new strategy, with periodic self-evaluations, can overcome its water stress problem.

12.2   MOROCCO'S WATER RESOURCES

The capacity of natural water resources in Morocco is estimated at an average of 21 billion cubic meters (surface water, 16 billion cubic meters per year; and groundwater, 5 billion m3 per year, and with the presence of 130 aquifers) (Laouina 2006). This potential is marked with a great territorial disparity: more than 70% of surface waters are concentrated in the extreme northwest of the kingdom—that is, on less than 15% of the national territory (Ministre de l'Equipement et de l'Eau, a).

Morocco currently has a heritage of more than 149 major dams of full capacity of over 19 billion cubic meter (m3) (ELGhomari 2015). This infrastructure has enabled the country to generalize:

- Availability of drinking water in urban areas.
- Availability of drinking water at a rate of about 96% in rural areas.
- The irrigation of nearly 1.5 million hectares.
- Protecting vast territories from flooding.
- The coverage of nearly 10% of electricity needs through the production of hydroelectric power.

12.3   DAMS: GENESIS OF SURFACE WATER MOBILIZATION POLICY

Between 1925 and 1956, Morocco embarked on the first phase of a policy to build dams with high surface water potential to meet the needs of the urban center for drinking water and to ensure energy production.

In 1929, the first dam in Morocco intended to provide potable water to the city of Casablanca was the Sidi Mâachou dam. In 1953, the Bin El Ouidane dam was built with 1,384 million m^3 in capacity.

Around 1956, the kingdom saw the construction of 13 dams, with a mobilized volume of 1.8 billion m^3.

During the second phase, between 1956 and 1966, Morocco entered a transitional phase, where it began to evaluate its available water resources and to define objectives. During these ten years, it built three dams for water's drinking supply, with a mobilized volume of 0.4 billion m^3.

In 1967, the third phase, the former Moroccan King Hassan II gave a new impetus to the policy of dams for the irrigation of a million hectares by 2000; this policy is continued by HM Mohamed VI with the realization of two to three large dams per year.

Due to the dam's policy, the kingdom has more than 149 major dams of full capacity of more than 19.1 billion m^3, 15 large dams are under construction of full capacity

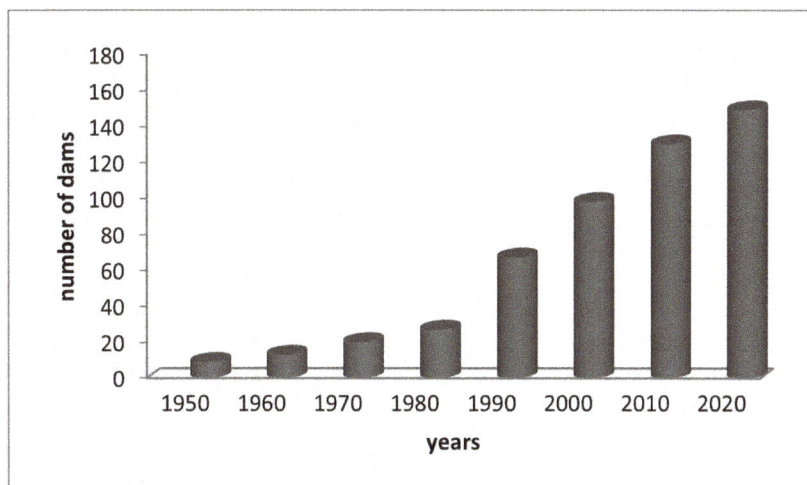

FIGURE 12.3 Evolution of dam numbers from 1929 to 2020.

Source: Based on data from Minister of Equipment and Water.

of 4.8 billion m³, there are 13 hydraulic works of water transfer with a volume of 2.5 billion m³/year, and also more than a hundred small dams and hill lakes, without forgetting the development of water courses (Ministre de l'Equipement et de l'Eau, b). As illustrated in Figure 12.3, this describes the evolution of dam numbers from 1929 to 2020.

12.4 THE PRINCIPAL STEPS OF WATER POLICY IN MOROCCO

In Morocco, the first text relating to water dates from 1914, completed by the Dahirs of 1919 and 1925, and includes all waters, whatever their form, in the public hydraulic domain. Later, essential water-related legislation was developed to meet new needs (Agence Bassin Hydraulique De Souss Massa 1995). In the sixtiess and seventies, Morocco launched major hydraulic schemes and sectorial projects and began the first water resources planning studies. From the 1980s onwards, it generalized integrated and anticipatory planning on the level of several hydrological basins in the country (Agence Bassin Hydraulique De Souss Massa 1995).

To ensure this planning, in the 1990s, it institutionalized the management, integrated planning, and decentralization of water resources by Law 10–95. This law marked a clear progress in its time by offering a unifying text that specifies, as shown in Table 12.1.

Despite this, the achievement of the objectives of true integrated water resources management (IWRM) (after several years of implementation) (Arrifi 2009) has encountered significant constraints. For example, the new institutional structures in many cases lack sufficient authority to fulfill their functions (Houdret 2008).

TABLE 12.1

Basic Principles and Limits of Law 10–95

Principal Basis of Law 10–95	Limits of Law 10–95
• Public ownership of water • The uniqueness of water management • The consecration for the first time to the fight against water pollution and its regeneration • The adoption of the principle of the taxpayer pays and the polluter pays • Consultation in water management • Decentralized management by river basin agency • Recognition of the social, economic, and environmental value of water • Solidarity between users, sectors, and regions	• Lack of legal rules on seawater desalination, wastewater reuse, and rainwater harvesting • The lack of provisions for flood protection • The complexity of the procedures for the delimitation and use of the public hydraulic domain • The lack of definitions for certain expressions, such as "direct or indirect discharge," which has delayed the implementation of the "polluter pays" principle • The difficulties encountered by the boards of directors in the management and control of water basin agencies

Source: Based on data from Agence Bassin Hydraulique De Souss Massa (1995).

Between 2000 and 2015, the country recognized a renewal of the water policy, especially in the following areas: water demand management, diversification of supply, and the environmental and sustainable development dimension.

In 2015, through a new regulatory framework, Water Law 36–15 (Agence Bassin Hydraulique De Souss Massa 2015), Morocco strengthened governance and long-term planning (national and territorial) of these water resources through:

• The creation of water basin agency councils
• The establishment of a legal framework for seawater desalination
• The obligation of liquid sanitation master plans, how to take into account stormwater and wastewater use requirements
• The definition of a number of terms and the deletion of others that are unclear for the implementation of the "polluter pays" principle

According to Molle et al. (2019), about 20 billion dirhams per year is the overall amount made available by the government to develop basic infrastructure.

In 2020, the national program for drinking water supply and irrigation (2020–2027) was prepared, which aims to accelerate investments in the water sector to strengthen the supply of drinking water and irrigation and to reinforce the resilience of our country to climatic hazards and disturbances (Ministre de l'Equipement et de l'Eau, c).

This program is based on five axes: improvement of water supply, demand management and water valuation, reinforcement of drinking water supply in rural areas, reuse of treated wastewater, especially in agriculture (Bielorai et al. 1984; Brister and

Schultz 1981; Neilson et al. 1989; Day et al. 1962; Rhallabi 1990; Seydou 1996), and finally, communication and awareness raising (Diao 2021).

This program constitutes the first phase of the national water plan's draft (2020–2050), which was presented to the interministerial water commission on December 25, 2019. This national water plan's draft outlines the actions that different stakeholders have taken over the next 30 years to ensure the country's water security (Ministre de l'Equipement et de l'Eau, c).

12.5 THE SITUATION IN MOROCCO WITH REGARD TO WATER STRESS AND PLANNED SOLUTIONS

Morocco is classified as a high-water-stress country, having experienced seven periods of widespread drought during the period of 1955–2004 (Bennouna 2020b). The most severe were in the 1980s, 1995, and 2000. The first drought inaugurated the installation of wells (Ameur et al. 2013). The second and third droughts were marked by a significant innovation, namely, the deep drilling technique (Chaudhry 1990).

Currently, its water resources are estimated at less than 620 m^3 per person per year, compared to 3,500 m^3 in 1960, which means that the kingdom is well below the "water poverty level" of 1,000 m^3 per person per year (Mseffer 2021), as shown in Figure 12.4, on the water stress level in Morocco.

According to the United Nations report, this allocation is expected to fall to less than 600 m^3 per capita per year by 2030, and by 2050, the country will be below the level of "extreme water stress."

While population is growing and the water need of key sectors (agriculture, industry) makes water demand management an imperative, it is also due to inadequate supply infrastructure (e.g., silting of dams and reservoirs) and inefficient water management.

Faced with this water stress, Morocco intends to secure the mobilization of water, according to the Ministre de l'Equipement et de l'Eau, a:

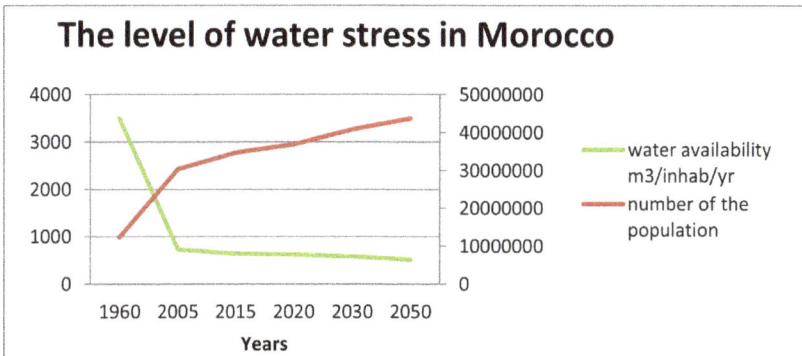

FIGURE 12.4   The level of water stress in Morocco.

Source: Based on data from Minister of Equipment and Water.

1. The construction of over 21 dams of full capacity of 5.6 billion m³.
2. The installation of the seawater desalination system for the water's drinking supply, such as the one in Al Hoceima, with a capacity of 55,080 m3/d after commissioning:

   - Al Hoceima with a capacity of 55,080 m³/d, after start-up.
   - From the city of Laâyoune and that of Agadir (under construction), to reinforce the supply of drinking water and the irrigation of Chtouka, with a production capacity of 275,000 m³/d at the start, and will eventually reach 400,000 m³/d.
   - The Office Chérifien des Phosphates, with a current production capacity of 72 329 m³/d.
   - From Jorf Lasfar and Laâyoune, with an additional production capacity of 68,000 m³/d.
   - Casablanca's system is the largest, with a production capacity of 822,000 m³/d at start-up.

3. The demineralization of brine water; the current production capacity is of the order of 90,400 m³/d, which will be reinforced by the projects in progress with a production capacity of 5,830 m³/d.
4. Reuse of treated wastewater, especially as 500 million m³ of wastewater is discharged from urban agglomerations in Morocco, according to the Conseil supérieur de l'Eau et du Climat (CSEC 1994), and 900 million m³ by the year 2020. Currently, 46 organized and controlled reuse projects have been or are being implemented nationwide for different uses (green areas, golf courses, and agriculture).

   - In 2019, the wastewater treated and reused volume is around 65 mm³, 51% for watering golf courses and green spaces, and 17% for industry.
   - In 2021, the wastewater treated mobilized volume will reach 100 mm³/year (Ministre de l'Equipement et de l'Eau, a).

12.6 CONCLUSION

This chapter contributes to a better understanding of the water policy in Morocco, which was instituted in the 1960s and waited 20 years to establish its modern version. This policy, which was launched in the mid-1960s as a dam policy, had as its main objective the irrigation of 1 million hectares and relied on water resources estimated on the basis of the water cycle by annual mobilizable volumes of 22 billion m³: 18 in surface water and 4 in groundwater. The regulatory and institutional reforms around water on the one hand and the measures of the new strategy for mobilizing nonconventional water and the solutions of the circular water economy on the other hand were at the center of this work. In this chapter we were able to answer the question in the title, namely, "Circular Economy Strategic Approaches: Morocco Case Study?" Concerning this question, in addition to the construction of new high-scapacity dams, Morocco is planning to set up a total of 20 desalination plants, of which we have cited the operational desalination plants with their treatment capacities, as well as those in the study phase. These stations would allow the supply of drinking water and the irrigation of large agricultural areas. We also discussed the projects of reuse of operational wastewater with their capacities of saved water, and the projects of demineralization of brackish water, which Morocco also based its calculations on to mobilize these nonconventional waters. Currently, the kingdom's water policy requires the implementation of movements that require changes in attitudes and approaches, so while waiting for desalination to free Morocco totally or partially from its dependence on rainwater, it is necessary to implement actions to recover what can be lost, in addition to the critical analysis at the level of each water basin, to have an idea of the problem of adequacy between demand and resources.

REFERENCES

Agence Bassin Hydraulique De Souss Massa, 1995, *La Loi n° 10–95*. www.abhsm.ma/index.php/sliders/loi-n-10-95

Agence Bassin Hydraulique De Souss Massa, 2015, *La Loi n° 36–15*. www.abhsm.ma/index.php/sliders/loi-n-36-15

Ameur, F., Hamamouche, M.F., Kuper, M., Benouniche, M., 2013, *La domestication d'une innovation technique: introduction et diffusion du goutte-àgoutte dans deux douars au Maroc*. Cahiers Agricultures 22: 311–318.

Arrifi, EL M., 2009, *La gestion intégrée en eau au Maroc: ressources, contraintes et implications sur l'économie d'eau*, Revue HTE N°140 • Septembre 2008.

Banque Mondiale, 2007. *Obtenir le meilleur parti des ressources rares Une meilleure gouvernance pour une meilleure gestion de l'eau au Moyen-Orient et en Afrique du Nord*, Rapport Sur Le Développement Région Mena, ISBN: 0-8213-7094-4, Washington DC 20433

Bennouna, A., 2020 a, *Gestion de l'eau au Maroc et changement climatique*, page 251, Revue No 32, Revue Espace Géographique et Société Marocaine, Laboratoire de Recherches, d'Etudes Géographiques, Fès.

Bennouna, A., 2020 b, *Gestion de l'eau au Maroc et changement climatique*, page 253, Revue No 32, Revue Espace Géographique et Société Marocaine, Laboratoire de Recherches, d'Etudes Géographiques, Fès.

Bielorai H, Vaisman I & Fegin A., 1984, *Drip irrigation of cotton with treated municipal effluents I. Yield response*. Journal of Environmental Quality 13: 231–234

Boukhari, T. R, Naïmi, M., Chikhaoui, M., Raclot, D., Sabir, M, 2019, *Evaluation Des Performances Du Modele Agro-Hydrologique SWAT à Reproduire. Le fonctionnement Hydrologique Du Bassin Versant Nakhla (Rif occidental, Maroc)*. European Scientific Journal, ESJ: Vol. 15(5) (2019), ESJ February Edition.

Brister, Gh., Schultz, RC,.1981. *The response of a southern applachian forest to wastewater irrigation*. Journal of Environmental Quality 10: 148–153

Chaudhry, J,. 1990. *The Adoption of Tubewell Technology in Pakistan*, The Pakistan Development Review, Pakistan Institute of Development Economics, vol. 29(3 and 4), p. 291–303.

Conseil Economique, Social et Environnemental, 2019, *Economies Circulaires Au Maroc: Traitements Des Déchets Ménagers Et Des Eaux Usées*, p. 3.

Conseil Supérieur de l'Eau et du Climat (CSEC) (Maroc), 1994, *Réutilisation des eaux usées en agriculture*. Rapport 8 ème Session.

Day, A.D., Tucker, T.C., Vavich, M.G., 1962, *Effect of city sewage effluent on the yield and quality of grain from barley, oats, and wheat*. Agronmy Journal 54(2), 133 à 135

Diao, M., 2021, *Gestion de l'eau: "Le Maroc a opté pour une planification anticipative et à long terme"*, Finance New. https://fnh.ma/article/developpement-durable/gestion-de-l-eau-le-maroc-a-opte-pour-une-planification-anticipative-et-a-long-terme-1

ELGhomari, K., 2015, *Bilan De La Politique De L'eau Au Maroc.* www.barrages-cfbr.eu/IMG/pdf/07_-_el_ghomari politique_des_barrages_au_maroc.pdf

Hinrichsen, D., Robey, B. and Upadhyay, U. 1997, *Solutions for a Water-Short World.* Report No. 14, Johns Hopkins School of Public Health, Population Information Program, Baltimore.

Houdret, A., 2008, *Les conflits autour de l'eau au Maroc: Origines sociopolitiques et écologiques et perspectives pour une transformation des conflits.* Phd thesis, Université Paris VIII Vincennes-Saint Denis; Universität Duisburg-Essen.

Kirby, A., 2004, *Energy: Meeting soaring demand*, BBC NEWS. http://news.bbc.co.uk/2/hi/science/nature/3995135.stm

Kuper, M., Hammani, A., Chohin, A., Garin, P., Saaf, M., 2012. *When groundwater takes over: Linking 40 years of agricultural and groundwater dynamics in a large-scale irrigation scheme in Morocco.* Irrigation and Drainage, Wiley, 2012, 61 (suppl.1), pp. 45–53.

Laouina, A., 2006, *Prospective " Maroc 2030 ", Gestion Durable Des Ressources naturelles Et De La Biodiversité au Maroc*, Rapport pour le compte du Haut Commissariat au Plan.

Llamas, MR., Martinez-Santos, P.,2005, *Intensive groundwater use: Silent revolution and potential source of social conflicts.* Journal of Water Resources Planning and Management 131: 337–41.

Molle, F., Berkoff, J., 2006, *Cities versus agriculture: Revisiting intersectoral water transfers, potential gains, and conflicts.* Comprehensive assessment of water management in agriculture. Research Report 10. IWMI, Colombo.

Molle F., Tanouti O., Faysse N., 2019, *Morocco.* In: François, Molle., Sanchis-Ibor, C., Avellà-Reus, L. Irrigation in the Mediterranean: Technologies, Institutions and Policies. Cham: Springer, pp. 51–88. (Global Issues in Water Policy; 22). ISBN 978-3-03-003696-6. ISSN 2211–0631.

Ministre de l'Equipement et de l'Eau, a, *Planification des ressources en Eau*, La Direction Générale de l'eau. http://81.192.10.228/ressources-en-eau/planification-des-ressources-en-eau

Ministre de l'Equipement et de l'Eau, b, *Ressources Eau en chiffres*, La Direction Générale de l'eau. http://81.192.10.228/ressources-en-eau/chiffre-de-leau/

Ministre de l'Equipement et de l'Eau, c, *Approvisionnement en eau potable et l'irrigation (2020–2027)*, La Direction Générale de l'eau. http://81.192.10.228/ressources-en-eau/lapprovisionnement-en-eau-potable-et-lirrigation/

Ministry of Energy Transition and Sustainable Development, 2009, *La Stratégie Nationale De L'eau*, Ministère Délégué chargé de l'Eau, p. 7.

Mseffer, D.Z., 2021, *État des lieux des ressources en eau au Maroc*, p. 26., N° 1035, Conjoncture.www.cfcim.org/wp-content/uploads/2021/04/1035-avril-2021-Politique-de-leau1.pdf

Neilson, G., Stevensen, D.S., Fitzpatrick, J.J., Brownlec, C.H., 1989, *Nutrition and yield of yong apple tree irrigated with municipal wastewater.* Journal of the American Society for Horticultural Science (USA) ISSN: 0003–1062114 (3): 377–383.

OECD, 2008, *Freshwater, OECD Environmental Outlook to 2030*, OECD-ilibrary, p. 219–220. https://read.oecd-ilibrary.org/environment/oecd-environmental-outlook-to-2030_9789264040519-en#page523

Rhallabi, N., Moundib, R., Maaroufy, M., Marhich, M., 1990, *Effet des irrigations avec des eaux usées brutes et épurées sur le sol, le rendement d'une culture de tomates et la qualité hygiénique de récolte.* Actes Institut Agronomique et Vétérinaire 10 (2): 57–66.

Seydou, N., 1996, *Utilisation des eaux usées domestiques en maraîchage périurbain à Dakar (Sénégal).* John Libby Journal, Science et Changements Planétaires/Sécheresse 7 (3): 217–22

Shawiza, V., 2018, *85 Percent of World's Population live in the Driest Half of the Planet*, Soko Directory.

Stour, L., Agoumi, A., 2008, *Sécheresse climatique au Maroc durant les dernières décennies*. Hydroécologie Appliquée 16: 215–32.

Rosegrant, M.W., Cai, X., Cline, S., 2002, *World Water and Food to 2025: Dealing with Scarcity*. Report at International Food Policy Research Institute/IWMI, Washington, D.C./ Colombo.

United Nations, 2007, *The Millennium Development Goals*, the United Nations- Department of Economic and Social Affairs. www.un.org/millenniumgoals/pdf/mdg2007.pdf

United Nations, 2021a, *Water in a Changing the World*, The United Nations World Water Development Report 2021: Valuing water—UNESCO Digital Library, p. 11. https:// unesdoc.unesco.org/ark:/48223/pf0000375724

United Nations, 2021b, *Water in a Changing the World*, The United Nations world water Development Report 2021: Valuing Water—UNESCO Digital Library, p. 8. https://unesdoc.unesco.org/ark:/48223/pf0000375724

United Nations, 2021c, *Water in a Changing the World*, The United Nations World Water Development Report 2021: Valuing Water—UNESCO Digital Library. https://unesdoc.unesco.org/ark:/48223/pf0000375724

United Nations News, 2021, *Billions Risk Being Without Access to Water and Sanitation Services by 2030*, UN News. https://news.un.org/en/story/2021/07/1095202

World Bank, 1995, *Une stratégie pour la gestion de l'eau au Moyen-Orient et en Afrique du Nord*. Banque internationale pour la reconstruction. Directions in Development, Washington: World Bank, Washington, D.C. 20433

World Bank, 2007, *Making the Most of Scarcity, Accountability for Better Water Management in the Middle East and North Africa, Mena Development Report*. Washington, DC: The International Bank for Reconstruction and Development, The World Bank.

World Bank Group's Water Global Practice and water and sanitation services,2015, *Water and Sanitation Program End of Year Report, Fiscal Year 2015*, World Bank Water.

Index

Page locators in **bold** indicate a table.
Page locators in *italics* indicate a figure.

For Product Safety Concerns and Information please contact our EU
representative GPSR@taylorandfrancis.com
Taylor & Francis Verlag GmbH, Kaufingerstraße 24, 80331 München, Germany